How Science Works

John D. Norton
Department of History and
Philosophy of Science
University of Pittsburgh

The McGraw-Hill Companies, Inc.
Primis Custom Publishing

New York St. Louis San Francisco Auckland Bogotá
Caracas Lisbon London Madrid Mexico Milan Montreal
New Delhi Paris San Juan Singapore Sydney Tokyo Toronto

How Science Works

McGraw-Hill's Primis Custom Publishing Series consists of products that are produced from camera-ready copy. Peer review, class testing, and accuracy are primarily the responsibility of the author(s).

12 XRC XRC 9 0 9

ISBN 0-07-230800-1

Editor: M. A. Hollander
Cover Design: William C. Whitman Jr.
Printer/Binder: Xerographic Reproduction Center

With thanks to Heather Douglas, Ofer Gal, Chris Martin,
Elizabeth Paris, David Raikow and Ted Richards

Version 1.1a 1998

Introduction

Contents

This course is intended to provide an introduction to science and scientific thinking for students who have not had much contact with science. Its goal is to explain what is distinctive about the scientific approach and its product, scientific theories. The emphasis will be on quantitative approaches and on showing how the use of number in science greatly extends the reach of our investigative tools. The course is divided into three parts:

Chapters 1 to 3 address general material. They ask, what are the problems scientists face in their investigations of nature? What are the general techniques commonly used to overcome them?

Chapters 4 to 8 introduce the science of thermodynamics. They show through a concrete example how theories are constructed: How does one develop the deep theoretical notion of energy? How can one develop a theory of heat engines that applies to all possible heat engines? It also shows how the theory can be applied to answer very practical questions in real life. Even though gasoline powered automobile engines have been the target of massive complaints for decades, why is an alternative so hard to find?

Chapters 9 to 13 develop one of the most important, general techniques of science: statistical analysis. Randomness appears in many of the systems investigated by science. On average we expect 50 girls in a group of 100 children. In practice the number of girls will deviate from this average value. Statistical analysis depends on the fact that these deviations form very orderly patterns, encoded in the so-called "bell curve". It will give us the means for taming randomness.

The layout of this text is not standard. Its presentation is intended to capture something of the flavor of a classroom lecture. Throughout, the page on the left hand side will provide a blackboard style summary of the material. A text explaining the material will appear on the right hand side.

12345678901234567890123456789012345678901234567890

1. 1.23 x 1,000 =

2. 1/3 + 1/6 =

3. 3/4 x 2/3 =

4. 3/4 as a decimal =

5. 0.60 as a fraction =

6. 0.5 x 0.1 =

7. If there are 3.8 liters in a gallon,
how many liters are there in 100 gallons?

8. If there are 3.8 liters in a gallon,
how many gallons are there in 76 liters?

12345678901234567890123456789012345678901234567890

12345678901234567890123456789012345678901234567890

This class is intended for students who do not normally take classes that involve quantitative matters. The mathematical content has been kept at an absolute minimum. No advanced techniques are presumed. All that is required are elementary arithmetical skills. The questions opposite will help you decide if you have these skills. You should be able to answer all these questions without undue effort or labor. You may need a calculator for 8.

12345678901234567890123456789012345678901234567890

Recitation 0. A Warm-Up Exercise

In science, we have to worry about two things: *what* we know and *how* we know it. Generally most attention is focussed on the *what*. In this course our emphasis will be on the *how*. In a sense the *how* is more important, for it is what makes the *what*special. The content of science is not supposed to be good guesses or revealed mysteries. It is supposed to be the reliable results of painstaking inquiries. They assure us that the Earth does orbit the sun, that everything is made of atoms and that our own physical characteristics are encoded in immense DNA molecules. We are going to try to understand how such inquiry works.

According to astrology, knowledge of the positions of heavenly bodies at the instant of one's birth allows quite detailed prediction of the course of our lives. Today we will not be concerned with whether astrology is true or false. Most of you will already have opinions on the question. Rather we will worry about *how* we could go about finding out if it is true or false. To get us started, consider the astrological predictions for today. Doubtlessly some of you will find striking matches between the prediction for today and what actually happened to you today. Others will find striking failures. This is the sort of data that can be used to evaluation astrology.

How might we decide the truth or falsity of astrology?

In trying to answer the question, consider to following difficulties:

•There is an inherent vagueness in what counts as a successful prediction. What a supporter of astrology counts as a success, a detractor may discount. (e.g. The prediction says: "Be prepared for unexpected good fortune." Which of the following count: You find a dollar in the street. You find a piece of jewelry you thought you had lost. You forget to put money in your parking meter but do not get ticketed?)

•If an astrologer makes enough predictions, some will come true by chance alone. (e.g. On any day of our lives, something good happens unexpectedly, even if it is trivial. So by chance alone, the prediction "Be prepared for unexpected good fortune." will always have some successes.) The detractors say these are the *only* predictions that work. The supporters say the successes are more than mere chance.

The method used to decide astrology must be acceptable to both detractor and supporter. How can it resolve these and other similar problems?

Chapter 1

Sir Cyril Burt and the Inheritance of Intelligence

Do our children inherit their intelligence from us, their parents?

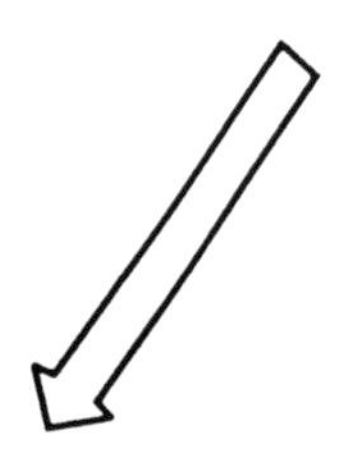

Possible answers

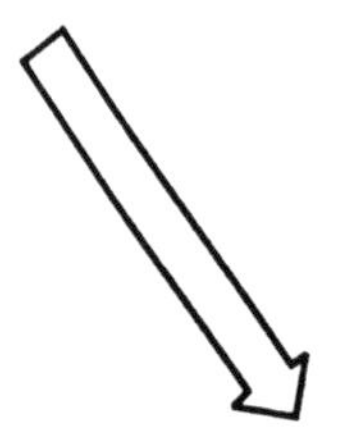

YES. Intelligence is genetically determined.

i.e. there is an upper limit to "smarts" set by parentage.

NO. Intelligence is fixed by environment.

e.g. quality of schooling, degree of stimulation in home life, etc.

Or some combination?

Nature versus Nurture

The purpose of this chapter is to give you a preview of the types of issues that will be addressed in this class and the methods that will be used to cover them. In a short case study of the work of Cyril Burt we will illustrate two themes. First, science strives to find out the truth about things that may seem quite beyond the grasp of ordinary inquiry. With ingenuity and diligence, however, we will see that it is possible to decide these issues. Second we shall see the surprising and far-reaching power of statistical methods.

The question under investigation is whether children inherit their intelligence from their parents. Clearly children inherit some things from their parents. The ones we notice most easily are simple things like skin color, hair color and eye color. But how far reaching is this inheritance. Does it extend to intelligence as well? Will smarter people have smarter children in the same way that red-haired people are more likely to have red-haired children?

If you answer yes, then you believe there is a range for the intelligence of children set by the intelligence of the parents. If you want to be smart, choose your parents well! Intelligence is fixed by your nature and this nature cannot be altered. If you answer no, then you look to other factors to set the intelligence of children. The most obvious candidates are environmental factors like the quality of schooling, the quality of home life--how much mental stimulation is provided, *etc.* A "no" answer would imply that intelligence is fixed by nurturing. Of course this is not a question that is likely to allow a simple yes or no answer. Presumably the truth lies somewhere between.

Our emphasis here will not be on the question of whether or to what degree nature or nurture fixes intelligence. The material we look at in this lecture will leave the issue undecided. Rather we are interested in the issue of how one might go about finding out. For—as we shall see—it is not at all so clear how one can decide between nature and nurture. Nevertheless opinions on what determines intelligence have exercised an extraordinary influence on matters of social policy in various countries. Knowing how to decide between nature and nurture really matters.

Qualifications:

(1) The Question has profound political and moral implications.

Not just, "How cute, little baby Johnnie is just as smart as his Daddy."

Intelligence is inherited.

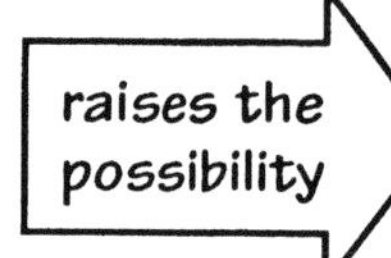

Different races may have different average intelligence

Translate into social policy: ⇒ Use it to justify school segregated by race. Put most resources into schools with smartest kids. In the U.S., that means schools with white kids.

Translation fails, even if there are differences in average intelligence of races!

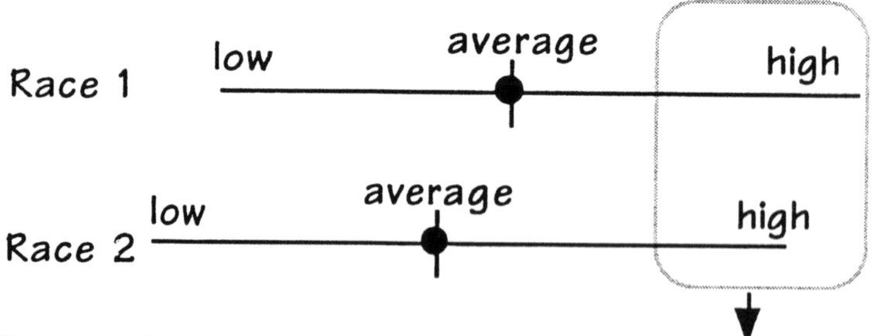

Differences in averages are slight.
Differences in individuals are great.

Averages do not matter, individuals matter.

If you want to assign more educational resources to smarter kids, throw them at these.

Before proceeding, we should take a little time to consider two very important qualifications. The first is that the question is far less innocent than it might seem. Whether intelligence is inherited is a matter of biology and we cannot wish it to be one way or another. What is up to us is how such facts of biology are used in matters of social policy. And here the record is not such a happy one.

If intelligence is inherited—and the "if" must be emphasized since we do not assume here that it is the case— then different groups that share a common gene pool will have similar intelligences. More specifically it is assumed by people who hold that intelligence is inherited that different racial groups will differ in average intelligence. In the U.S., people who think on these lines have tended to accord higher intelligence to white children.

This is now translated into an instrument of social policy. The supposition is that one would want to assign more educational resources to smarter children. The result is that white children are sent to better funded schools.

There are many problems here. In translating what might be a result from biology into social policy, a number of non-scientific judgments have been added. Is it obviously the best allocation of resources to direct more towards smarter kids? Might they not need them less? Or, if we decide to allocate more resources towards smarter kids, is division along racial grounds the most effective? Would it not make more sense to select out the smarter kids, whatever their race, and give them better resources?

The inheritance of intelligence in Great Britain, home of Cyril Burt.

Burt and others concluded that intelligence is inherited.

Different social classes may have different average intelligence

Burt and others concluded the higher class children are genetically smarter.

Choose kids that are genetically smart early and give them better schooling.

The "11+ exam"

taken by all children at age 10-11 years

The outcome irrevocably decides a child's secondary school future.

Top scoring 20% → Grammar school → ✔ University

Remaining 80% → Technical or "secondary modern" school → ✘ University (Unfit)

The educational future of 80% of children was sealed at age 11.

In Great Britain, where Cyril Burt did his work, the great social divisions are not along the lines of race as they tend to be in the U.S. Rather they tend to be along the lines of class. The social classes range from the lower classes, beginning with simple manual laborers, and ascends through the professions (doctors, lawyers) to the upper classes, in whose heights we find the nobility. Burt believed that intelligence is inherited and this was translated into the expectation that there are different average intelligences across the different social classes, with the smarter residing in the higher classes.

For many years the British education system was so set up as to perpetuate differences in education. All children at the ages 10-11 years took an exam, the "11+ exam", designed to measure their innate or inherited intelligence. Their performance on the exam decided their educational fate. The best 20% were given better schooling at a grammar school which would then allow passage to university. The remainder were sent to lesser schools that would not lead to university admission.

If intelligence is inherited, this system ensures that those blessed with "smart" genes end up in better education. But if Burt is wrong and the quality of education is the decisive factor, it cuts off children who have had a slow start from recovering through better schooling. It also directs those children into a lesser career thus making it more likely that their families will also perform less well educationally.

This examination system was as a result of a long series of government reports written in 1926, 1931, 1938 and 1944. Burt's research played a part in the evidence that these reports relied on. By the 1960s the approach has fallen from favor and it began to be dismantled.

See Stephen Jay Gould, *Mismeasure of Man*. New York: Norton, 1981, p.281+.

(II) Is there a single simple quality "intelligence"?

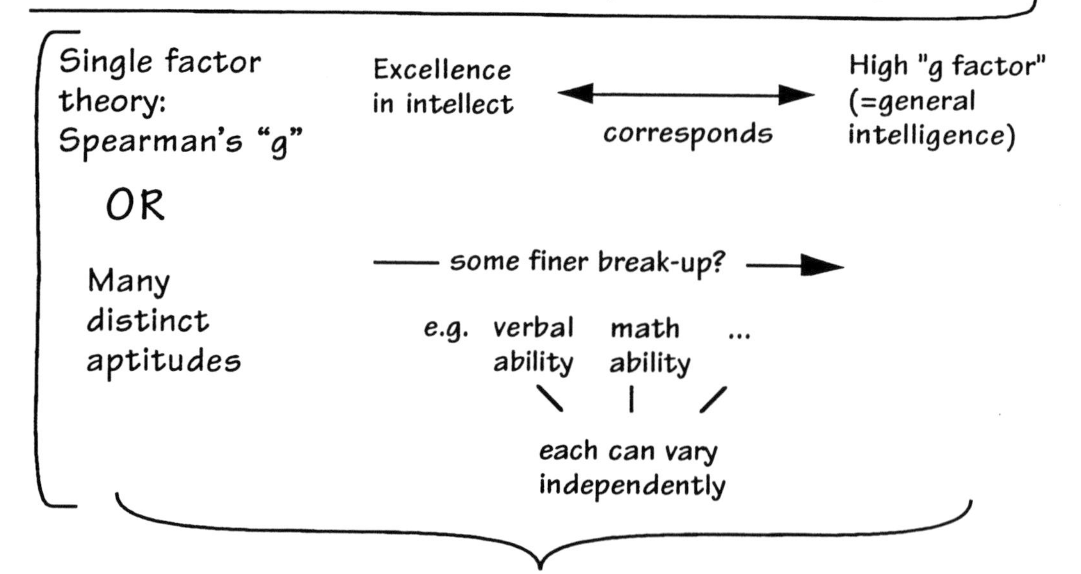

The answer is independent of whether mental abilities are inherited.

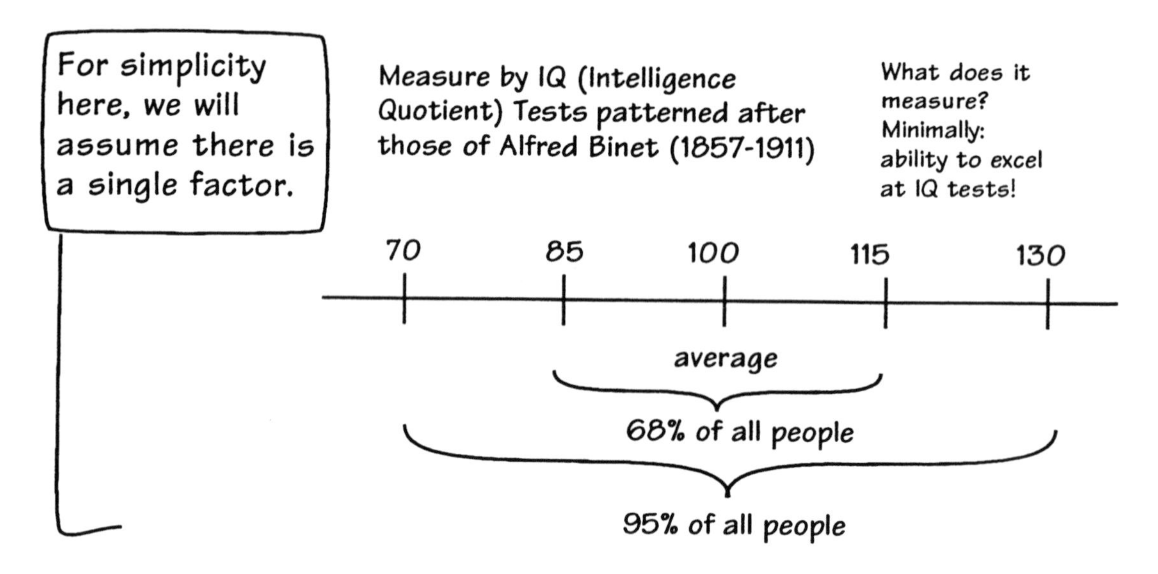

Is intelligence inherited?

(1) Final result: I don't know.
(And this answer is irrelevant to what we will do in this chapter.)

(2) Wait out the story
...it has a surprise ending!

The second qualification is that it is not so obvious that there is a simple quality "intelligence" that can be measured by a single number in humans.

Those who say there is think of intelligence much like the height of a person. Everyone has a definite height and it can be measured directly and reported as a single number. An older school in psychology holds that intelligence is, to a large extent, like height. It can be measured on a simple scale. This was the view of Cyril Burt's mentor, Charles Spearman. He sought to measure intellectual abilities through a single index he called the "g factor". Both Spearman and Burt were Professor of Psychology at University College, London, with Burt succeeding Spearman in 1931.

An alternative view holds that intelligence is like body shape. No single number captures it. It must be represented by an array of attributes--short legs or long legs, large belly or small belly, fine boned or heavy boned, and so on. While the various attributes may be connected, the connections may be loose.

Whether intelligence is inherited is actually independent of whether a single number can measure it. In the following, we shall set this new question aside and assume, with Burt, that intelligence can be measured to significant degree by one number. To serve this purpose Burt's work employed the familiar "IQ" test.

The results of IQ tests are scaled routinely in a special way that will become very important in this chapter. The scores of a group are adjusted so that the average score comes out at 100. The scale is expanded or contracted so that most people cluster around the 100. More specifically, they are adjusted so that 68% of people fall in the 85 to 115 range and 95% in the 70 to 130 range.

In this chapter we will not be able to decide whether intelligence is inherited. But that is not necessary for our concern: how could we go about finding out if intelligence is inherited. Also, be patient and let this story play out--it has a surprise ending!

How can we know if intelligence is inherited? The simple approaches fail.

#1 Do Genetic factors cause intelligence like yeast causes dough to rise?

Simple to check this.
Dough without yeast does not rise.
Dough with more yeast rises more and with less yeast less.

No similar test here!
How do we make person without genetic factor for intelligence?
Or a person with a weaker factor?
If we could, would we want to!

#2 Look at individual case histories:

Tom is smart and Tom's parents are smart ————????————> Tom's smarts are inherited

BUT

Does this one case prove inheritance of intelligence?

- Could it be an accident that smart Tom has smart parents?
- Could it be that the environment provided by Tom's smart parents makes him smart (and that he inherited nothing)?

Now we can turn to the basic question. How can we find out whether intelligence is inherited? We can see quite quickly that the obvious and simple approaches are not going to give us a clean answer.

We can know that yeast causes bread dough to rise through a simple experiment. We prepares batches of dough with yeast. They all rise. Next, we prepare batches of dough without yeast. We find they do not rise. Through more tests, we find that if we put more or less yeast in, the dough rises more or less. Clearly yeast causes dough to rise.

Can we decide whether genetic factors cause intelligence through similar experiments? We cannot easily do comparable experiments. What would it be to have a person without the genetic factor for intelligence? How might we prepare a person with more or less of the genetic factor for intelligence? Is this sort of manipulation of children something we could do morally or that any parent would allow?

We can conceive experiments that do not require such meddling. We can merely observe individual case histories without interfering. We may notice in a particular family that the child is smart and so are the parents. Does this prove that the child's intelligence is inherited? Certainly not. The agreement in one case could be coincidental, so that many cases would be required. Even if we had many such cases at hand, how would we know that it was the genes provided by the parents that made the child smart? Presumably smart parents will have a home life to match with plenty of stimulation to direct the child towards intellectual pursuits. They may also be more likely to educate their child better. How could we know that this environment is not responsible for the greater intelligence of the child?

Cyril Burt (+ many others) conducted series of experiments designed to test inheritance of intelligence.

Early Studies

Problem	How do we measure intelligence objectively?		How do we rule out accidental pairings smart parents / smart ? kids
	⇩		⇩
Solution	Evaluate intelligence by objective intelligence tests.	in	many pairs of parent/child

Results

In 40,000 parent/child pairs tested, drawn from the population of London:

Smarter child ⟷ (goes with) Smarter parent ⟷ (goes with) Higher class

Conclusion

(1) Children inherit smarts from parents.
(2) Higher classes are where the genetically smarter are.

Sample data

Class	Average IQ of parent	Average IQ of child
I. Higher Professional	139.7	120.8
II. Lower Professional	130.6	114.7
III. Clerical	115.9	107.8
IV. Skilled	108.2	104.6
V. Semi-skilled	97.8	98.9
VI. Unskilled	84.9	92.6

Data reported in D. Dorfman, "The Cyril Burt Question: New Findings," *Science*, 201 (1978), 1177-86.

Starting roughly in 1900, Cyril Burt was one of many who began experiments to discover whether intelligence is inherited. The design of the early experiments avoided some of the simple pitfalls. There is an inherent ambiguity in deciding how intelligent someone is. Rough impressions can be very unreliable. Burt tended to use objective tests like the now familiar IQ test. Also he could rule out the possibility that relationships he uncovered were due to accident by looking at many pairs of parent and child. By the 1960s he was able to report on an astonishingly large number of pairs of parent and child—40,000—drawn from his research over many years starting in 1913. (C. Burt, "Intelligence and Social Mobility," *Br. J. Stat. Psychol.* 14 (1961) 3.) Burt was in a good position to assemble this many cases since he was involved in the administration of school education in London and and, as a result, had access to a lot of data about children in London.

His results were that smarter children tend to have smarter parents and the smarter families were the higher class families. The result suggested is obvious: children inherit their intelligence from their parents and these smarter genes are to be found more in higher class families.

The table reproduces a sample of Burt's data. The patterns Burt found are easily read from it. Notice how the higher class families have the higher IQs and how there is a close connection between the IQs of parent and child. You might want to take a moment to look at the numbers in this table. They will prove to be rather interesting.

Problem

Smarter children have smarter parents.

Smarter children inherit their smarts from smarter parents?

or

Smarter children learn their smarts from better environment provided by smarter parents?

How can we determine the relative contribution of each?

Study Twins !!

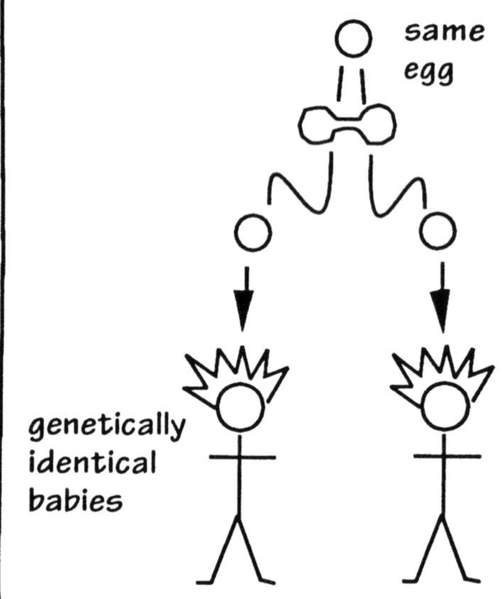

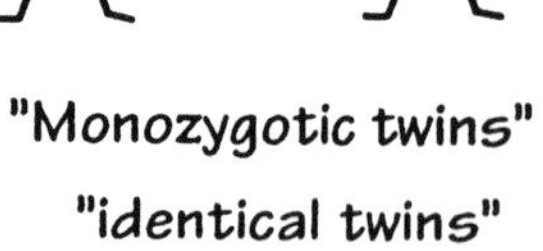

"Monozygotic twins"
"identical twins"

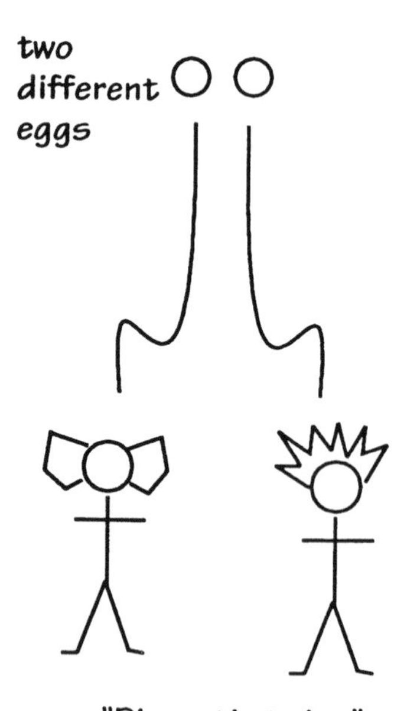

"Dizygotic twins"

The number of pairs reported as studied is impressively large. But a significant problem remains and it is not helped by increasing the number of pairs. Demonstrating that smarter children have smarter parents does not show that the children inherited their intelligence from the parents. It leaves open the possibility that the children are smarter not because of any genetic factor, but because of the better environment the smarter parents supply for the children. This last possibility has to be taken very seriously since we see directly from data that the smarter parents tend to be in the higher classes. One would expect that higher class parents would provide better opportunities and education for their children merely by virtue of wealth and privilege.

To solve this problem, Burt drew on a brilliant idea suggested by Francis Galton in 1883. In order to check the degree to which something is inherited, our ideal subjects would be a pair of individuals that are identical genetically. Galton noted that nature provides exactly these cases as twins. Twins could be used as a test bed to check for the relative importance of genetic and environmental factors in humans.

There are two types of twins. "Dizygotic" twins result from the fertilization of two separate eggs in the mother. The ensuing twins will have distinct genetic material. They will be no more closely related genetically than siblings who are not twins. The case that interests us is that of "monozygotic" or "identical" twins. These twins result when a single fertilized egg in the mother splits and the two parts each grow into an individual. The resulting twins will be identical genetically. They will share exactly the same genetic disposition to intelligence—to the extent that such a thing exists.

Burt's studies of twins 1940s-60s

Find

Monozygotic twins
- separated at birth
- reared apart.

⇨ Any agreement in intelligence must be due to sameness of genes.

Results

High correlation in intelligence of separated twins.

Conclusion

There is a strong inherited factor in intelligence.

Burt's Data for monozygotic twins reared apart:

	Reared in			
	Own home		Foster home	
Case number	"IQ"	Social class	"IQ"	Social class
1	68	6	63	6
2	71	4	76	5
3	73	5	77	5
...	...	...	...	...
25	95	6	96	5
26	96	2	93	R
27	96	4	109	1
...	...	...	...	...
51	125	4	128	5
52	129	2	117	5
53	131	1	132	4

smarter twin ⟷ smarter twin
duller twin ⟷ duller twin

even though social classes of paired twins varies widely

Key
6=highest class
1=lowest class
R=residential institution

Data reported in A. R. Jensen, "Kinship Relations Reported by Sir Cyril Burt," *Behavior Genetics*, 4 (1974), 1-28 on p. 15.

Burt undertook a series of studies in the 1940s-1960s devoted to exploiting the cases of identical twins available to him. The cases of special interest were those in which a pair of identical twins were unable to remain with their parents after birth and they were brought up in separate families. These cases are rare, but if we can find them, these unfortunate families have provided precisely the test arrangement we need. The twins are genetically identical but in different environments.

If genetic factors are the dominant determinant of intelligence, then we would expect the intelligence of the twins to tend to agree whatever may be the difference in their environment.

If the environment determines intelligence, then we would expect the intelligence of the twins to be essentially unconnected; instead the intelligence would follow the environment.

By 1966, Burt was able to report 53 cases of identical twins separated at birth. (C. Burt, "The Genetic Determination of Differences in Intelligence: A Study of Monozygotic Twins Reared together and Apart," *Brit. J. Psychol.*, 57 (1966), 137-153.) The results indicated strong agreement between the intelligence of the twins separated at birth. This suggests that the genes they shared played the major role in determining their intelligence.

Correlation coefficient
measures degree of agreement.

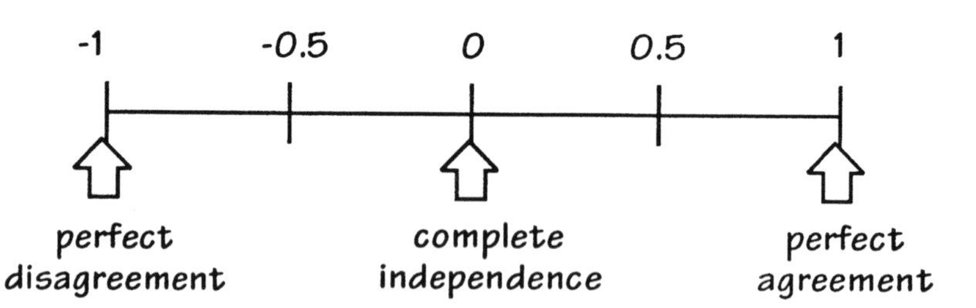

Burt's 53 monozygotic twins reared apart: Correlation = 0.877

Later reported as 0.874 after correcting for the possibility that each twin could be in either column.

Simplified illustration of how the correlation coefficient is computed

Assume intelligence is not measured on continuous numerical scale.

Values are high "H" or low "L" and nothing else.

$$\text{correlation coefficient} = \frac{\text{number of twin pairs that are both H or both L} - \text{number of twin pairs that are different (one H, the other L)}}{\text{total number of twin pairs}}$$

Perfect correlation

Twin 1	H	L	L	H	L	H	H	H
Twin 2	H	L	L	H	L	H	H	H

$$\text{Correlation} = \frac{\text{8 pairs same} - \text{0 pairs different}}{\text{8 pairs total}} = 1$$

Partial correlation

Twin 1	H	L	L	H	L	H	H	H
Twin 2	H	H	L	H	H	H	H	H

different

$$\text{Correlation} = \frac{\text{6 pairs same} - \text{2 pairs different}}{\text{8 pairs total}} = 0.5$$

No correlation

Twin 1	H	L	L	H	L	H	H	H
Twin 2	H	H	L	H	H	L	L	H

$$\text{Correlation} = \frac{\text{4 pairs same} - \text{4 pairs different}}{\text{8 pairs total}} = 0$$

One cannot just look at Burt's data on the previous page and see that the intelligences of the twins are closely connected. Such subjective impressions are too vague for serious work. One person's impression of "strong" connection may be another's "moderate". Who can say which is right? The solution is to compute a numerical measure of the degree of agreement. That is something everyone can agree on. The number is the "correlation coefficient". It varies between -1 and +1, with +1 signaling perfect agreement between the twins' intelligence, 0 no connection and -1 perfect disagreement.

Burt found a correlation coefficient of 0.874 for the intelligences of his 53 monozygotic twins reared apart. This is a strong correlation.

The calculation of the correlation coefficient is done with a simple formula. To get a sense of how it works consider this simplified version of the correlation coefficient. It is how the coefficient would be calculated if IQ measurements just yielded two values, high "H" or low "L", instead of a number. The formula operates by a scoring system. Every time a pair of twins agree in intelligence (both are H or both are L), score +1. Every time they disagree (one is H, one is L) score -1. The correlation coefficient is just the total score divided by the number of pairs overall.

In the case of perfect correlation, all twins agree. Whenever one twin is an H, so in the other; whenever one twin is an L, so is the other. In the case shown opposite, we score +8, which gives a correlation coefficient of 8/8 = 1.

In the case of partial correlation, most of the pairs agree. That is, whenever one twin is an H , more often than not so is the other; and correspondingly for L. In the example shown, they agree in 6 cases of eight (score +6) and disagree in 2 (score -2). The total score is 6-2=4. The correlation coefficient is 4/8=0.5.

In the case of no correlation or independence, if one twin is an H, then the other is just as likely to be an H as an L; and correspondingly for L. In the case shown, 4 pairs agree (score +4) and 4 disagree (score -4). The total score is 4-4=0 and the correlation coefficient is 0/8=0.

Burt's data in summary:

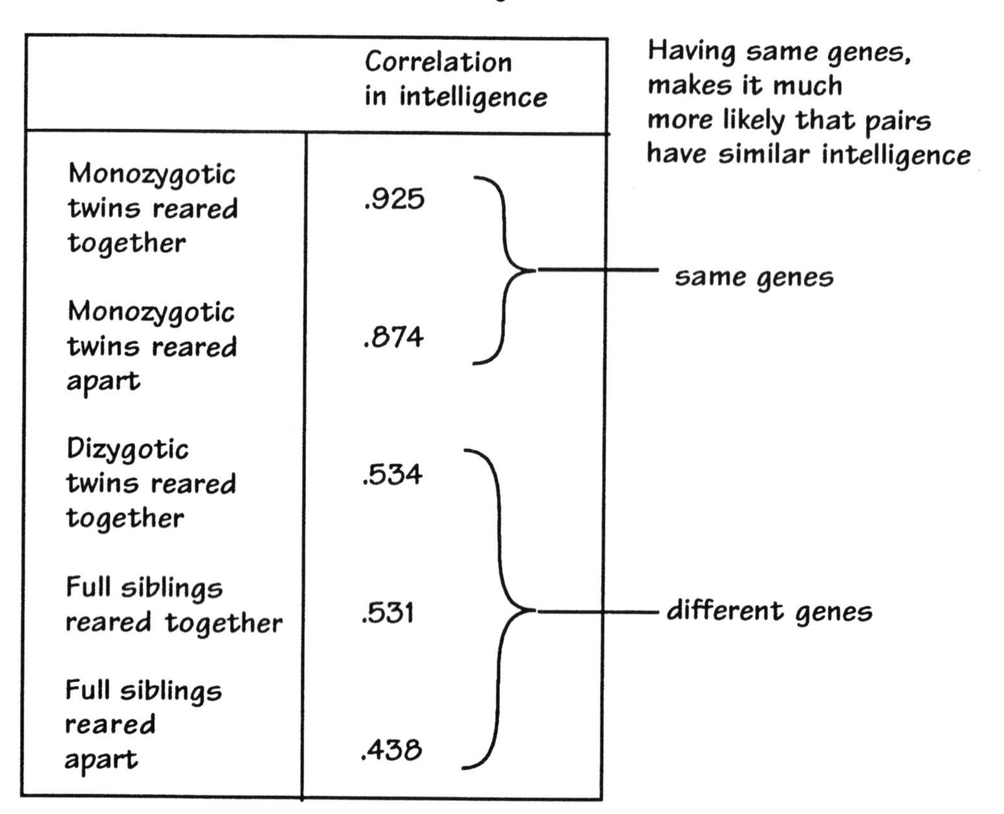

	Correlation in intelligence
Monozygotic twins reared together	.925
Monozygotic twins reared apart	.874
Dizygotic twins reared together	.534
Full siblings reared together	.531
Full siblings reared apart	.438

As reported in Jensen, *op. cit.*

Once a correlation coefficient has been established, we now need to interpret it. Burt reported a final correlation coefficient of 0.874 for the intelligences of twins separated at birth. This is clearly a high correlation.

The remainder of Burt's work gives an even clearer perspective on the significance of this value. Burt also reported on the correlation in intelligence of many other combinations of siblings and twins, as shown in the table. The lowest correlation (0.438) appears for siblings reared apart. Both their environment and genes differ the most of all cases considered. Rearing the siblings together improved the correlation only slightly, even if the siblings were dizygotic (non-identical) twins. So the environment seems to be of less importance in determining intelligence. If the twins are monozygotic (identical), the correlations jumps to 0.874, even if they are reared apart, suggesting that the identity of genes is decisive. The correlation increases a little more if the identical twins are reared together, so that the environments become the same.

BUT

Burt died October 10, 1971

Scandal O. Gillie *Sunday Times*, London 24 October, 1976, p.1.

Accusation Burt fabricated his results!

Basis of accusation:

(1) Failure of Gillie to find any trace of existence of Burt's assistants: Jane Conway, Margaret Howard

Shortly (1) proved less problematic. John Cohen of University of Manchester remembered Howard from when he was a PhD student under Burt.

(2) **Burt's data was too good!!**

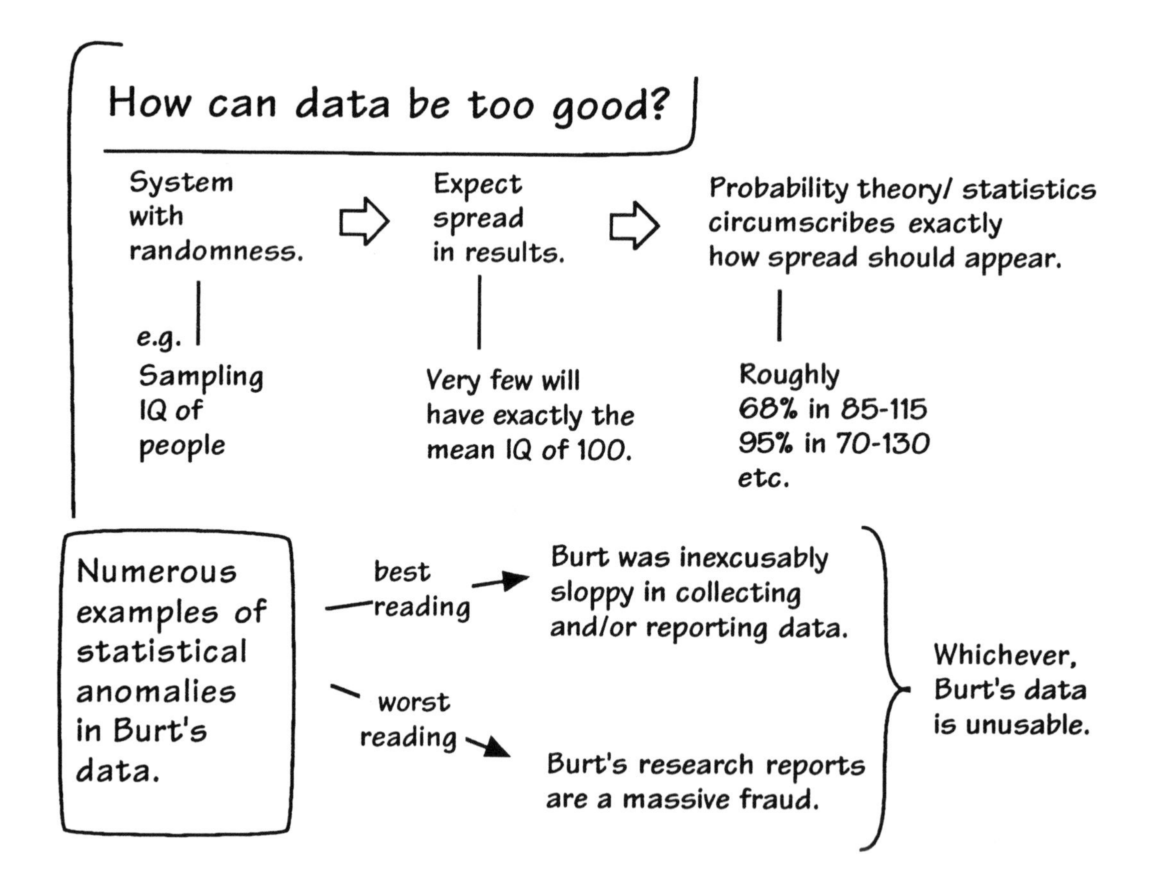

Burt died in 1971. His lifetime of work on the inheritance of intelligence stood proudly as one of the major discoveries of modern psychology and as an example of how careful study can cause Nature to reveal her secrets.

That is until the scandal broke. In 1976 a journalist, O. Gillie, made public what had been suspected for a while. There was something seriously wrong with Burt's studies. The problem had emerged in various forms. Gillie had been unable to track down the assistants Burt spoke of in his study. And there was an odd omission of essential methodological details from Burt's reports and the citations for the details to sources that proved elusive. But the most serious and inescapable indication of defects in the studies was a surprising one. It was not that Burt's data failed to show the regularities claimed. Rather they did it too well!

In any system with randomness, one does not expect to get perfect results. The results will vary around average values and the way in which they vary will be characteristically regular. For example if we test a large group of people with an IQ test, on average we expect a score of 100. That does not mean that all the people will score 100. If we read a report that claimed that every single member of the group tested at 100 exactly, we would have good reason to doubt the report. The result reported is too good. While the group tested may be ordinary, they will not all have exactly the average score of 100. We would expect the scores to be spread around 100. The way they are spread would also be characteristically regular. On scaling we ought to find 68% in the 85-115 range and 95% in the 70-130 range.

Burt's data was too good in exactly this way. We may have to guess how it came to be that way. But no reading allows us to continue to take Burt's data seriously. At the kindest, we might just conclude that Burt did the studies, but he was sloppy in reporting. There is evidence of incomplete reporting of how his data was collected and processed. Perhaps also he was too lazy to calculate his figures, reporting instead what he expected them to be. Or, less generously, we may suspect a massive fraud. How the defects of Burt's data arose continues to be a subject of debate. (See, for example, N. J. Mackintosh ed.), *Cyril Burt: Fraud or Framed?* Oxford Univ. Press, 1995.)

Examples of Anomalies

(1) Correlation coefficients reported for twin studies.

Many remain exactly the same as the number of twins under study increases. This is virtually impossible for real data.

Kamin's case:

IQ as measured in group IQ test.

Monozygotic twins reared apart.

Report of 1955: 21 pairs Correlation 0.771 → Report of 1958: "Over 30" pairs Correlation 0.771 → Report of 1966: 53 pairs Correlation 0.771

Because the study collects random occurrences, we expect the correlation to fluctuate as the number of pairs is increased.

Jensen's *op. cit.* 1974 review of Burt's data.

165 correlation coefficients reported with number of pairs in study. ⇨ 20 suspicious instances of the same correlation coefficient to 3 decimal places.

(But not all are independent, since some are based on repeated data.)

As summarized in A. R. Jensen, "Sir Cyril Burt in Perspective," *American Psychologist*, 33 (1978), 499-503

The anomalies in Burt's data are a little more complicated that just giving everyone an IQ of 100. But they are only just slightly more complicated, so that we may wonder why they were not noticed earlier.

The first anomaly to draw attention was pointed out by a psychologist, Leon Kamin, in a colloquium at the University of Pennsylvania on September 19, 1972. In a series of reports, Burt had increased the number of identical twins separated at birth under study, presumably as a result of further searching for such twins. Yet the correlation between the intelligence of the twins stays the same as the group grew in size. It did not just stay roughly the same. That would be quite all right. If the correlation measures some stable relationship, then altering the group studied ought not to alter the correlation coefficient that much. The problem was that it stayed *exactly* the same. The result on the group IQ test held constant at 0.771—no change at all in the three decimal places reported.
(Notice that the group IQ test is just one component of the figures used to determine the final IQ assessed, which is the .87 mentioned earlier. For the reports of 1955, 1958 and 1966, these final assessments were also nearly constant: 0.876, 0.876 and 0.874 respectively.)

This constancy is extremely unlikely to happen in real data. The studies treat random occurrences. Each correlation coefficient reported is only an estimate of the true degree of correlation and is very unlikely to coincide exactly with it. As the group changes in size, we would expect small fluctuations in the quantity we compute as an estimate. In an analogous case, we might try to find the average intelligence of people in the US by measuring the IQ of 10 people. We expect an average result around 100 in the sample group, but not exactly 100. It may be 95 or 103. If we add 5 more people, we expect the average of the group to remain around 100, but not to be exactly what we found in the smaller group.

Kamin's anomaly was not the only one found. Even Burt's sympathizers had to agree that something was seriously amiss with the correlation coefficients Burt reported. Arthur Jensen was a student of Hans Eysenck, who was a student of Cyril Burt. Jensen insisted there was no fraud in Burt's work, only sloppiness. Nonetheless, Jensen's later review (*op. cit.* 1974) of Burt's data revealed 20 suspicious instances of constant correlations coefficients out of 165 correlation coefficients reported.

Why we expect the correlation coefficient to fluctuate as sample size increases:

Earlier example:

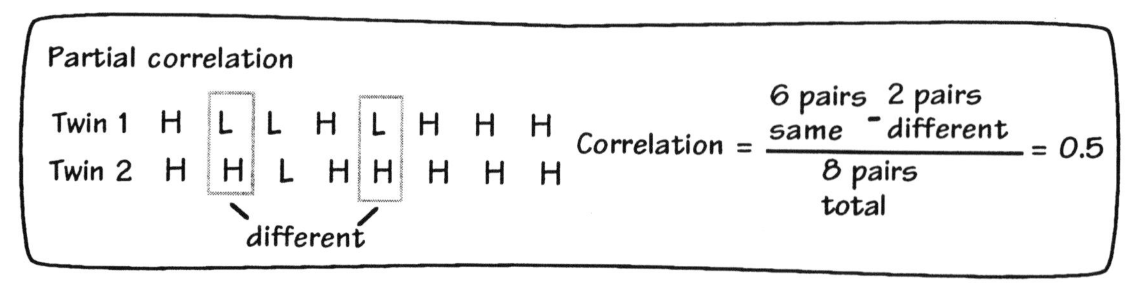

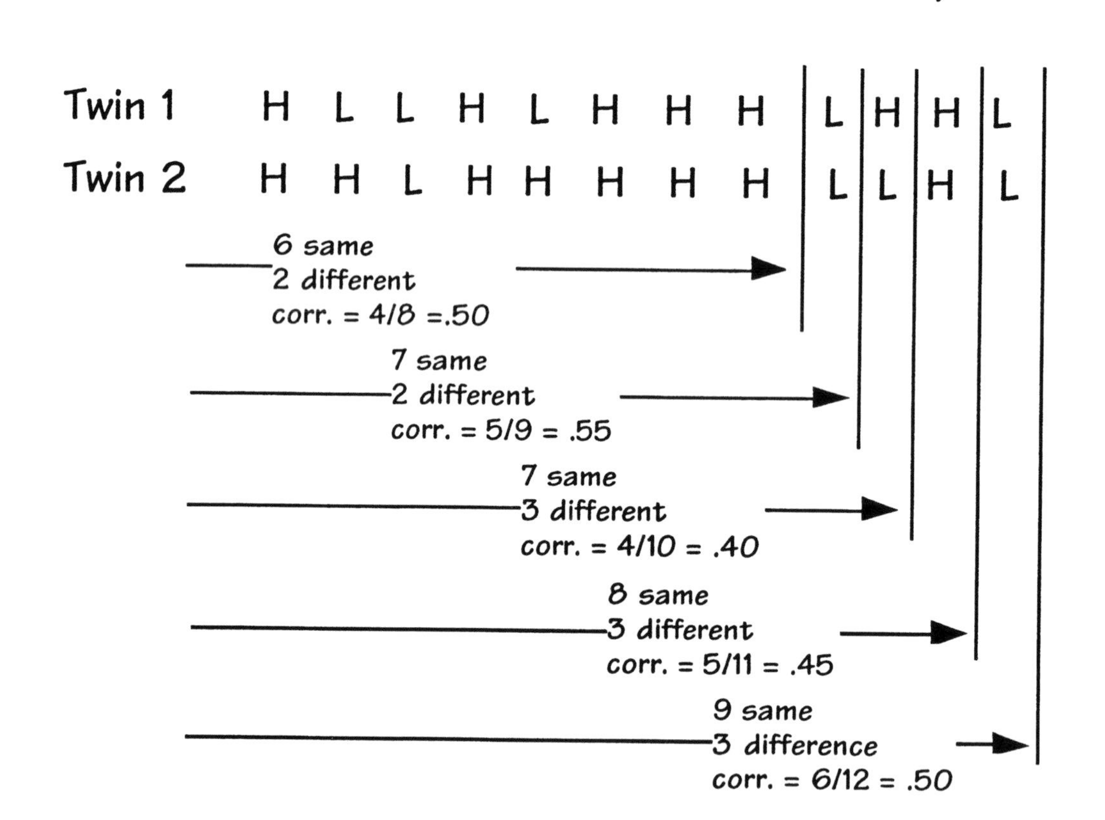

We can see very quickly why the correlation coefficient ought to alter as Burt adds to the number of twins in his study. Consider the example of partial correlation discussed earlier. In our 8 pairs of twins, we had 6 cases of agreement and 2 of disagreement, yielding a correlation coefficient of 0.5.

Now imagine that we add more pairs while trying to keep the correlation coefficient constant at 0.5. Each new pair that we add is a pair in which the intelligences either agree or disagree. Each adds 1 or subtracts 1 from the total score used to compute the correlation coefficient. Thus there is no way to add individual pairs without the correlation coefficient altering in value as new pair is added. The calculation shows one such attempt. Only after 4 new pairs are added are we able to bring the correlation coefficient back to 0.5.

Study of the psychology of intelligence developed hand in hand with new and powerful statistical techniques tailored to solving the problems peculiar to the study. Burt himself was an expert in statistical analysis. Therefore it is incomprehensible that Burt failed to see the significance of his perfect data and that its perfection immediately and irrevocably tainted his work.

(II) Burt's 1962 study: Fathers and children were spread over the range of IQ's **exactly** according to the expectations of the normal distribution.

Statistical theory tells us it is very UNLIKELY to get a perfect spread!

This suggests the data was invented or "corrected" to fit to normal distribution.

In real data we do not expect to get EXACTLY these perfect figures.
(68% in 85-115)
(95% in 70-130)
We expect figures to be close but not perfect e.g. 67%, 69.5% etc. is more realistic.

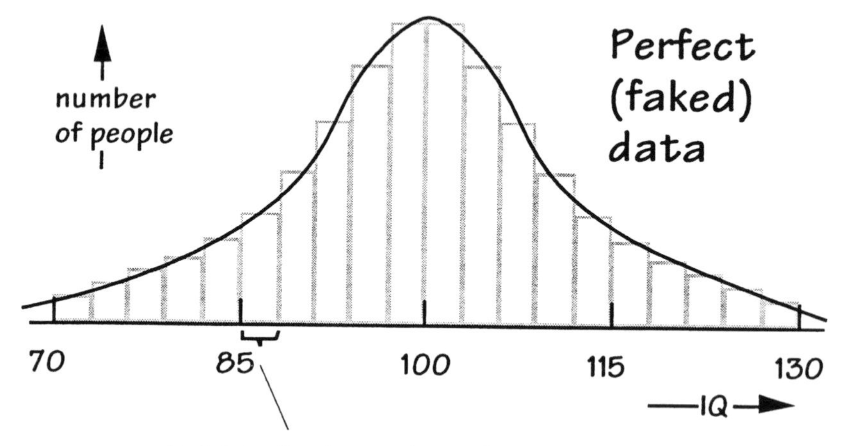

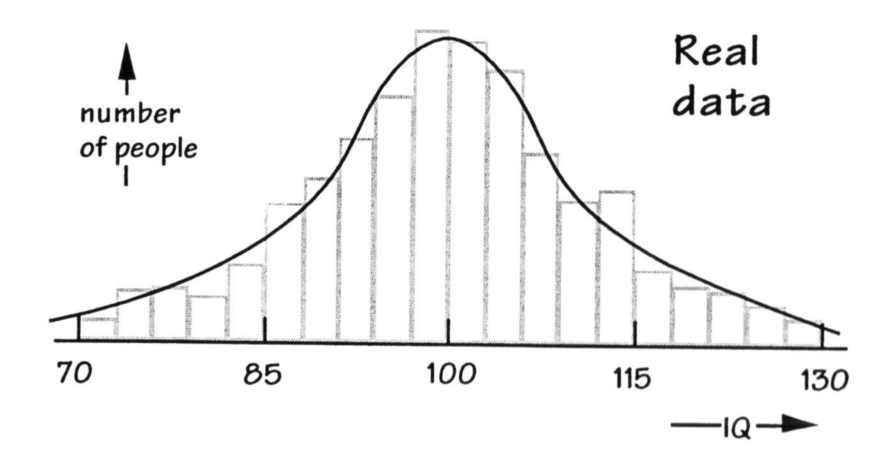

A second anomaly in Burt's statistics was pointed out by Dorfman, *op. cit.*, 1978. If we measure the IQs of a large group of people, we expect the average IQ to be 100 and the actual IQs to be spread over a range of values around 100. We have already said that they should be spread in a very regular way: 68% falling in the range 85-115, 95% falling in 70-130, and so on. Technically this is known as a "normal distribution." But this description of the spread is itself a description of what would happen on average were we to repeat the test on the whole group many times.

On one run, for example, we should not be surprised to find a little more than 68% falling in the 85-115 range; on another we may find less than 68%. What would raise suspicion is if we found exactly 68% in the 85-115 range every time—that is too perfect a distribution, just like finding an IQ of exactly 100 for every person in the group.

We can visualize this graphically. We imagine the number of people with IQ's of 70, 71, 72, ... represented by columns, with the height of the column set by the number of people in the group with an IQ at the indicated value. The first figure shows what perfect data would look like. It is impeccably regular with a bell shaped curve fitting it perfectly. These are data that match exactly the normal distribution. Real data will differ from it in various random ways as illustrated in the second figure. The columns will be irregular and the fit of the bell curve imperfect—although this fit will improve as the number of subjects in the group increases.

Burt's 1962 study of the IQs of fathers and children suffers exactly this defect. The IQs are spread normally—but perfectly so. In principle such perfect data can arise by chance. But it is fantastically unlikely, just as it is possible but very unlikely that everyone in a group should happen to have an IQ of exactly 100. We are left to conjecture how Burt arrived at such perfect data. That he fabricated the data is a very real possibility.

(III) Burt found perfect regression of child's IQ towards average.

Clarke and Clarke (1974) as reported by Dorfman, *op. cit.*, 1978.

Burt's rule predicted that a child's IQ regresses toward the average of 100 away from the parent's IQ.

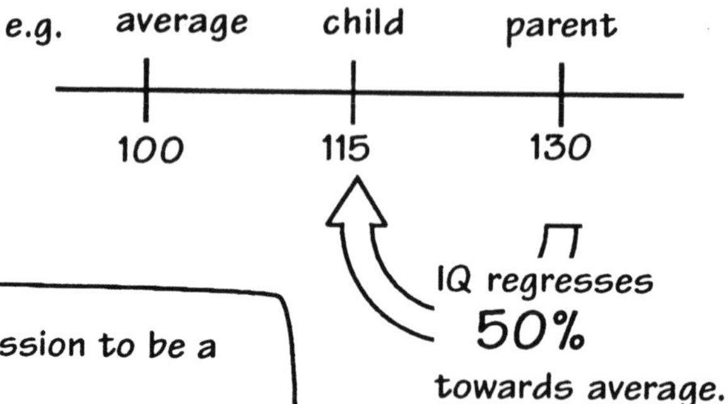

At best we expect this 50% regression to be a rough rule of thumb.

Burt's 1961 data has **exactly** 50% regression on every line with an error < 1% in every case.

Burt's 1961 data (again)

Class	Average IQ of parent	Average IQ of child	IQ of child based on assumption of 50% regression	Difference between reported and calculated IQ of child
I. Higher Professional	139.7	120.8	119.85	-0.8%
II. Lower Professional	130.6	114.7	115.3	+0.5%
III. Clerical	115.9	107.8	107.95	+0.1%
IV. Skilled	108.2	104.6	104.1	-0.5%
V. Semi-skilled	97.8	98.9	98.9	0%
VI. Unskilled	84.9	92.6	92.45	0.2%

N.B. Because of large number of pairs in study, Burt can estimate the average intelligences with an error of less than 1%.
What is suspicious is that the resulting estimates return precisely the figure of 50% regression in all cases.

+ MORE

The third anomaly that we will look at arises in the context of what is known as regression. Burt's studies reported that a child's IQ follows that of the parent. But the child's IQ does not exactly match the parent's. A parent with IQ of 130 would not on average have a child with IQ 130. Rather the IQ of the child would lie between the high value of 130 of the parent and the average value of 100. Similarly If the parent has a low IQ of, say, 70, then the child's IQ would be higher, on average, lying between the 70 and the average value of 100. Burt investigated a simple rule for governing this regression. The child's IQ would on average have regressed 50% towards 100. In the case of the parent with IQ 130, the parent is 30 points above 100. Under the regression rule, the child will, on average, be just 50% of this (=15 points) above 100. That is, the child will, on average, be at an IQ of 115.

Burt's data once again proved to be untrustworthy, not because it was incompatible with this regression. Rather, it incorporated the regression all too perfectly. The table shows the average IQs from Burt's data for parent and child in the various classes. The children's IQs have regressed 50% towards the average from the parents' in each of the class groups--they have regressed by EXACTLY 50%. The columns added on the right of the table show what the children's IQs would be if we just assumed the 50% regression rule exactly correct and calculated the children's IQs from the parents'. They differ from the numbers Burt reported by less than 1%.

What raises suspicion is that Burt should find the 50% regression so perfectly. The rule is more a rule of thumb than an exact law. What we should really expect is that the children's IQs would be roughly half way regressed towards the average, not exactly 50%. The fact that Burt's results differ from the calculated results in all classes by an error in the range 0-1% suggest that these errors may have arisen as a round-off error in calculating figures of size 100 when a decimal was rounded off at some point in the calculation.

These three anomalies do not exhaust the catalog of statistical anomalies in Burt's data. They do suffice to show just how flawed his data was and how easy it is to see the flaws once one looks.

Morals

- Science seeks to know extraordinary things about the world.

  How could we know
  if intelligence is inherited?

- They can be known by ordinary means, if we approach the problems with diligence and ingenuity.

  Studies with identical twins if done properly should allow us to disentangle inherited and environmental contributions.

- Science is a human endeavor and prone to all human weaknesses.

  Scientists like all humans can be led astray by a failure of critical judgment. But Burt was caught.

- Statistical methods are very powerful.

  What other method could reveal potential fraud solely by looking at perfect data?

This review of Cyril Burt's work in the inheritance of intelligence illustrates some very important themes which will be pursued. Science seeks to know so many things about the world that, at first glance, seemed buried beyond recovery in the chaos of experience. Yet there are ways to recover them. We can disentangle the different contributions of environment and inheritance to intelligence once we see that nature has kindly provided us just the opportunity we need through identical twins. That Burt bungled these studies does not mean that the method itself is defective. In the hands of a more honest or more careful researcher it ought to yield reliable results—assuming the other problems of the research can be resolved.

The more subtle, but, in the long run, equally important moral is to do with statistics. One may think that when there is randomness in a system we will be unable to infer much about it. Does not randomness obliterate all patterns? Statistical analysis exploits the fact that this is not so. Even in systems with randomness, the variations occur with a regularity that can be measured and exploited. Precisely this proved to be the undoing of Burt. The random variation of his data failed to exhibit the regularities we expect in statistical analysis. Even though the results were perfect in a non-statistical sense and exhibited precisely the regularities Burt claimed, we can be sure that they are spurious.

In this example one senses the immense power of statistical methods. What other tool can reveal data as decisively flawed even though the data seems perfect?

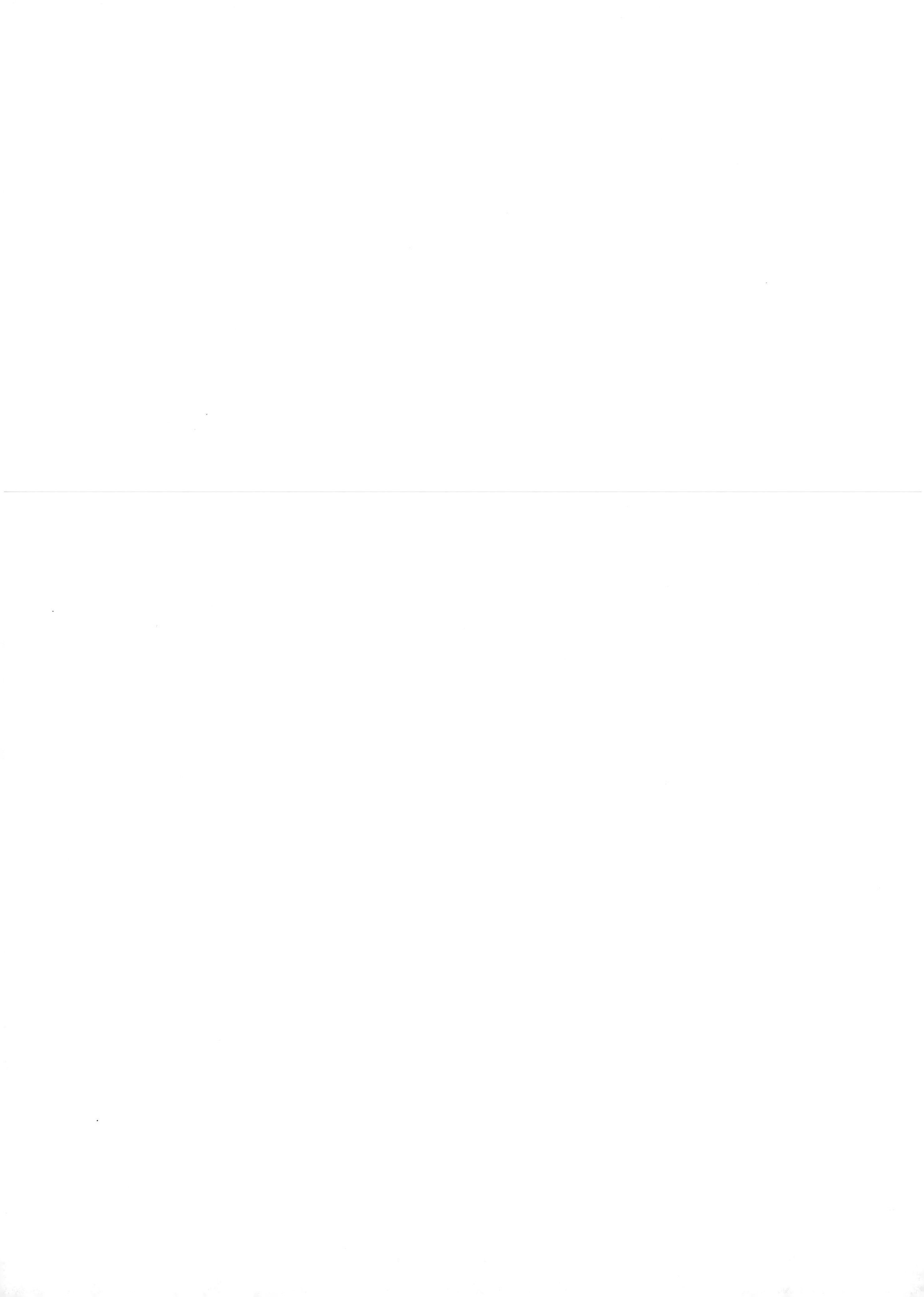

Assignment 1 Sir Cyril Burt: What Went Right and What Went Wrong

1. Burt's investigation of the inheritance of intelligence by means of twin studies was an ingenious solution to a difficult problem. To get a sense of how clever it was, you need to develop a sense of why the obvious, simple approaches are unreliable. A simple survey of the people you know and their children would be an unreliable way to decide if intelligence is inherited. To get a sense of why more elaborate methods are needed, consider the following simpler example.

To what extent do people resemble their dogs?

Your assignment is to observe one or more dog owners and compare them with their dogs. If you find the sort of resemblance in question, you will know it--it is often so striking as to be quite funny. Bouncy, energetic people often have bouncy energetic dogs. Prim and neat people often have prim and neat dogs.

(a) Enter your observation into the table attached for submission. Consider both behavior and appearance.

While we can all find cases where the resemblance is very striking, is it a real effect or just something we have talked ourselves into? If there were no systematic resemblance of owner and dog and we sift through enough examples, we would eventually find some striking cases of resemblance, purely now as an accident. In real life, there are, of course, many cases where the resemblance fails. How do we know that the remainder are not just accidental resemblances?

(b) How could we check that the resemblance are part of systematic regularity and not merely accidents?

2. Burt was caught because his data were too good. Data that is *too* good--that is an odd idea! The following exercise will help you see how data can be too good. Consider the tossing of a coin. We know that on average it will come up heads ("H") one in two times and tails ("T") one in two times. So if I throw it ten times. I expect it to come up H in*about* five of ten throws . The *about* is the crucial qualification. In fact, I will get exactly 5H only about 25% of the time. The remainder will be 4H, 6H, *etc.*

Take any coin and toss it ten times recording how many H's and T's there were. Do it ten more times and ten more times again, recording results in the table attached. (You might find it faster to take ten pennies, shake them *well* --perhaps in a large jar so they can turn over many times--and toss them out.)

(a) Record your results in the table attached.

Think of the sequence of three runs of ten throws as a long experiment designed to estimate whether the coin being tossed is fair. A fair coin ought not to favor H or T. But that doesn't mean that *exactly* half of all outcomes are H or T. The chances that you will get 5 H out of 10, then 10 H out of 20 and finally 15 H out of 30 are very small--about 1.5%

(b) If someone reports the outcome of this experiment and tells you they got exactly one half H at each stage (5H/10 and 10H/20 and 15H/30), what would you suspect?

(c) How is this connected with the suspicious "invariant" correlations reported by Burt, where he increased the number of twins in the trials but the correlations stayed exactly the same.

Assignment 1

Name:______________

1(a).

	Degree of similarity in behavior	Degree of similarity in appearance	Comments
First human/dog pair			
Second human/dog pair			
Third human/dog pair			
Fourth human/dog pair			

2. (a)

First ten throws	number H's out of 10 =	
Second ten throws	number H's out of 10 =	cumulative number H's out of 20 =
Third ten throws	number H's out of 10 =	cumulative number H's out of 30 =

(Don't forget to submit answers to 1(b) and 2 (b), (c) as well.)

Assignment 1 Name:_______________

1(a).

	Degree of similarity in behavior	Degree of similarity in appearance	Comments
First human/dog pair			
Second human/dog pair			
Third human/dog pair			
Fourth human/dog pair			

2. (a)

First ten throws	number H's out of 10 =	
Second ten throws	number H's out of 10 =	cumulative number H's out of 20 =
Third ten throws	number H's out of 10 =	cumulative number H's out of 30 =

(Don't forget to submit answers to 1(b) and 2 (b), (c) as well.)

Chapter 2

To Err is Human: Scientists are Human

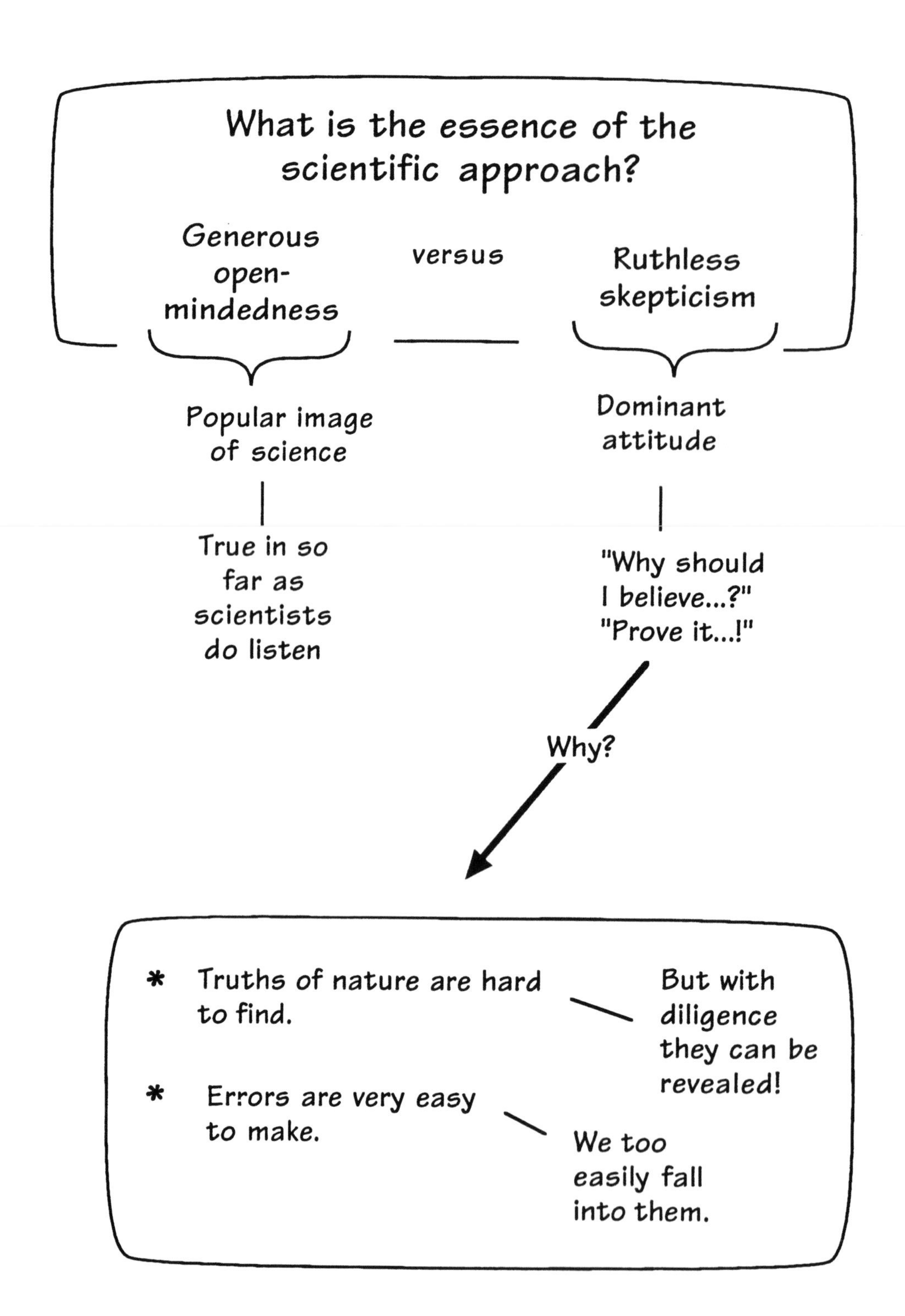
What is the essence of the scientific approach?
Generous open-mindedness
versus
Ruthless skepticism
Popular image of science
Dominant attitude
True in so far as scientists do listen
"Why should I believe...?"
"Prove it...!"
Why?
* Truths of nature are hard to find.
But with diligence they can be revealed!
* Errors are very easy to make.
We too easily fall into them.

In tackling many problems, we feel the need to adopt a "scientific approach" for we expect that this approach will yield the most reliable results. But what is this approach? At its core, what do we do when we adopt a scientific approach?

There is a popular image, fostered by movies, of the scientist as a benevolent misfit, open to all new ideas, no matter how bizarre. There is some truth in this image. Good scientists cannot afford to ignore new ideas. But a generous open-mindedness is not and cannot be the dominant turn of mind of scientists. Most new ideas that seem outrageous are just that--outrageous fantasies. Very, very few turn out to be sound. While a scientist may listen to new ideas, a scientist cannot afford to accept every new idea that floats past. Otherwise the scientist's understanding of nature would become thoroughly contaminated with errors and fantasies.

To guard against this problem, scientists adopt a rather different stance: ruthless skepticism. When confronting new ideas--even those originating with the scientist him or herself--the questions must always be "Why should I believe this new proposal?", "What evidence is there for this proposal?", "Can you prove it to me?" Nothing can be believed until a good case has been made for it. And if that case begins to fragment, the scientist must be prepared to reject the proposal.

There are excellent reasons for this type of skepticism. The truths of nature are hard to find--but they can be found if we are prepared to search diligently for them. In this search, we have learned, time and time again, that errors are very easy to make. We protect ourselves from them by demanding that nothing be accepted unless we have strong evidence for it. Until then our new proposal must be regarded with suspicion. We may remind ourselves of this suspicion by labeling the proposal an "hypothesis"--something that has not yet been confirmed sufficiently to enter the body of accepted science.

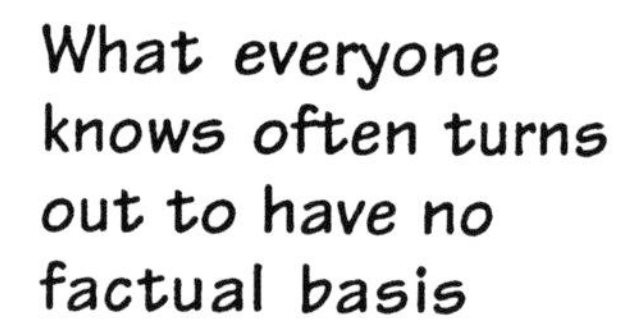

Skepticism is essential in investigating nature.

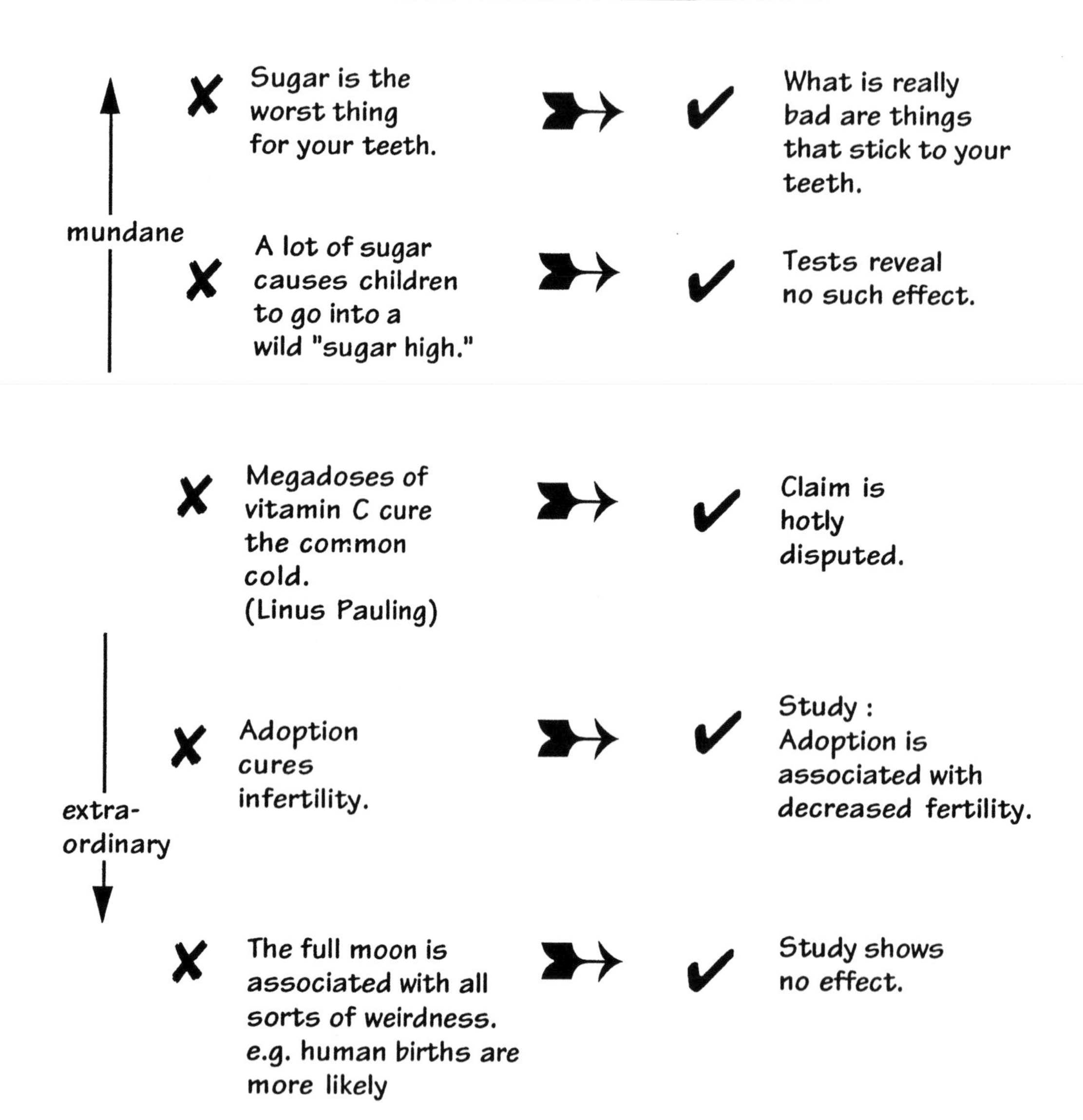

To get a sense of just how easily erroneous belief can gain currency, consider this collection of popular misconceptions that span from the mundane to the extraordinary.

Parents have long felt the need to protect their children from sugar. But it turns out that sugar is far from the worst of villains when it comes to tooth decay. The worst foods are sticky foods that adhere to the teeth and allow acids to accumulate on the tooth's surface.

Giving children sugar does not lead to a sugar high with wild behavior. In a controlled study on children identified by their parents as sensitive to sugar, the behavior was the same if the children were given a sugar rich food or one without sugar but artificially sweetened. Perhaps the wild behavior derives from the exciting circumstances surrounding the eating of very sweet food--a rewarding treat. (See M.L.Wolraich et al., *New England Journal of Medicine*, 330 (1994), 301-7)

Linus Pauling, a Nobel Laureate in both Chemistry and Peace, has long promoted the idea that huge doses of vitamin C either cure or alleviate the common cold. It turns out the suggestion has dubious foundation and that Pauling himself is regarded as a crank by the mainstream medical community. (See S. Barrett, "The Dark Side of Linus Pauling's Legacy," *Skeptical Inquirer*, Vol. 19 (Jan./Feb. 1995), pp.18-20.)

Many of us have had the striking experience of watching an infertile couple adopt a child and then conceive a child. As a result, many believe that adopting a child can increase the chance of subsequent fertility. Studies actually show the reverse! Couples who adopt are more likely to remain infertile subsequent to the adoption than other infertile couples. There is no real mystery in this. Couples with the least hope of subsequent fertility are more likely to adopt a child! (See E.J.Lamb and S. Leurgrans, "Does Adoption affect subsequent fertility," *Am. J. Obstet. Gynecol.* 134 (1979), 138-44.)

The eerie spectacle of the luminous full moon hanging ominously in the sky has led to all sorts of speculation of mysterious lunar influences. One story popular in maternity wards is that a full moon makes births more likely. Examination of hospital records reveals this is a myth. We can understand how it comes about, however. If there is a hectic night in the maternity ward and no one notices a full moon, the experience is forgotten. However, if, by chance, it is the night of the full moon and this is noticed, it becomes a striking confirmation of the belief and long remembered! (See G. O. Abell and B. Greenspan, *New England Journal of Medicine*,Jan. 11, 1979, p. 96)

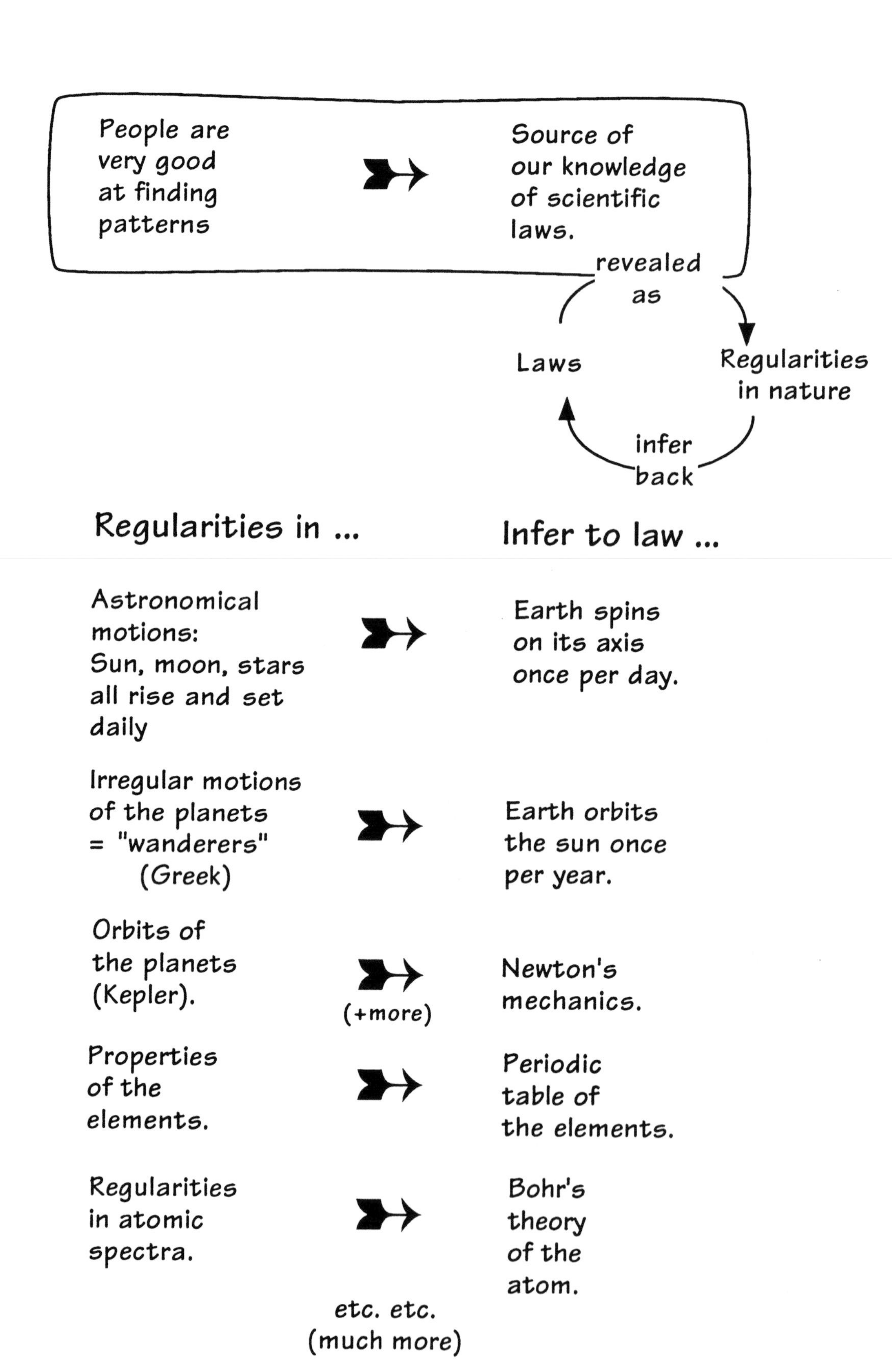
People are very good at finding patterns
Source of our knowledge of scientific laws.
revealed as
Laws
Regularities in nature
infer back
Regularities in ...
Infer to law ...
Astronomical motions: Sun, moon, stars all rise and set daily
Earth spins on its axis once per day.
Irregular motions of the planets = "wanderers" (Greek)
Earth orbits the sun once per year.
Orbits of the planets (Kepler).
(+more)
Newton's mechanics.
Properties of the elements.
Periodic table of the elements.
Regularities in atomic spectra.
Bohr's theory of the atom.
etc. etc. (much more)

Human ingenuity may never exhaust the many ways that mistakes can be made. Fortunately some types of errors are more common than others. In particular there are a few types of errors that tend to be made over and over. If we can identify these common types of errors, then we can develop general strategies for avoiding them. In this chapter, we will describe two types of error commonly committed and in the next chapter we indicate the general strategies used to control these types of error.

The first type of error derives from a human ability that is extremely important in science. Experience confronts us with a bewildering array of sights, sounds and sensations. People are very good at looking within this confusion and finding patterns. This ability has proven, time and time again, to be the key to our discovery of central laws in all sciences. Opposite you will see a number of these listed. For example, from the simple observation that the sun, moon and stars move, not randomly but uniformly about us, it is possible to infer that the earth spins on its axis. More recently, it was noticed that there are distinctive patterns in the colors present in light emitted by excited atoms. These patterns provided the key to understanding how an atom is built from its constituent particles.

The details of these examples need not detain us here. Each requires and deserves considerable explanation. All that concerns us is that our ability to find patterns--even where there may first seem to be none--has proven to be of decisive importance in scientific discovery.

Problem We are too good at finding patterns!

We find patterns where there are none.

...and hold on to them tenaciously.

Examples

Crank Pyramidology

19th. century tradition finds all sorts of highly "significant" regularities in the size and proportion of the the Great Pyramid of Egypt.

John Taylor, *The Great Pyramid: Why Was It Built and Who Built It.* 1859

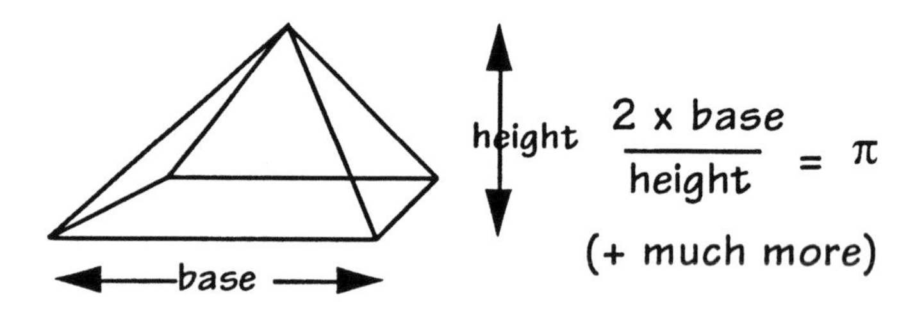

Charles Piazzi Smyth, Astronomer Royal of Scotland; Professor, University of Edinburgh

Our Inheritance in the Great Pyramid. 1864 (+more)

$$\frac{\text{Pyramid base}}{\text{Width casing stone}} = 365 \quad \text{number of days in a year}$$

"Pyramid inch" (Approx. modern inch) = $\frac{1}{25}$ width casing stone = $\frac{1}{10{,}000}$ Earth's polar radius

Height pyramid × 1,000,000 = Earth-Sun distance

... also find earth's mean density, period of precession of earth's orbit, mean temperature of earth's surface all encoded in pyramid dimensions!

The problem with our remarkable pattern finding abilities is that we can sometimes be far too good at finding patterns. We delude ourselves into thinking there are significant patterns when in fact there are none. People who fall into this trap can hold on to these spurious pattern with remarkable tenacity.

We see this pattern most acutely in crank literatures. We shall review some examples because they supply an especially vivid illustration of a very common type of error.

The nineteenth century fascination with the Egyptian pyramids spawned a literature devoted to showing that huge amounts of highly sophisticated scientific and other knowledge were encoded by the ancient Egyptians in the dimensions of the pyramids. The literature begins innocently enough. If one compares the base and height of the Great Pyramid, one can quickly recover a good approximation for π. This may have happened by chance; or it may be that the ancient Egyptians did have a sufficiently sophisticated knowledge of geometry to encode π into the dimensions of the pyramid.

Once one such result is found, the exercise reduces to systematic examination of all the dimensions in the hope of finding further significant results. Charles Piazzi Smyth was the undisputed champion of this enterprise. He measured the pyramid in units of the pyramid's casing stone and the "pyramid inch." He found an astonishing list of physical parameters of the earth encoded in the pyramid. Each new parameter indicates that the ancient Egyptians had a surprisingly modern grasp of another aspect of science. We may want to concede that ancient Egyptians did know more than we usually allow. But Smyth's list grows so rapidly and in such bizarre directions that it strains all belief.

As the list grows one becomes more and more convinced that there are no such parameters encoded within the pyramid by its builders. The results are just those of a fanatic determined to recover significant number patterns. Given the huge number of dimensions that can be compared within the pyramid, it is just a matter of patience before pure chance supplies numbers that can be paraded spuriously as significant.

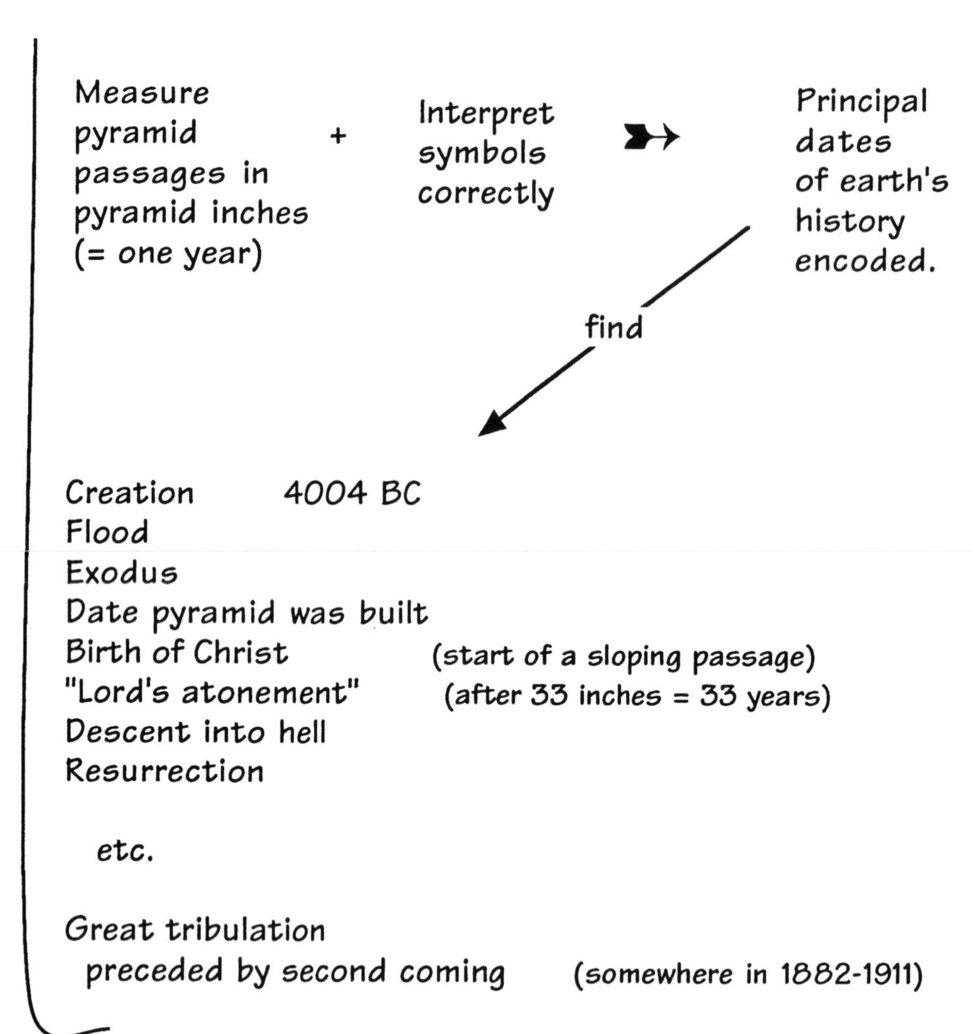
Measure pyramid passages in pyramid inches (= one year)
+
Interpret symbols correctly
Principal dates of earth's history encoded.
find
Creation 4004 BC
Flood
Exodus
Date pyramid was built
Birth of Christ
(start of a sloping passage)
"Lord's atonement"
(after 33 inches = 33 years)
Descent into hell
Resurrection
etc.
Great tribulation
preceded by second coming
(somewhere in 1882-1911)

Piazzi Smyth was prepared to carry his investigations to quite absurd lengths. He went so far as to claim that a certain passage in the pyramid had encoded within it the principal dates of biblical history and prophecy--as he saw them. How do we explain that Smyth just happens to be decoding a chronology prepared thousands of years before him and finds the prediction of the "Great tribulation" within a few decades of the time of his writing? It would seem that whoever prepared the chronology had a special obsession with the late 19th century! An amazing coincidence! Or is Piazzi Smyth just reading his own special interests into the pyramid?

M. Gardner, *Fads and Fallacies in the Name of Science*. Dover, 1957, Ch. 15.

Who wrote Shakespeare's plays?

A huge literature starting in 18th century is devoted to showing they were not written by William Shakespeare. The favorite alternative candidate is Francis Bacon.

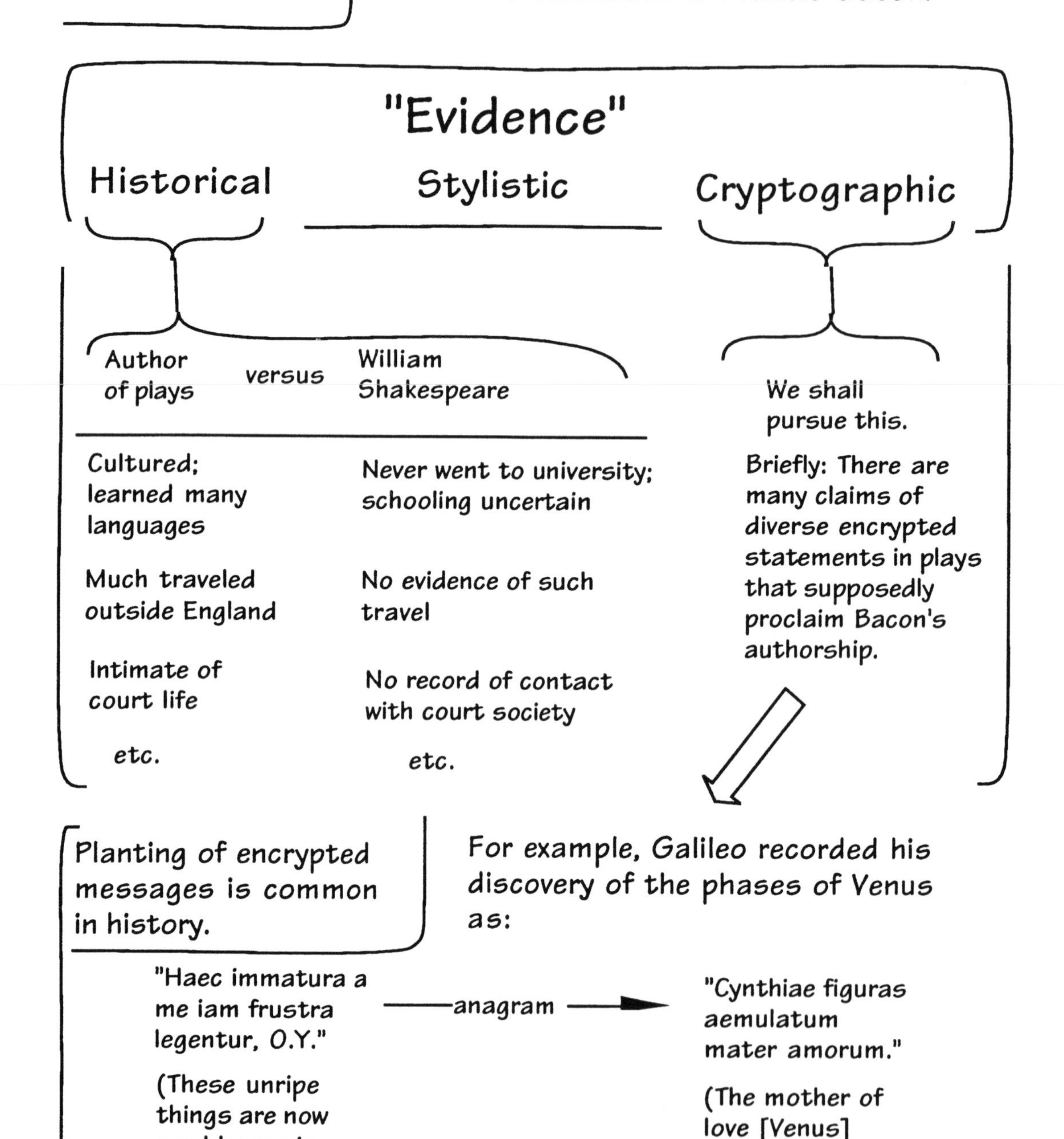

Planting of encrypted messages is common in history.

For example, Galileo recorded his discovery of the phases of Venus as:

"Haec immatura a me iam frustra legentur, O.Y." —anagram→ "Cynthiae figuras aemulatum mater amorum."

(These unripe things are now read by me in vain, O. Y.)

(The mother of love [Venus] imitates the phases of Cynthia [Moon].)

These extra letters indicate something more than meets the eye!

There is another example of our ability to find and defend entirely spurious patterns. It lies in a huge literature devoted to deciding who wrote Shakespeare's plays. It energetically defends the claim that Shakespeare was not the author of the celebrated plays. There are many other candidates. One outstrips them all: Francis Bacon, a notable philosopher, politician and writer. The literature on this question grew to an enormous size. A 1947 bibliography by Joseph S. Galland of Northwestern University--just a list of the works in the field--was 1500 pages long. Devotees formed learned societies with journals across the world, including: *Journal of Bacon Society* (England 1885), renamed in 1891 as *Baconiana*; the Bacon Society of America (1922) with its journal, *American Baconiana* (1930-32) and even a German journal, *Deutsche Baconiana: Zeitschrift für Bacon-Shakespeare Forschung* (Frankfurt 1930-32).

There is some room for discussion of these claims. Shakespeare did, apparently, co-author some plays without wide acknowledgement of the co-author. Also there are some anomalies between what we know of the author of the plays and what we do know about William Shakespeare. (Here we must be careful since we know very little about him.) The type of evidence that we will consider is cryptographic. The claim is that Francis Bacon, the true author of the plays, left encoded messages sprinkled throughout he plays announcing himself as the true author.

The use of secret messages was not uncommon in Shakespeare's time. When Galileo observed Venus through his telescope, he found that the planet went through phases just like the moon (new moon--half moon--full moon). In one of many examples of the use of secret messages, Galileo recorded this discovery in an anagram. That way his priority in the discovery could be assured without actually revealing what he had discovered! Someone who held the jumbled message would be very unlikely to figure out how to unscramble it. But once the unscrambled message is revealed its presence becomes clear.

The finding of these encoded messages is an exercise in pattern finding. We shall see that this literature has gone to extraordinary lengths to force patterns onto unwilling texts. Once one is determined to find a pattern, it can be wrestled into existence--although the outcome may be dramatically (!) implausible.

Earliest cryptographic "evidence" circa 1880

Especially Ignatius Donnelly, *The Great Cryptogram*

Based on exploiting the "biliteral" cypher described by Francis Bacon.

Text of message everyone sees....

More text of message everyone sees....

Switch between two fonts

Use switches in font to carry to real message in secret

But do it with two fonts that are more alike than these so it is not at all obvious unless you know to look for it!

ggggg → a

ggggg → c

ggggg → b

etc.

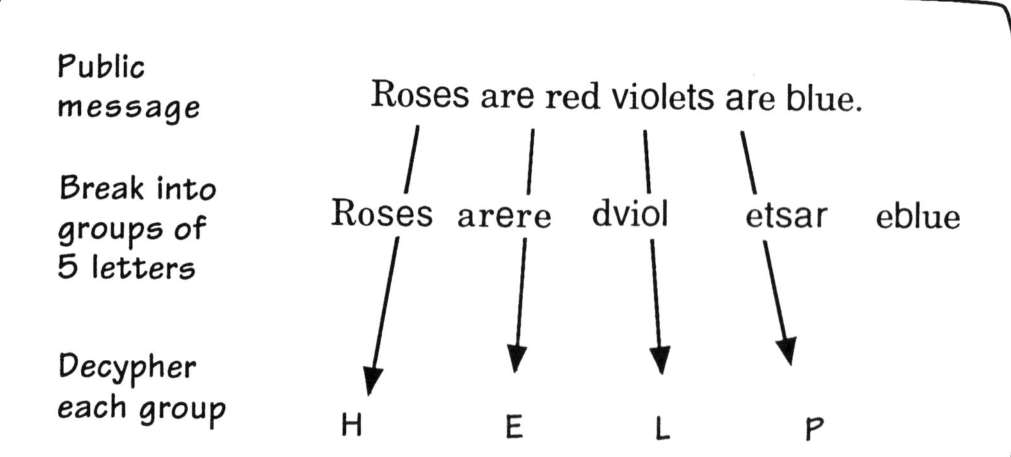

What makes Bacon an attractive candidate for the cryptographically inclined is that he had a strong interest in codes and cyphers. This interest derived from his political life in which secure, secret communication is very important.

Bacon described an ingenious "biliteral" cypher. In it, an apparently innocent piece of text is used to pass a secret message. The trick is to use different fonts to carry and hide the message. These two t's come from different fonts:

T T

The first is from a "serif" font with little feet on the letters; the second is from a "sans serif" font, which lacks the little feet. Different letters can be encoded by the pattern of switches between the two fonts. Overall 5 letters in the public message are needed to encode each letter of secret message. For example ggggg encyphers the letter "b" because it uses the pattern

1st font - 1st font - 1st font - 1st font - 2nd font

Since ggggg uses the same letters, the trickery is obvious. The same pattern of font changes is better hidden in normal text, such as:

Dear Mary

The first five letters have the same pattern of font changes. So they also encypher the letter "b".

In the examples shown, the presence of the secret message is still obvious since the two fonts used are very different. If the two fonts are similar, however, it may take an expert to pick them apart. If you were not alerted to the fact that two fonts have been used, would you notice that these two letters come from different fonts?

T T

Can you disentangle the mix in

Now is the winter of our discontent.

Can we find a message encoded in Shakespeare's writings or those associated with him?

No, all attempts that yield any result work by being fantastically contrived.

e.g. Hugh Black, *North American Review*, 1887

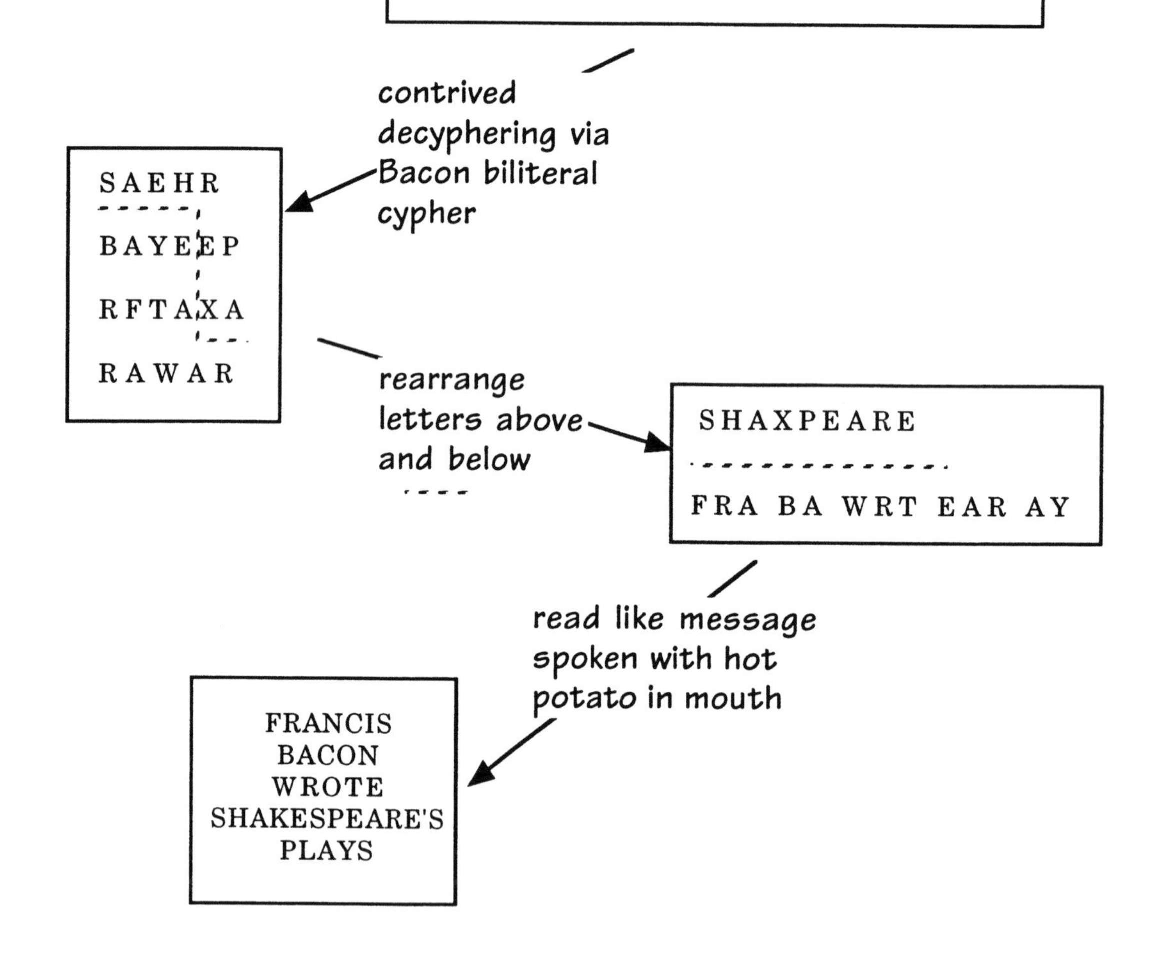

In the case of Galileo's anagram, when the correct unscrambling is revealed, we have little doubt that we have found a message hidden by the author. It is unlikely that the letters of a randomly chosen message can be rearranged perfectly to make a sensible message. It is all the more unlikely that the message recovered should describe one of the latest scientific discoveries--unless it was planned that way.

Attempts to find secret messages in Shakespeare's works or associated writing have not met with any comparable success. One text that attracted a lot of attention is an inscription beneath a bust of Shakespeare. The use of odd lettering looks promising, at first, since it calls to mind the Baconian biliteral cypher. But attempts to find the hidden message just do not work. One attempt is shown opposite. Huge liberties are taken in trying to recover the message. First a biliteral cypher is invented and used. The resulting letters are then scrambled. Even after all this, the resulting message is little more than garbled nonsense. But, we are told, it attributes the plays to Shakespeare if we read it as though we had hot potato in the mouth!

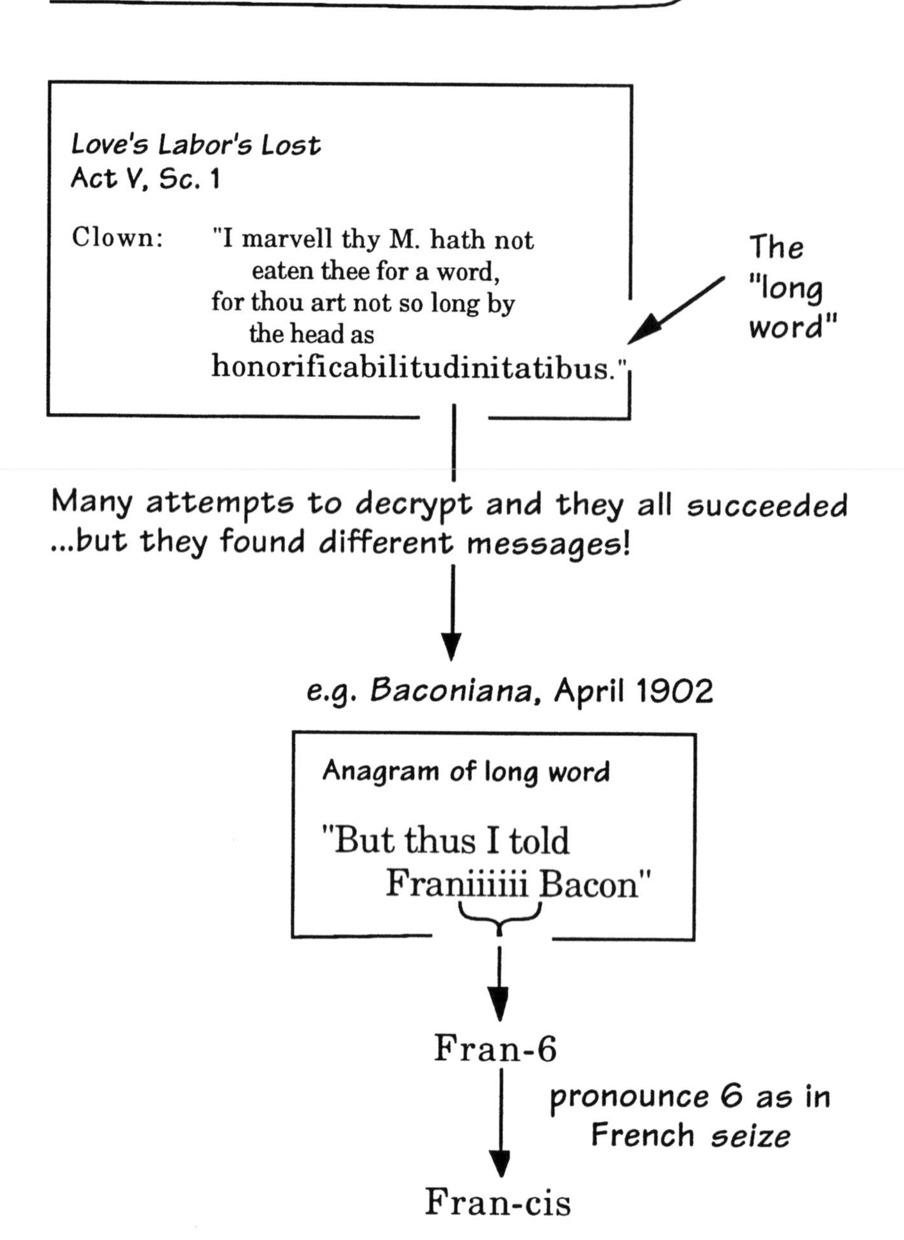
Attempts to decrypt the "long word"
Love's Labor's Lost
Act V, Sc. 1
Clown: "I marvell thy M. hath not eaten thee for a word, for thou art not so long by the head as honorificabilitudinitatibus."
The "long word"
Many attempts to decrypt and they all succeeded ...but they found different messages!
e.g. Baconiana, April 1902
Anagram of long word
"But thus I told Franiiiiii Bacon"
Fran-6
pronounce 6 as in French seize
Fran-cis

Another popular target of this literature is the "long word"--honorificabilitudinitatibus--which appears as a nonsense word in *Love's Labor's Lost*. There have been no shortage of messages found hidden in the long word. But since they are all different messages, it is hard to take any seriously. Many of the messages do not make much sense and can only be recovered if all sorts of odd liberties are allowed. See one example opposite. The anagram of the long word is only able to produce "Francis" after a very odd maneuver on "iiiiii" and the final message is hardy a decisive claim of authorship.

In any case, it is inconceivable that the long word could encrypt Bacon's authorship of the plays. It has been shown that the long word was already a popular nonsense word of the time. It had appeared in print as early as 1460--well before Shakespeare.

W. F. Friedman and E. S. Friedman, *The Shakespearean Ciphers Examined.* Cambridge U. P., 1957.

A simple scheme for hiding secret messages:

THERE IS NO MESSAGE
ENCLOSED HERE PRIVATELY!

Look at every ninth letter after first H.

tHEREISNOMESSAGEENCLOSEDHEREPRIVATELY!

Put the text in a table to make identification easier.

T	**H**	E	R	E	I	S	N	O
M	**E**	S	S	A	G	E	E	N
C	**L**	O	S	E	D	H	E	R
E	**P**	R	I	V	A	T	E	L
Y	!							

Witztum, Rips, Rosenberg; Michael Drosnin: *The Bible Code:* The same technique reveals hidden messages in Hebrew letters in the Hebrew bible.

Mock example using English letters of King James bible by skeptic David Thomas.

S	A	N	D	M	Y	D	A	U	G	H	T	E	R	S	T	H	O	U	H
M	Y	D	A	U	G	H	T	E	**R**	S	T	H	O	U	H	A	S	T	N
U	G	H	T	E	R	S	T	H	**O**	**U**	H	A	S	T	N	O	W	D	O
E	R	S	T	H	O	U	H	A	**S**	T	N	O	W	D	O	N	E	F	O
H	O	U	H	A	S	T	N	O	**W**	D	O	N	E	F	O	O	L	I	S
A	S	T	N	O	W	D	O	N	**E**	**F**	O	O	L	I	S	H	L	Y	I
O	W	D	O	N	E	F	O	O	**L**	I	S	H	Y	L	I	N	S	O	D
N	E	F	O	O	L	I	S	H	**L**	Y	I	N	S	O	D	O	I	N	G
O	L	I	S	H	L	Y	I	N	S	**O**	D	O	I	N	G	I	T	I	S
H	L	Y	I	N	S	O	D	O	I	N	G	I	T	I	S	I	N	T	H

Efforts to find secret messages and evidence of the "true" author of Shakespeare's plays continue today. There is a yet older tradition of secret message seeking that has flourished recently. This is the tradition of seeking hidden messages in the bible.

The technique now proposed as used for hiding these messages is extremely simple. We take a text such as "There is no message enclosed here privately!" We start at some letter and look at letters that follow at uniform distance. For example, we start with the first H and look at letters and characters 9 positions along. We recover the secret message "help!" What greatly simplifies the quest for hidden messages is to lay out the text in a table. Then the secret messages can be read off from the table's columns as shown opposite.

The recent attempts to find messages encoded in this way in the bible were brought to wide public notice with Michael Drosnin's *New York Times* bestseller, *The Bible Code* (Simon and Schuster, 1997). He in turn drew on the efforts of Doron Witztum, Eliyahu Rips and Yoav Rosenberg, " Equidistant Letter Sequences in the Book of Genesis," *Statistical Science*, 9, 1994, pp.429-439.

Drosnin's cases all dealt with the bible passages in their original Hebrew. He was able to report many striking messages as visible once one laid the text out in a table, suppressing all spaces and punctuation. For example, the analysis reveals the name of the former Israeli prime minister, Yitzhak Rabin, vertically in a table and it is crossed horizontally and ominously by the Hebrew for "assassin that will assassinate." This appears to be an astonishing, ancient prediction of Rabin's assassination 1995.

Shown opposite is the extraction of a hidden message from the English translation of Genesis in the King James version of the bible. This example is presented since most readers of this text cannot read Hebrew. The message was found by skeptic David Thomas ("Hidden Messages and the Bible Code," *Skeptical Inquirer*, Nov./Dec. 1997, pp.30-36.) It alludes to a 1947 incident at Roswell, New Mexico, that UFO buffs believe to be the crashing of an alien flying saucer. (The incident is more likely merely to be the crashing of balloon supported radar reflectors within the government's secret "Project Mogul.")

Finding messages does not assure us they are placed there by design and not merely by chance:

Enough text to look in. + Willingness to use very large step sizes (in thousands) + Computer to conduct search. = Guarantee of finding some striking message.

Skeptic David Thomas set out to find "Nazi" and "Hitler" connected in the English translation of *War and Peace* and he found it:

W I T H Q U I C K S H O R T S W A Y I N G S T E P S H E
E S S S A T D O W **N** O N A S O F A N E A R T H E S I L V
H E R S E L F A N D T O A L L A R O U N D H E R I H A V
R B A G A N D A D D R E S S I N G A L L P R E S E N T M
C K O N M E S H E **A** D D E D T U R N I N G T O H E R H O
P T I O N A N D J U S T S E E H O W B A D L Y I A M D R
W A I S T E D L A C E T R I M M E D D A I N T Y G R A Y
R E A S T S O Y E **Z** T R A N Q U I L L E L I S E Y O U W
N N A P A V L O V N A Y O U K N O W S A I D T H E P R I
T U R N I N G T O A G E N E R A L M Y H U S B A N D I S
L M E W H A T T H **I** S W R E T C H E D W A R I S F O R S
I T I N G F O R A N A N S W E R S H E T U R N E D T O S
E L I G H T F U L W O M A N T H I S L I T T L E P R I N
T H E N E X T A R R I V A L S W A S A S T O U T H E A V
A C L E S T **H** E L **I** G H **T** C O **L** O R **E** D B **R** E E C H E S
R O W N D R E S S C O A T T H I S S T O U T Y O U N G M
L K N O W N G R A N D E E O F C A T H E R I N E S T I M

BUT

Witztum, Rips and Rosenberg: sound statistical analysis proves the biblical mesages are more than chance alignment of letters.

Names are coordinated with correct dates far more often than chance allows.

The volume of messages recoverable from the bible is astonishing. "...every major figure, every major event in world history can be found..." writes Drosnin (p.46). They include "Roosevelt," "Churchill," "Stalin" and "Hitler"; "America," "revolution" and 1776 encoded together; "Napoleon," "Waterloo" and "Elba"; "revolution" and "Russia" encoded with 1917; "Shakespeare"--"presented on stage"--"Hamlet"--"Macbeth" encoded in one sequence; "Wright brothers" with "airplane"; even Newton and Einstein are named with details of their discoveries; and so on in unyielding flood.

This flood by itself proves very little. The techniques employed are extremely powerful. The message seekers are willing to look at letter spacings of not just 5 or 10 letters, but hundreds and thousands. The name Yitzhak Rabin was found with letters 4,772 spaces apart. They use computers to search though the text checking a huge array of possibilities. In a text as large as the bible, we are assured that eventually this technique must recover a purely chance alignment of letters that looks like a secret message. There is little doubt; one just needs enough patience to check enough possibilities in enough text.

The inevitablility of success has been demonstrated by skeptics who have succeeded in finding comparable hidden messages in texts which no one supposes harbors them by design. Searching the Supreme Court's 1987 *Edwards* ruling on creationism, David Thomas (*op.cit.*) found thousands of hidden names, including politicians, scientists and talk show hosts! Other texts, such as *War and Peace* and *Moby Dick*, harbor more messages, including paired words (see opposite).

There is a small complication. While such messages are assured in all texts, Witztum *et al.* (*op.cit.*) presented a simple proof that the bible's messages were more than mere chance alignment of letters. They compared how often names of designated personalities were encoded with appropriate dates. They claimed to find that each name was far more likely to be associated with the correct date than would be the case if the names and dates appeared by chance. The study was published in a refereed statistics journal. However their claim did not survive scrutiny. Later examination of their study found sufficient deficiencies to nullify the result claimed. (See M. Bar-Hillel *et al.*, *Chance*, Vol.11, 1998, pp.13-19; B. McKay, *Statistical Science*, May 1999, forthcoming.)

We are incorrigible pattern finders.

We find them even when we know they are not there.

e. g. I threw a die 6 times to get

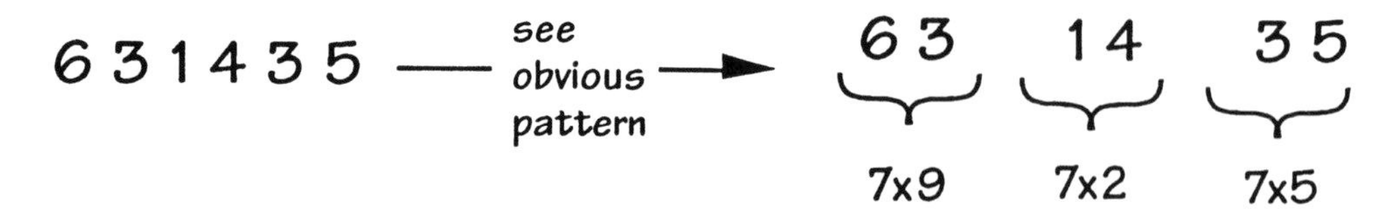

Could this happen by chance?
NO

Chance all three pairs divisible by 7 = Chance first pair divisible by 7 x Chance second pair divisible by 7 x Chance third pair divisible by 7

$1/216 = 1/6 \times 1/6 \times 1/6$

Would anyone do this seriously?

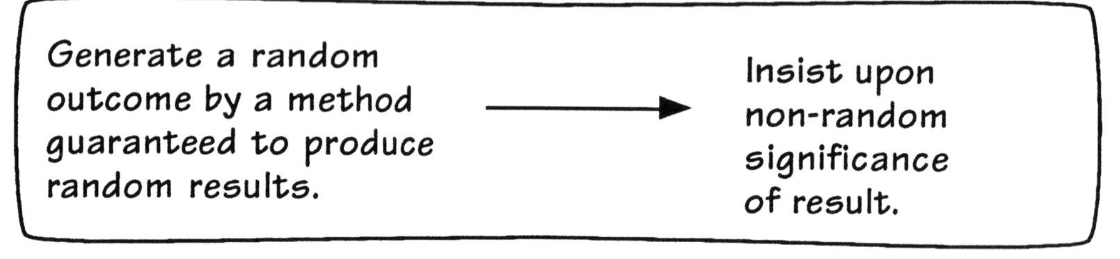

YES

Tarot cards: Subject sees cards shuffled: Order guaranteed random → Then told order is significant

I Ching: Subject sees sticks scattered: Arrangement guaranteed random → Then told arrangement is significant

Also tea leaves, Jung's synchronicity

It is hard not to chuckle at the obsession of the pattern finders we have seen so far. However we should not dismiss their obsession as something we are free from. Unless we are vigilant, we too can walk the same path. Here is a little experiment:

I threw a die six times the die showed 6,3,1,4,3,5. You will immediately see the pattern. Each pair of numbers just happens to be divisible by seven. Now that is striking! Could this happen by mere chance? A quick calculation suggests that it is unlikely to happen by chance. There is a one in six chance that any pair of digits thrown by the dice is divisible by seven. (Of the 36 possible combinations, there are six divisible by seven: 14, 21, 35, 42, 56, 63.) So overall there is a one in 216 chance that the outcomes of the die throws have the pattern they show. It is very unlikely that this pattern would arise by chance!

Notice what we have done in the last paragraph. We started with a process *guaranteed* to produce a random outcome. (I really did throw a die six times!) We then convinced ourselves that the pattern could not have arisen merely by chance. We are drawn into the conclusion that there is a significant pattern there--although we may well balk at the task of saying just what it signifies!

Here is how the trap operates. In any set of six outcomes, there will be some pattern evident. If the pairs are not divisible by seven, it may be by three or five or eight; or the numbers may be prime; or they may all be odd; and so on. The chance is small that we find a *particular* pattern that we chose in advance. The chance is great that *some* pattern will arise. We may well wonder if there is any set of six outcomes for which we cannot *retrospectively* find some pattern.

Would anyone fall into this simple trap in real life? It turns out that this trap has spawned great industries. In reading Tarot cards, I Ching, tea leaves and many other systems of divination, one starts with a process whose outcome is guaranteed to be random: card shuffling, stick throwing etc. One then insists on deep significance in the random arrangement. One defense of hard core practitioners is that they really do not believe that random patterns have intrinsic significance. They are merely aids to trigger the mental powers of the practitioners. You may like to ponder whether this is an acceptable escape.

Correlation and Causation

Correlation: This TENDS TO GO WITH that.

Causation: This CAUSES that.

Examples

* Smokers and former smokers tend to get lung cancer much more than non-smokers. since Smoking causes lung cancer.

Four of five lung cancer victims are smokers or former smokers but only one in four of the US population smokes.

Smoking does not inevitably lead to lung cancer. Only about one in 250 smokers or former smokers will contract lung cancer.

* Workers exposed to asbestos (e.g. asbestos miners, asbestos weavers) are far more likely to develop lung disease (asbestosis, lung cancer). since Prolonged inhalation of asbestos fibers causes lung disease.

Subject	Relative risk of lung cancer
Non-smoker, no asbestos exposure	1
Smoker, no asbestos exposure	11
Non-smoker, asbestos exposure	5
Smoker, asbestos exposure	53

So far the errors involved in the examples of overzealous finding of patterns have been easy to see. The confusion of correlation and causation is an example of erroneously fitting patterns that can be quite hard to detect and have devastating effects in terms of mistaken information.

Two things are correlated if they tend to go together. Certain habits and occupations tend to be correlated with lung cancer and other diseases, for example. That means that people who smoke or work with asbestos are more likely to contract these diseases than people who do not. The table gives estimates of the relative risks. Someone who both smokes and works with asbestos is 53 times more likely to develop lung cancer, for example.

The reason for this correlation is that smoking and asbestos exposure cause cancer. That is, they bring about the cancer.

Notice the causal claim is far stronger and more useful. Once we know what causes the lung cancer, we know what has to be done to reduce its likelihood. We need to reduce our exposure to these agents. As we shall now see, knowledge of a correlation alone cannot be turned so directly into actions. Therefore mistakenly identifying mere correlation as causation can lead to improper action.

Correlation —	does NOT imply	— Causation
This TENDS TO GO WITH that.	does NOT imply	This CAUSES that.

Examples

*	Children with larger shoe sizes tend to read better.	but	Having bigger feet does not make you read better. Why not?
*	Most car accidents occur close tc the driver's home.	but	Driving close to home is not more dangerous than driving far from home. Why not?
*	Patients receiving treatment in the most sophisticated hospitals are more likely to die than those in underequipped hospitals.	but	Treatment in the most sophisticated hospitals is not more dangerous to your health. Why not?

---

*	People exposed to weak electromagnetic fields have slightly higher rates of leukemia and brain cancer.	does NOT imply	Exposure to weak electromagnetic fields causes leukemia and brain cancer.

Now comes the catch: that two things are correlated does not mean that one causes the other. Opposite are listed a few examples in which two things are correlated but one does not cause the other. They are connected in a more subtle way. Can you figure out the connection?

To see a causation when there is only correlation is a classic example of seeing more pattern than the evidence warrants. Notice how it immediately turns into erroneous policy decisions. We do not get children to read better by making their feet swell. We do not minimize the chance of car accidents by never driving close to home. We do not maximize our chance of surviving a hospital visit by insisting that we be treated in the worst hospitals.

Important, current debates in public health sometimes reduce to a determination of whether a correlation is actually supported by the causation suspected. It has been found that exposure to electromagnetic fields correlate with slightly greater rates of various cancers. The presence of the correlation is clear. Does this mean that these fields cause the cancers in the same way smoking causes cancer?

That remains a subject of considerable debate. Many other factors come into play. If there is a higher rate of cancer in houses located near power lines, that may merely reflect that such houses tend to be in poorer or more heavily industrialized parts of town where there are many other potential causes of cancer. While electromagnetic fields of the power lines may correlate with higher cancer rates, they may not cause them. If we mistakenly believe they do, then we may end up making hugely wasteful decisions about the location or relocation of power lines. (Conversely if we mistakenly reject the fields as causes, the consequences are needless illness and death.)

A greater rate of cancer deaths could even reflect the *greater* health and welfare of these workers. At first glance this seems impossible. But it is not. If for some other reason these workers do have better health and welfare, they are more likely to survive to older ages. Older people are more prone to cancer, which tends to develop in later years. So their cancer rate would be higher, not because the industry caused more cancers, but because other causes of death are suppressed.

Problem We are poor at assessing the import of evidence:

We overestimate the importance of positive evidence.

We undervalue the force of negative evidence.

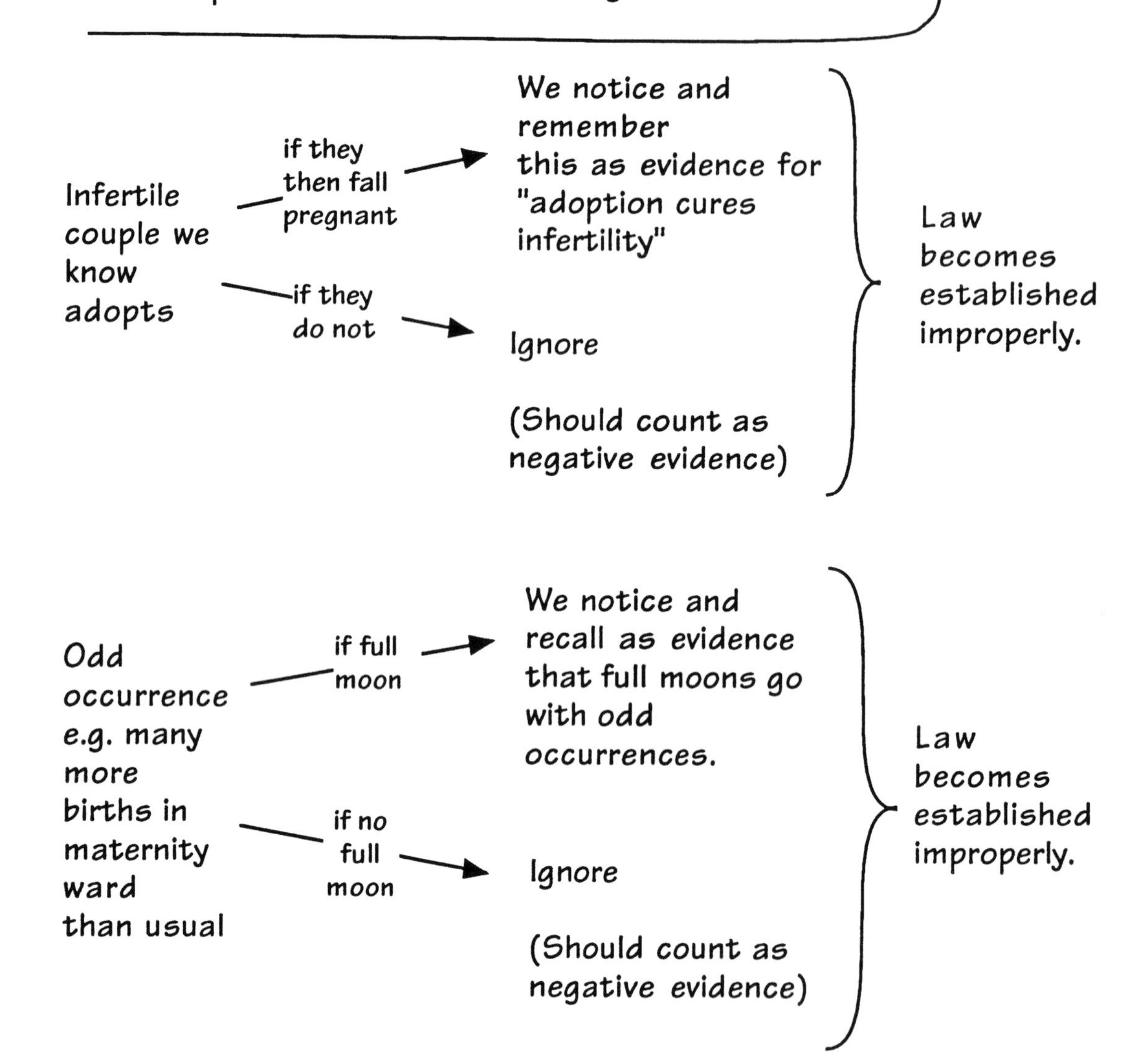

There are two errors we commonly commit in seeking to discover the secrets of nature. The first pertained to our tendency to find patterns spuriously. We now turn to the second. It pertains to positive and negative evidence. Positive evidence supports an hypothesis; negative evidence undermines it. Unless we are very systematic in our treatment of evidence, it turns out that we do not treat the two types equally. If we have some favored hypothesis, we tend to be very impressed by evidence favorable to it. We may even let a small number of positive items of evidence dictate our belief. At the same time, we tend to undervalue negative evidence. The overall effect is that we may come to believe that a body of evidence strongly supports some hypothesis when it may not do so.

We have seen some examples of this earlier. It is a very striking experience to know an infertile couple who adopt a child and then falls pregnant. One tends to remember this. If an infertile couple adopts and then does not fall pregnant, this tends to leave no impression. As a result our communal memory accumulates an impressive body of cases which support the "adoption cures infertility" hypothesis. But the case has been prepared by inadvertently selecting the positive evidence and ignoring the negative evidence. A more complete study of cases shows that the hypothesis is not supported by the evidence of many cases of infertile couples that adopt.

Precisely the same fallacy has convinced many of the old fable that many weird things happen at the time of the full moon. Imagine you are in a maternity ward and the night is very hectic. If you happen to glance out the window and see the full moon hanging eerily and ominously in the sky, you cannot help but be struck by it. If you see no full moon you probably would not even recall that you looked in the sky. Once again you start to collect a body of cases all of which support the hypothesis of a connection between the full moon and odd occurrences. But the case is based on simply selecting and remembering positive instances and ignoring negative instances from a body of evidence that, in balance, provides no support for the hypothesis.

Anecdotal evidence.

Items of evidence that have not been collected systematically. They are often circulated as stories--"anecdotes"--about personal experiences.

Notoriously unreliable and misleading.

"I don't believe that tobacco and alcohol are so bad for you. My Great Uncle Bill smoked and drank every day of his life-- and he lived to be 93 years old."

What is NOT the problem with anecdotal evidence:

The truth of the anecdotes is not always in question. They may well be perfectly true stories.

But we may also have doubts about the reliability of the anecdote.

What IS the problem with anecdotal evidence:

The anecdotes are purported to represent faithfully a more complete body of evidence.
The anecdotes pretend to be "typical".

But they are not!

- Great Uncle Bill's health is supposed to be typical of those who smoke and drink heavily.

- Our friends who adopted and then fell pregnant are supposed to represent a subgroup of infertile couples who suddenly gain greater fertility on adopting a child

This problem of haphazard selection of evidence is known as the problem of "anecdotal evidence." Such evidence is labeled anecdotal since it often circulates as anecdotes about personal experiences or those of others we know. Anecdotal evidence is always suspect, to begin with, since we usually have no control or option to review the source of the anecdote. The anecdote may be perfectly true; but it may also be distorted in subtle or not so subtle ways.

This danger of distortion is not the deepest problem with anecdotal evidence. The real problem is that the anecdotal evidence purports to tell us what is typical of a broader body of evidence. This is where anecdotes are very likely to mislead. Even if we convince ourselves that the events related in the anecdote happened exactly as related, the anecdote may still mislead in so far as it fails to represent what it typical. We have seen the subtle pressures that lead us to overvalue certain pieces of evidence and undervalue others. Certain episodes are very striking or memorable precisely because they fit with some hypothesis we already entertain. Thus we let these episodes, circulating as anecdotes, speak louder than others. Anecdotal evidence is always at risk of being the product of this type of biased process of selection. In this case, anecdotal evidence becomes another example of positive evidence mistakenly given disproportionate emphasis in our assessments of the whole body of evidence.

Anecdotal evidence is responsible for the widespread currency of many unfounded beliefs. One notable class of examples is ineffective or even crank medical therapies. Someone who suffers long term illness--arthritis, for example--may be driven to try many different treatments. If the illness is one that spontaneously improves and worsens, then it is quite likely that the person is taking an entirely ineffective treatment at the same time as the illness spontaneously improves. One now has a spurious anecdote supporting the effectiveness of the treatment!

The same mechanism can work to enforce stereotypes for different groups of people: racial, gender, national etc. If one meets a member of the group that fits the stereotype, that is striking anecdotal evidence for the correctness of the stereotype. If one believes the stereotype, one is tempted to discount cases that do not fit, however, as atypical exceptions.

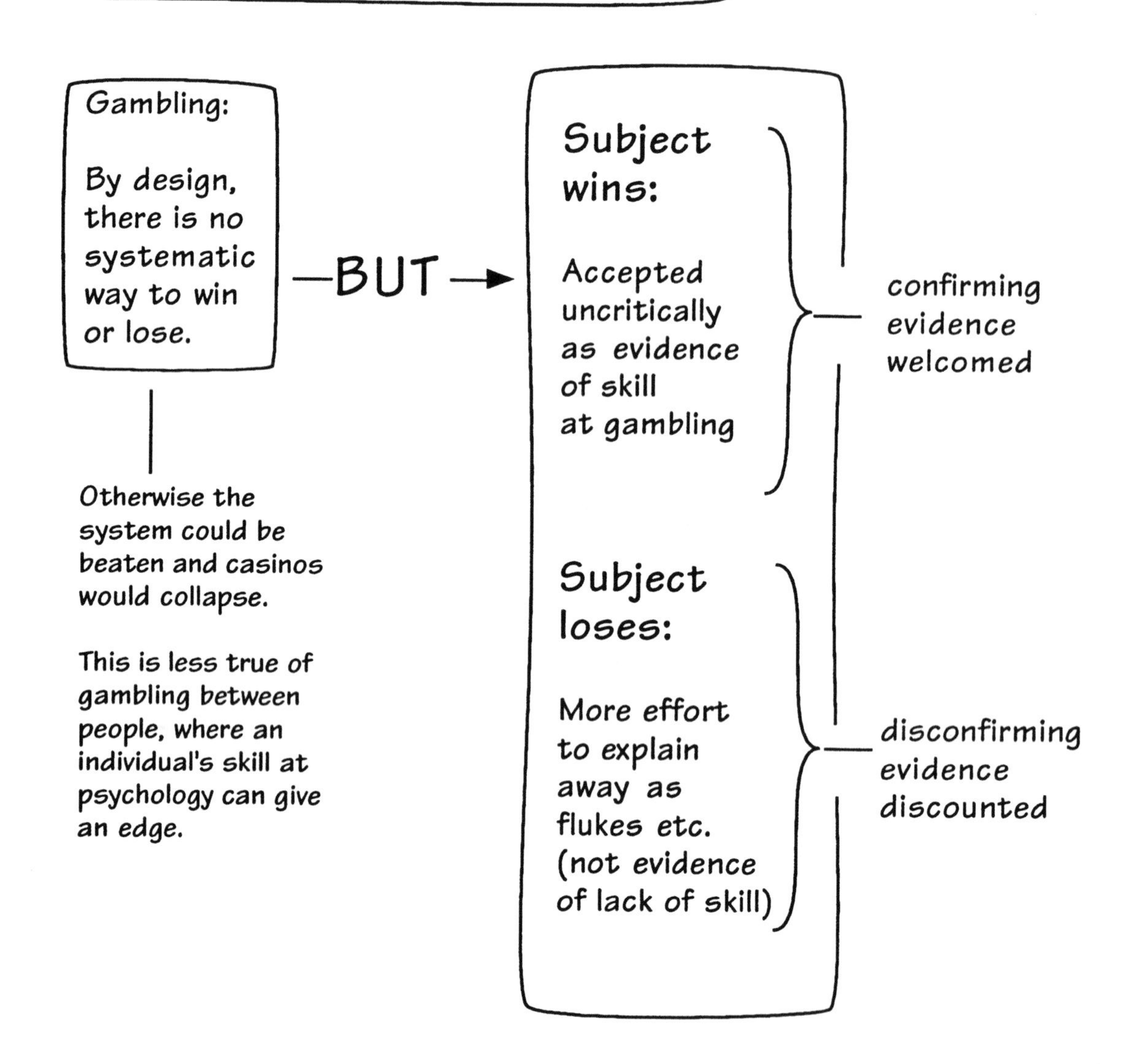
Studies in psychology show:
Same fallacy of faulty appraisal of evidence explains why people continue to gamble.
Gambling:
By design, there is no systematic way to win or lose.
—BUT→
Otherwise the system could be beaten and casinos would collapse.
This is less true of gambling between people, where an individual's skill at psychology can give an edge.
Subject wins:
Accepted uncritically as evidence of skill at gambling
confirming evidence welcomed
Subject loses:
More effort to explain away as flukes etc. (not evidence of lack of skill)
disconfirming evidence discounted

Our tendency to misunderstand the import of evidence has quite broad explanatory power. It is used in psychology to help explain why so many people continue to gamble even though they lose and, at the same time, believe that they can beat the odds. Many insist that they have a system for winning or that they are in the final stages of developing one.

This sort of belief is almost guaranteed to be ill-founded. One can be virtually certain that this is the case in casino gambling. Most of this type of gambling is so set up that there is no systematic method of winning. The casinos are there to make a profit. It is simply good business for them to set up their games that way. It is also good business to promote circulation of stories--anecdotal evidence--of players that "break the bank."

When a subject plays, the subject wins and loses randomly, for gambling is, by design, a random process. But studies have shown that gambling subjects interpret the wins and losses very differently. A win is viewed as evidence for the skill of the gambler. If the gambler is testing a system of gambling, it is evidence for the effectiveness of the system. A loss, however, is regarded as a momentary set-back, a random fluctuation in a record of success. If a system is being used, the loss is simply an indication that the system is fallible or that it needs some fine tuning. Thus the gambler is confronted by a record of wins and losses that gives no evidence overall for special skill at winning. Yet, by selectively interpreting the wins and losses, the gambler can become convinced that it confirms an ability to win. (T. Gilovich, "Biased Evaluation and Persistence in Gambling," *Journal of Personality and Social Psychology*, 44 (1983), 1110-1162.)

Gamblers are also prone to the fallacy of identifying spurious patterns. Gamblers can see that a randomizing machine, for example a roulette wheel or throw of dice, is creating the sequence of numbers on which they bet. Yet they still may convince themselves that they have detected a pattern in the outcomes that can be exploited to predict the winning bets. There is some suspicion that a similar mistake is made in stock market analysis. Critics complain that there is not sufficient regularity in the patterns of stock prices to allow any profitable system of stock purchase based on tracking time charts of stock prices--although many analysts claim otherwise. (See B.G. Malkiel, *A Random Walk Down Wall Street*. Norton, 1990.)

The story of "Bridey Murphy":

When our aversion for negative evidence is so great that we may even avoid seeking it.

Hypothesis
- slender evidence for hypothesis — This was sought eagerly.
- strong evidence against — This was not looked for at all.

Mrs. Virginia Tighe Pueblo, Colorado 1952	Under hypnosis, she began to talk in an Irish brogue, supposedly the voice of a previous incarnation of a 19th century Irish woman, "Bridey Murphy"

Sensationalized in press!	W. J. Marker, "The Strange Story of Bridey Murphy," *Denver Post*, Sunday Supplement, Sept. 12, 19, 26. M. Bernstein, *The Search for Bridey Murphy*. Doubleday, 1956. + much more.	Quest for evidence FOR Mrs. Tighe as a reincarnated Bridey Murphy
Massive search in Ireland for traces of a Bridey Murphy in the 19th century.	... no significant successes	

THEN ↓

Bridey Murphy was "found". Articles in Chicago American 1956	Reporters looked in Mrs. Tighe's old home town--Chicago. As teenager, Mrs. Tighe lived opposite an Irish woman, Mrs. Corkell. Mrs. Tighe was fascinated by Mrs. Corkell's Irish lore and soaked up masses of it. Mrs. Corkell's maiden name = Bridie ! Strong evidence that Mrs. Tighe's hypnotic recollections were simply childhood memories reconstituted.	Quest for evidence AGAINST Mrs. Tighe as a reincarnated Bridey Murphy (neglected by supporters)

Our natural aversion toward negative evidence explains some of the more spectacular failures of critical analysis. One famous example is the "Bridey Murphy" case. In 1952, under hypnosis, Virginia Tighe spoke in an Irish brogue with recollections of a 19th century Irish woman called Bridey Murphy. The obvious hypothesis was that Tighe was the reincarnation of the dead Bridey Murphy--Murphy's soul had passed into Tighe's body. The case was welcomed as spectacular evidence of reincarnation.

It triggered a sensation in the press, marked by extensive searches for independent evidence for the story. In particular, could traces of the 19th century Bridey Murphy be found in Irish records? These searches failed to produce unambiguous evidence. The problem, in part, was the fact that Murphy is a very common name in Ireland! The hypothesis that Tighe was the reincarnation of a Bridey Murphy was widely believed in spite of the meager evidence collected for it.

What proved remarkable about the episode was the tenacity of the hypothesis' believers in seeking positive evidence for the hypothesis. Had they only overcome their aversion to negative evidence, they would have found strong evidence against the hypothesis. They would have found evidence strongly favoring the hypothesis that Tighe's Irish recollections were simply the reconstitution of memories of Irish folklore soaked up as a child from an Irish neighbor with the maiden name "Bridie". This simpler hypothesis explained how Tighe had amassed her knowledge of Ireland and the Irish brogue without the need for extraordinary hypotheses about reincarnation.

Morals

* The goal of science is the discovery of truth and the avoidance of error. Both are hard to do.

* The finding of patterns is essential to discovery in science, but it is also very easy to find spurious patterns.

  A common and insidious form of this error lies in the fallacious extension of correlation to causation.

* We naturally tend to overestimate positive evidence and undervalue negative evidence.

  Our overestimation of positive evidence leads to an exaggerated confidence in anecdotal evidence, which usually misrepresents the overall burden of a body of evidence.

Scientists must be skeptics, for it is the only way to guard against the pervasive errors that we otherwise fall into. The ways of misreading, misinterpreting and misunderstanding are only limited by human ingenuity--and that is near boundless. But we have seen two very widespread types of errors against which we must guard.

First, we have a natural ability to see patterns in chaos. This ability proves the key to unlocking the secrets of nature. But we must also guard against overusing it and finding patterns where none genuinely exist. In particular, when we find a correlation, we cannot automatically presuppose that we have seen the particular causal process that may have generated it.

Second, we do not naturally weigh conflicting evidence well. We tend to overvalue evidence that favors an hypothesis we like; we tend to discount evidence that speaks against it. A common example of this fallacy lies in anecdotal evidence. We take one piece of evidence, selected because we find it striking, and take it as representing a whole body of evidence. Since we are likely to find the evidence striking if it fits our preconceived notions, the item of evidence is unlikely to be a good representation of the whole body of evidence, especially if that body counts, in balance, against the hypothesis.

How can we guard against these errors? We shall see some devices for this in the next chapter.

Answers to Correlation and Causation Puzzles

Children with larger shoe sizes tend to be older. As a result they read better.

Most driving that people do is done close to home. Therefore it is entirely possible that most accidents occur close to home even if driving close to home were less dangerous. When the accident finally does happen, it is most likely that the driver is close to home.

If a patient is gravely ill, an ill-equipped hospital is likely to transfer the patient to a better hospital. These better hospitals tend to collect the most gravely ill patients--the ones that are most likely to die.

---oOo---

Examples of spurious belief in unfounded patterns abound. Consider the widespread belief among patients and physicians alike that arthritis pain is associated with the weather. It is worse, they believe, when the weather changes. This belief extends to antiquity and was given more precise formulation in the nineteenth century when a connection was suggested between arthritis pain and barometric pressure. Nonetheless, objective studies have been unable to confirm the existence of such a connection. Researchers who investigate the connection point to the difficulty of assessing whether the two time series are correlated. Consider the time series of weather conditions and of a patient's arthritis pain. How much agreement between their movements is needed for us affirm a connection? Some agreement must happen by chance. Our tendency is to be excessively impressed by these chance instances of agreement in the two series. But we undervalue the instances of disagreement that negate them. So our selective attention to the evidence once again assures us of a spurious pattern. See D. A. Redelmeier and A. Tversky, "On the Belief that Arthritis Pain is Related to the Weather," *Proc. Natl. Acad. Sci. U.S.A.* 93(1996), 2895-6.

Assignment 2: Can you believe everything you read?

Get the latest issue of the supermarket tabloid *Weekly World News*. Your assignment is to read through it looking for:

An article you believe to be true (or at least substantially true).
An article you believe to be false (or at least substantially false).

Submit in writing:

A citation to the article you believe (headline, page number, issue date) **as well as an account of your reasons for believing the article.**

A citation to the article you believe false (headline, page number, issue date) **as well as an account of your reasons for believing the article false.**

In assessing the articles, you may find the following principle useful. It was developed by people who investigate paranormal claims:

> Ordinary claims require ordinary evidence;
> extraordinary claims require extraordinary evidence.

For example, if you tell me that it is raining in your back yard, I am likely to believe you. The claim is ordinary. Rain is a common occurrence. The evidence that supports it is ordinary. The testimony of a single observer is usually sufficient evidence. Now imagine that you tell me there is a live dinosaur in your back yard. That is an extraordinary claim. If I am to believe it I need extraordinary evidence to support it. The testimony of a single observer is relevant, of course. But I would require something far stronger before I would believe it. Even a movie of the dinosaur may be insufficient. As watchers of the movie Jurassic Park know, it is entirely possible to fake a thoroughly convincing movie of a live dinosaur.

Chapter 3

How We Know

Obstacles to Knowledge

(1) Nature hides her secrets in the confusion of experience

(2) We too easily delude ourselves that we have disentangled these secrets

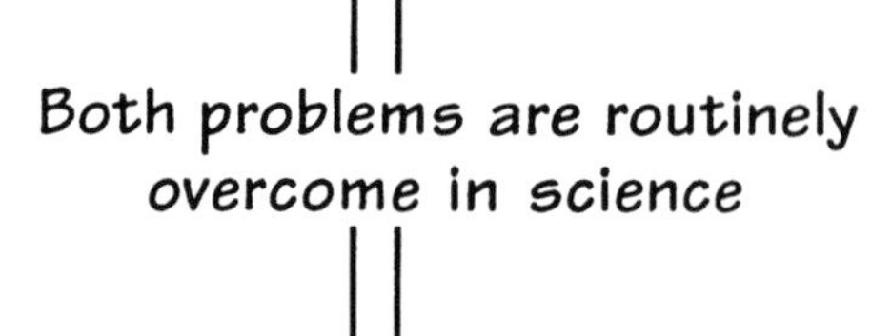

Techniques for finding Nature's secrets to be reviewed in this chapter.

1. Controlled Studies

2. Recast Claims in Objective Terms

3. Compare New Results against Well Established Theory

4. Use All Channels of Information

5. Use of Evidence: Overdetermination, Independence and Uniqueness

Scientists faces two obstacles in their quest for the secrets of nature. First, nature hides her secrets beneath the confusion of experience. The chemist who wants to discover which are the elements must organize a huge mass of scattered facts about this or that substance before a pattern forms in which the elements appear. Similar problems face those who work in all the sciences. Second, scientists must constantly resist error and self deception lest the results of careful investigation become corrupted with prejudged results and misconception.

It turns out that both obstacles can be overcome and have been overcome repeatedly; science has flourished with numerous episodes of such success. These successes result from ingenuity and opportunism by scientists, for we know of no simple recipe that ensures success for all the novel problems in science. In principle, each new problem may need an entirely novel approach for its solution. We routinely award our highest scientific honors to those who have devised new methods and approaches for discovery in science. However there are certain techniques that have proven effective in the past. It is possible to see that they address several of the common problems facing scientists. In this chapter we will review some of these techniques. They are listed opposite.

(1) Controlled Studies

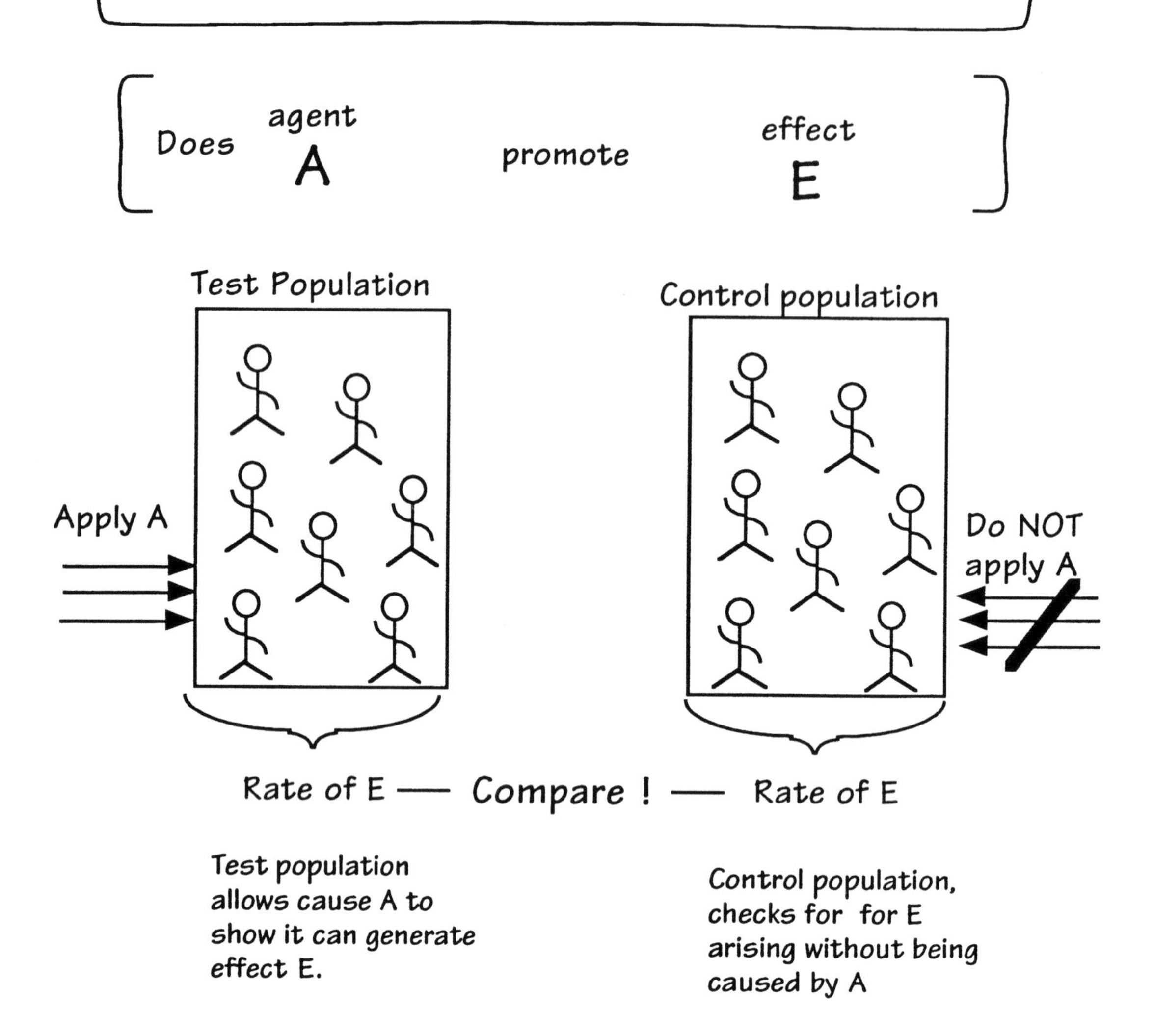

The controlled study is the most commonly used method for deciding whether some agent is capable of producing a designated effect. Many people believe that chewing garlic can cure the common cold. That is, eating garlic (the cause "A") leads to a cure of a cold (the effect "E"). Skeptics point out that if you chew garlic your cold will last about a week--and that it will drag on for about seven days if you don't!

The thinking behind the skeptics' quip is embodied in the design of a controlled experiment. To see if garlic (A) cures a cold, garlic is administered to a group of cold sufferers or potential cold sufferers, the "test population." The rate of cure (E) is then observed. To ensure that the garlic is responsible for the cure (E), we select a second group of cold sufferers as much like the test population as we can find. This second group is the "control population." They are given no garlic and the rate of cure is again observed. The efficacy of garlic can then be seen by comparing the two rates of cure. If the test population recovers faster or, perhaps, suffers fewer colds, then we have evidence for the efficacy of garlic. If there are no differences in the incidence or recovery rate between the groups, then we have evidence against the efficacy of garlic.

This technique of the controlled study is extremely powerful. Notice that we can discover whether an agent A does cause effect E without knowing anything about the mechanism that connects the two. For example, if the test shows that the garlic chewers did recover faster, we can then know that garlic works without having any idea of the way it acts within the body to bring about this cure.

Controlled studies provide an escape from two common errors.

(I) The problem of anecdotal evidence — In a controlled study one must consider the total evidence from all subjects in both test and control populations. Data may not discarded.

(II) The danger of confusing correlation and causation —

Correlation: This TENDS TO GO WITH that.

Causal claim: This BRINGS ABOUT that.

In setting up a controlled study, we strive to ensure that the only systematic difference between the test and control groups is A. Thus any systematic difference in E must be caused by A.

Controlled studies provide a means of escaping from two of the common errors described in the last chapter.

Anecdotal evidence may mislead us into thinking that garlic is an effective cold cure when it is not. We may hear spurious stories from believers who swear on the effectiveness of garlic. Or you may have your own anecdote in which you suspected that you were contracting a cold, chewed garlic and no cold came. (Of course the reason may merely have been that your sore throat was not the precursor of a cold after all!)

Since a controlled study forces us to evaluate an entire body of evidence, we cannot ignore the evidence against the hypothesis. The misleading anecdotes become overshadowed by the overall burden of the evidence. False cures--the sore throats that were not precursors of colds--would be just as frequent in test and control populations. They could no longer mislead us into concluding the effectiveness of garlic.

Our confidence in garlic may result from a confusion of correlation and causation. We may notice that people we know who chew garlic have fewer colds. This correlation by itself does not prove that garlic works against colds. It may be that the garlic chewers tend to be vegetarians or health food fanatics and that something else in their diet or lifestyle confers the immunity to colds. It may be as simple as a rural life, remote from the crowds of city, so that they are exposed to fewer cold viruses.

The controlled study corrects for this error. We strive to set up the test and control population so that the only systematic difference between them is the eating of garlic. Thus, if the test group has fewer colds or recovers faster, we know that garlic is the cause.

Early example of a controlled study

James Lind and the treatment of Scurvy

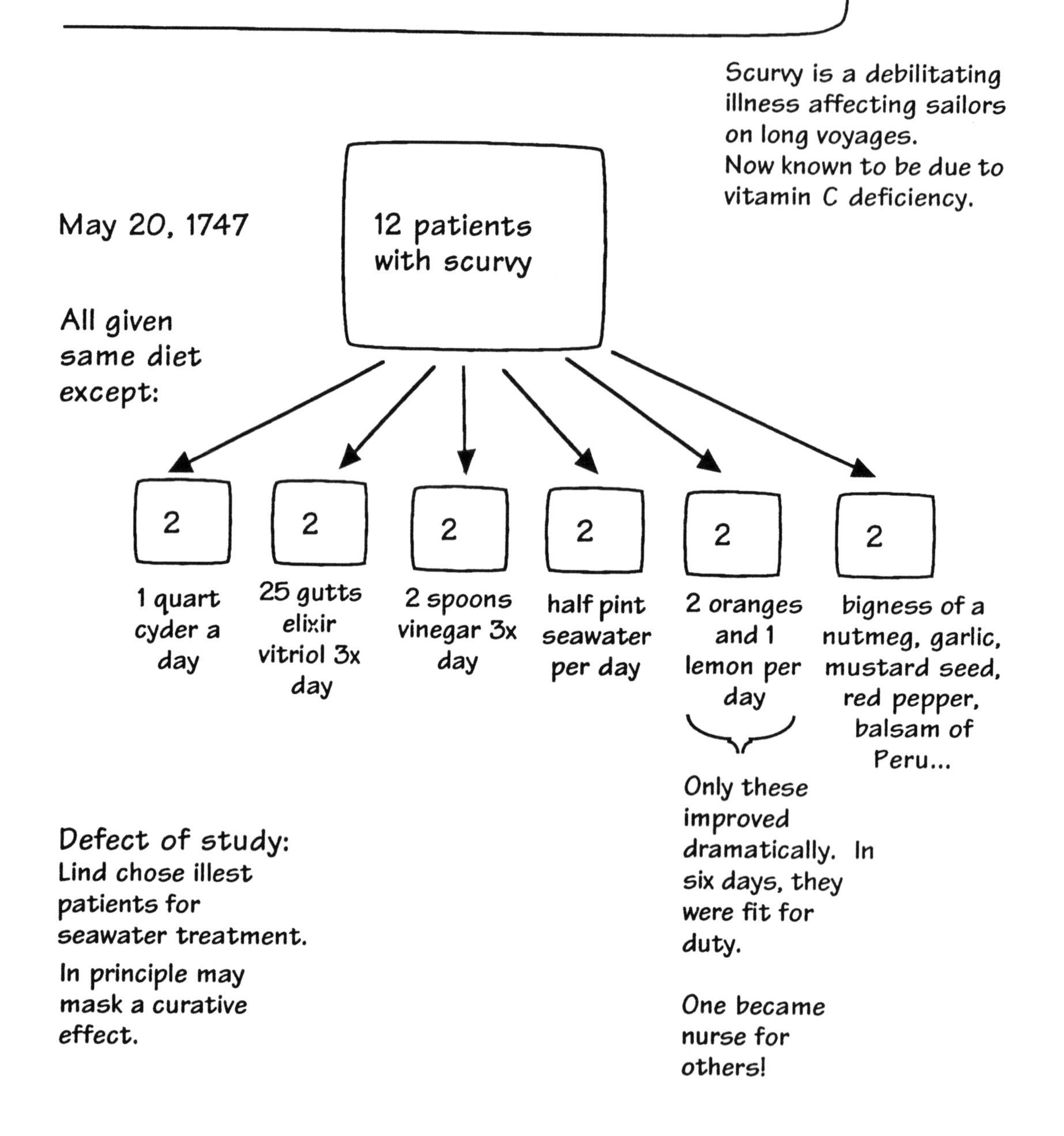

Early seafarers needed to cope with many difficulties as their voyages grew longer. We tend to think of the obvious problems of navigation and the supply of food and water. But there were others. The disease scurvy would strike sailors undertaking long journeys and could compromise the entire voyage. Thus finding a treatment of scurvy was at a premium; the viability of long distance sea travel was at stake. We now know that the illness resulted from a deficiency of vitamin C, a vitamin missing from the typical sailor's diet on ship. The simple cure was to include citrus fruit in the sailors' diets.

That scurvy could be cured by citrus fruit was demonstrated by James Lind in 1747. He took twelve patients ill with scurvy and tried the different treatments indicated opposite. Only the pair given citrus fruit recovered dramatically.

The general structure of Lind's experiment is that of a controlled experiment. In principle it fails to be a controlled experiment since there is no formal control population--scurvy patients who are left untreated. In practice, those patients given ineffective treatments stood in for the control group.

Notice that several of Lind's treatments involve acidic substances, the citrus fruit, vinegar and elixir vitriol--oil of vitriol is the old term for sulfuric acid. This suggests that something like citrus fruit was suspected as a cure. This diversity of acids would allow Lind to check whether it was citrus fruit in particular that effected the cure or whether any acidic substance would work. Thus patients treated with other acidic substances acted as a control for the the group treated with citrus fruit in this more limited regard.

One defect of the study was that Lind chose to give the seawater treatment to the illest patients. That risked compromising the study. If seawater had a mild curative effect, that may be masked by his choice of the illest patients for this treatment.

Many, many variations on form of study

Illustration: Does Vitamin C cure the common cold?

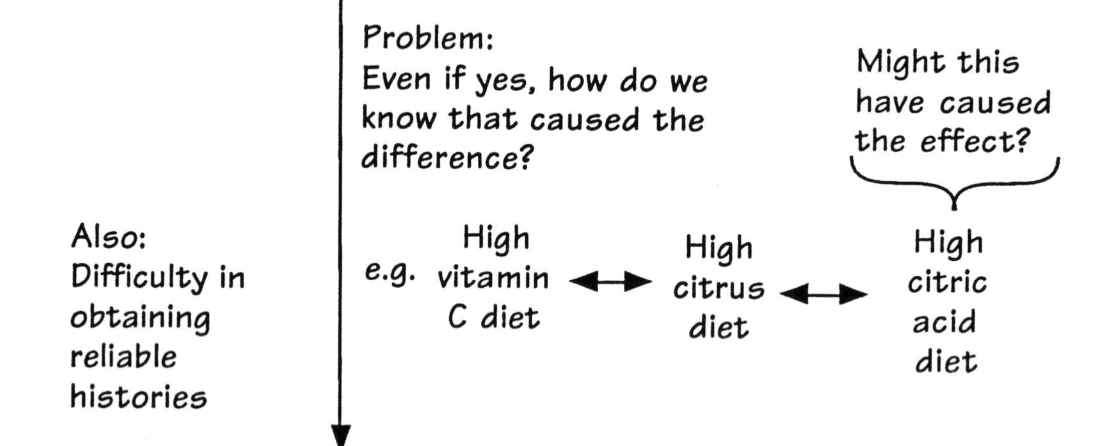

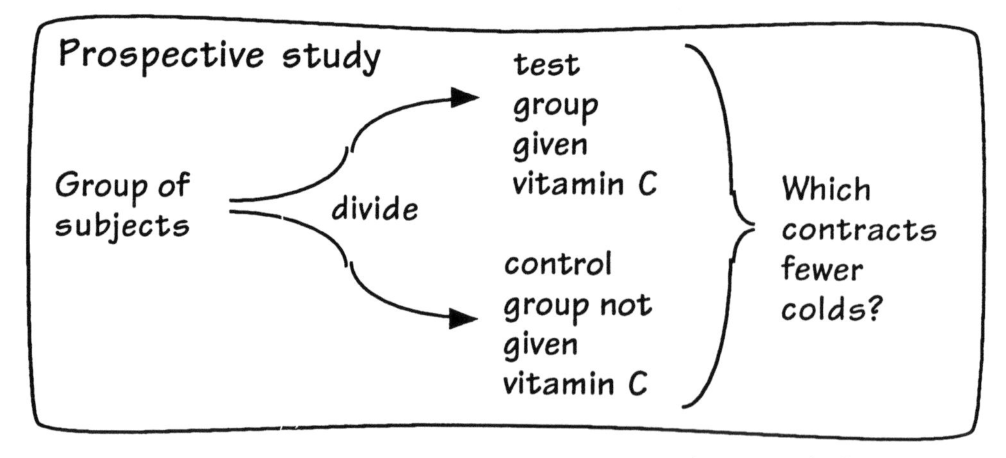

i.e. track future prospects.

So far we have seen the basic structure of a controlled study: the comparison of a test and control population. Many variations are possible, with each designed to address one or other shortcoming of the other forms. To illustrate some of these forms, we will imagine that we seek to investigate whether vitamin C can cure the common cold.

The simplest study we can do is a retrospective study. We collect records from as many patients as we can and compare the diets of those who had frequent colds with those who did not. It is "retrospective" since we are looking back in time to old records. If Vitamin C can cure colds, we may expect those with fewer colds to consume more vitamin C.

A retrospective study is attractive since it can be done rapidly, essentially by searching records and surveying patients. But such studies are rarely allowed to settle an investigation. They can be used at best to establish the plausibility or implausibility of the hypothesis in question. One major difficulty is that it is very hard to select a good control group. We may find that those prone to fewer colds did consume more Vitamin C. But that may be because they were mostly fruit lovers who consumed lots of citrus fruit. How can we rule out the possibility that some other constituent of the fruit conferred the beneficial effect? Citrus fruit has citric acid. Might that cure the common cold? Another difficulty is that it can be extremely difficult to obtain reliable patient histories. Unless we are fortunate, the standard records will not leave us enough information to determine the Vitamin C intake of patients. Patient interviews concerning past diet may also be unreliable. Memory can be very fallible. Patients may exaggerate how healthy their diet was to impress an interviewer.

To overcome these problems, we move to a prospective study. In it we divide a group of subjects into a test group, which is given Vitamin C, and a control group, which is not. It is "prospective" since we will track their future health. Notice that we are now in a much stronger position to eliminate other potential causes of fewer colds. We will give the test group pure vitamin C pills so we need not worry if some other component of the diet is responsible for any observed cure. We also have a better chance of tracking how much Vitamin C was in each subject's diet.

Problem:
How do we divide.
Ask for volunteers to take vitamin C?

→ No. Fruit lovers may all seek vitamin C. If test group does better, it may only be because they are fruit lovers!

Randomize

Use chance process to decide which subset is in which group

Randomization ensures no systematic difference between test & control group other than taking of vitamin C.

Moving to a prospective study does not automatically resolve all the problems. There are many places at which the study can still fail.

How do we divide the group of subjects into those that take vitamin C and those that do not? We must seek to divide the group so that the only systematic difference between the test and control group is the taking of vitamin C. If we merely ask for volunteers to go in each group, then we may find the fruit lovers tending towards the test group and the fruit averse placing themselves into the control group. In this case, we not know if a systematic difference in colds between the two groups was due to the vitamin C or to the fact that one group eats more fruit. Or we may place subjects who enter early into the test group and those who enter late into the control. Here we risk that the early subjects may be more eager and healthy than those we coerce to join later to fill out our numbers. This difference may explain any difference in susceptibility to colds and not the taking of vitamin C. In general any scheme we use for dividing the group risks introducing some systematic difference that will compromise the study.

The solution is to use a completely random procedure for the division. At its simplest, this means that we toss a coin for each subject. If we throw a head, they go into the test group; otherwise they go into the control group. In practice we may use a more elaborate way of introducing randomness. For example we may read numbers from a table of random numbers specifically prepared for the purpose.

The use of a completely random division ensures that there is no systematic difference between the two groups--at least none other than the one we choose to introduce; one group takes vitamin C, the other does not. Therefore any difference in susceptibility to colds between the two groups must be due to the only systematic difference, the taking of vitamin C.

This use of randomization is almost paradoxical. The key to finding regularity is randomization! Of course the paradox is only apparent. Randomization obliterates patterns. If we randomize everything, we obliterate all patterns. So we randomize everything excepting the particular factors of interest to us. Thus we obliterate all patterns other than those associated with the pattern that interests us.

Problem:
Deciding whether medicine has relieved cold is highly subjective.

e.g. Vitamin C takers may expect improvement → Report improvement since they anticipate it.

Blind study

Do not tell subjects who is in the test or control group. Control group gets a placebo that only looks like vitamin C.

"Placebo" =mock medicine that subjects cannot distinguish from real medicine.

There may also be problems in judging the results within the two groups. Whether a subject has a serious cold or not can be seen fairly unambiguously. But what if the study decides to test for severity of colds? That is more difficult to assess. What one person judges as a severe cold, another may not find so severe. Or other complaints may be confused with colds. A hay-fever attack may easily be mistaken for a cold.

Thus there is some latitude in deciding whether to report a subject ill with a cold or not and how severe that cold is. As long as this latitude introduces the same variation in both groups, we hope that it would not compromise the study overall. We would allow for the possibility that the cold rates in each group may not measure the true rate with perfect accuracy, but that it would not introduce a systematic difference between the two groups. Comparing the two rates would still reveal if one group was suffering fewer colds.

This problem may become severe if the latitude is exploited in another way. Observers tend to see what they expect to see. Assume subjects in the test group know they are taking vitamin C and that they believe in the efficacy of the vitamin. They will expect to catch fewer colds, or for those they catch to be less severe. Thus they may be strongly inclined to under-report the number and severity of colds they catch. ("It's just a little sniffle, not really a cold. Vitamin C has always helped me ward off serious colds.") Conversely those denied vitamin C expect to catch more colds and may over-report the number and severity of colds. If this effect is strong, the expectations of the subjects alone can give us a spurious positive result for the study.

The solution is to hide from the subjects which group they are in. To do this we give each pills, identical in look and taste. The pills for the test group contain vitamin C. The pills for the control group do not; their pills are a mock medicine, called a placebo. Thus the subjects in each group do not know which they are in. That may not prevent them from guessing. If they guess they are in the test group, for example, they may still under-report their colds. But that under-reporting will introduce no systematic differences between the two groups. Randomization ensures that the subject who guesses that he is in the test group is equally likely to be in the control group. The study has become a "blind study" in the sense that the subjects are blind to which group they are in.

Problem:

Experimenters know who are in test and which is control. Might this prejudice their evaluation?

Double-blind study Both subject and experimenters (who record results) do not know which subjects

- -take vitamin C (test group)
- -take placebo (control group)

...and so on!! e.g. Parallel design vs. cross-over design

cross-over design: test and control groups swap half way through

If the subjects' impressions may be prejudiced by their expectations, then we may expect the same problem to arise for experimenters who record the results. If an experimenter knows that a subject is in the test group, the experimenter may be inclined to expect fewer or less severe colds from that subject. Thus the experimenter may tend to discount reports from the subject about a cold or its severity. If the subject is in the control group, the experimenter may do the reverse. While experimenters may always make errors in assessing the health of the subjects, what is dangerous about this error is that it systematically leads to an overstatement of the health of the test group and an understatement of the health of the control group. Thus it may be sufficient by itself to yield a spurious, positive result.

The solution is to hide from the experimenters who record health data which subjects are in the test group and which are in the control group. The experimenters who have access to this data would not be those who conduct assessment and recording. Which subject is in which group and which pills are vitamin C and which placebo would be recorded by a code unknown to the record taking experimenters. This device does not preclude all errors in recording. But it will eliminate this particular bias.

Since both subjects and record taking experimenters are blind to which group each subject is in, this design of study is called a "double blind" study.

The devices we have seen here are only a few of the many used to eliminate spurious results in controlled studies. For example, the design used so far is a parallel design in which the test and control groups remain the same throughout the study. In a cross-over design, we switch half way through the study; the vitamin C takers become the placebo takers and *vice versa*. This design further reduces the chance of some inadvertent difference between the two groups. If a test group happens to have a resistance to colds not due to vitamin C, this resistance will not prejudice the results since these subjects will become the control group in the second half of the study.

Classic Controlled Study

Field trial of Salk Polio vaccine 1954

Does vaccine render immunity to polio virus?

Huge numbers of subjects needed since the incidence of polio very small (1/2000)

Over 1 million students participated!

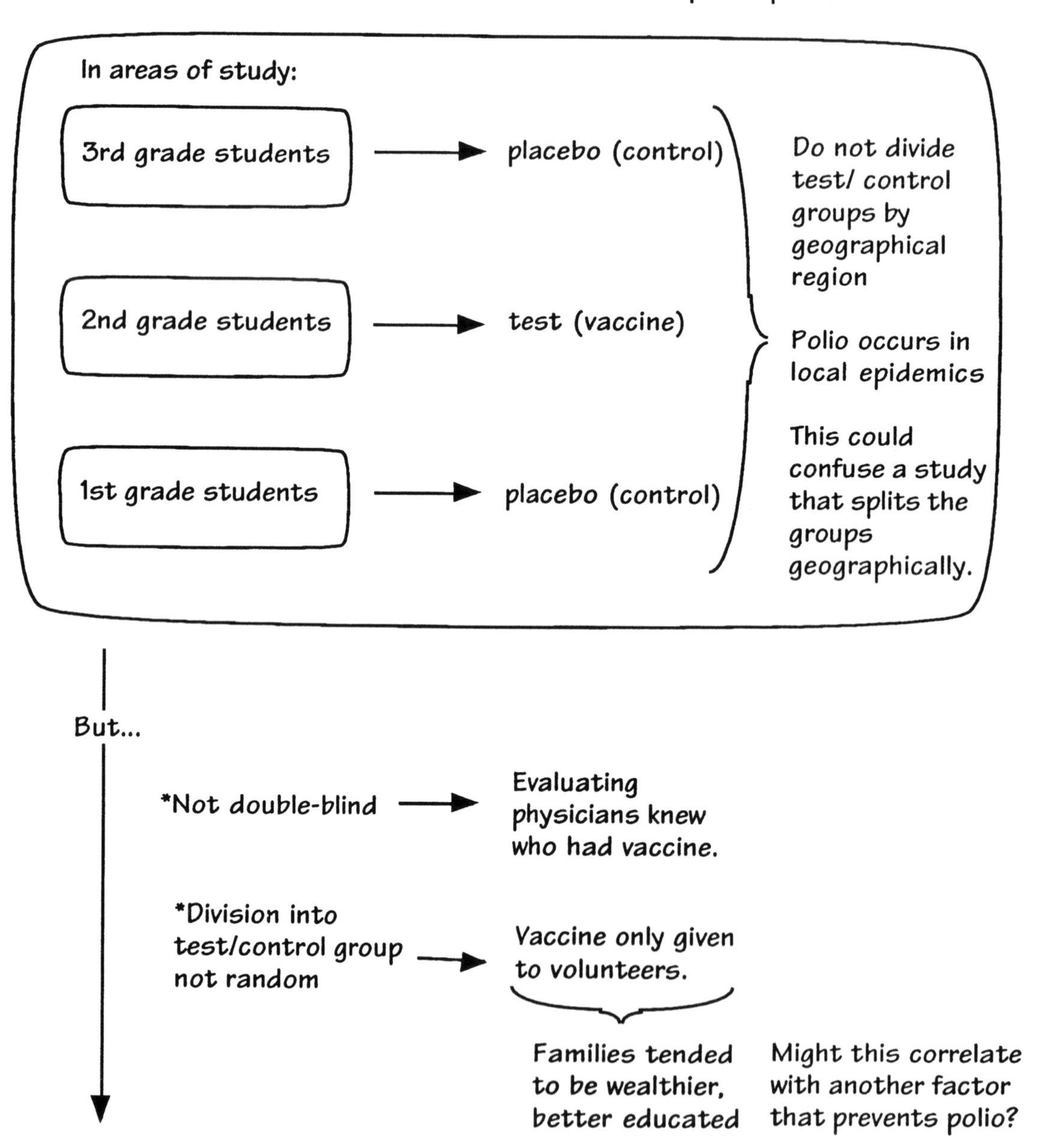

One of the most famous of all controlled studies was the testing in 1954 of the Salk vaccine against the infamous childhood disease, polio. The problem researchers faced was that the incidence of polio is quite small-- one in 2,000 children. So imagine that they give the vaccine to 100 children and none contract polio. It would prove nothing. Since the rate is one in 2,000, one would not expect any of the 100 children to contract polio, even if the vaccine does not work. Huge numbers of children are needed to in order for any systematic effect of the be obvious. The study became a massive effort. Over a million children participated in the study.

Selecting a suitable control group was also a problem. Polio tended to occur in local epidemics. If children from one city are used as the test group and those from another as the placebo, these epidemics may compromise the study. For example, if entirely by chance there is an epidemic of polio in the city with the control group, we may be led to think that the vaccine is effective. But all that happened was that the city with the test group happened just by chance not to be struck with a polio epidemic. The solution chosen was to use students in different grades in the same cities as test and control groups. There were still some systematic differences in the students with regard to age, but we could reasonably expect that they would not bias the study. The test group was controlled by students who were both older and younger. So any systematic differences associated with age should be visible in comparing the older and younger control groups.

The results of this first study showed the vaccine to be very effective. But some doubts about the study remained. The study was not a double blind study. The physicians who evaluated the students and recorded cases of polio knew which students had been given the vaccine. So there remained some chance that these record takers could prejudice the study. Further the vaccine was only given to volunteers. These volunteers tended to come from families that were wealthier and better educated. This meant that the division of students into the test and control group was not random. There was a systematic difference: the students in the test group tended to come from wealthier and better educated families. Might there be something about this lifestyle that decreased the chance of polio, so that the vaccine is not as effective as it seemed?

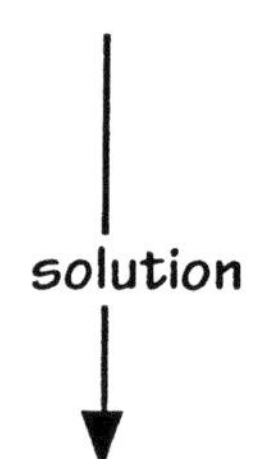

Further study

800,000 volunteers

- randomized assignment to test control group.
- double blind

Results: Rate of polio 50% less in test group

Rate of paralytic polio 70% less in test group

Only 4 deaths from polio -- all in control group

These results were compatible with the results of the other study.

These doubts were settled by a second study which corrected for both problems. This study took 800,000 volunteers. They were randomly assigned to the test and control group to eradicate systematic differences between the two groups. A double blind design was also followed.

The results were decisive and agreed with the results of the earlier test: the vaccine was very effective. The use of controlled studies enabled proof of the effectiveness of the polio vaccine. Its continued use has all but eliminated what was once a greatly feared childhood illness.

(2) Seek to cast claims in terms of objective, not subjective, judgments.

Anyone should be able to perform the same tests and obtain the same results.

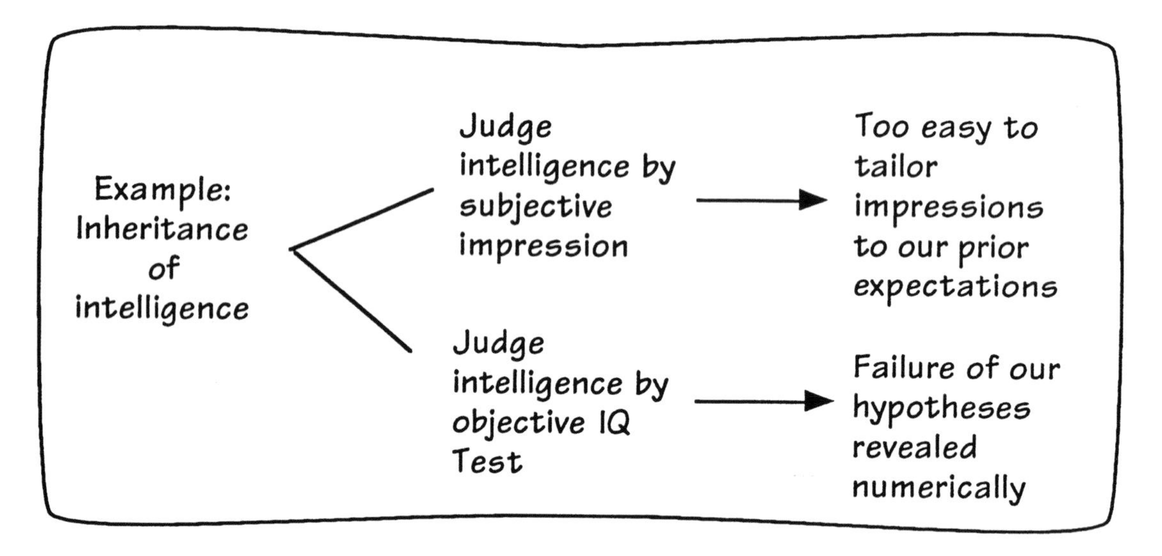

Problem: Does IQ measure what we want it to?

Perhaps. But we can test it. e.g. we can ask does a high IQ correlate with the ability to perform tasks requiring intelligence?

We can still try to force numbers to fit our expectations

But...

Burt tried and failed

One of the virtues of the controlled study is that it reduces the chance of our own expectations and prejudices compromising the study. There is a more general way that we try to do this. We try to design our experiments so that the results can be interpreted objectively. That is, what counts as a successful outcome can be seen so clearly that there is no real opportunity for the experimenters' own expectations to cloud the result. Very commonly we try to achieve this by replacing vague assessment with numerical assessments.

We have already seen an attempt to do this in the investigations over the inheritance of intelligence. To decide who is intelligent and who is not on the basis of the impressions of the experimenter clearly invites disaster. We routinely find that we differ from others in our impressions and evaluations of people. The solution was to reduce assessment of intelligence to the result of an IQ test. This greatly reduces the ambiguity in the assessment. What it takes to count as intelligent is now quite publicly available: one must score well on a test. Of course we may wonder if the test does accurately measure what we want it to. That, in turn, can be investigated. These investigations may convince us that the test is good; or they may convince us otherwise. In either case, we are not left to wonder exactly how different subjects were judged.

Once the judgment of intelligence is reduced to a numerical measure, the success or failure of the experiment is decided by the numbers. That still does not guarantee that our study is free of objection. That we learned from the case of Cyril Burt. But if the numerical results do not accord with our expectations, it is a very visible failure and one we must explain. The discovery that something was amiss with Burt's studies happened precisely because his studies relied so heavily on numerical assessment. Had he left his results in the form of vague impressions without numerical support, the deficiencies may never have come to light.

One may think that the need for vigilance over objectivity of results is localized in the human sciences, such as psychology. The case we will look at now shows this is not so. Even in fundamental research in physics, one must take care that results are given by objective methods lest we generate spurious results from wishful thinking.

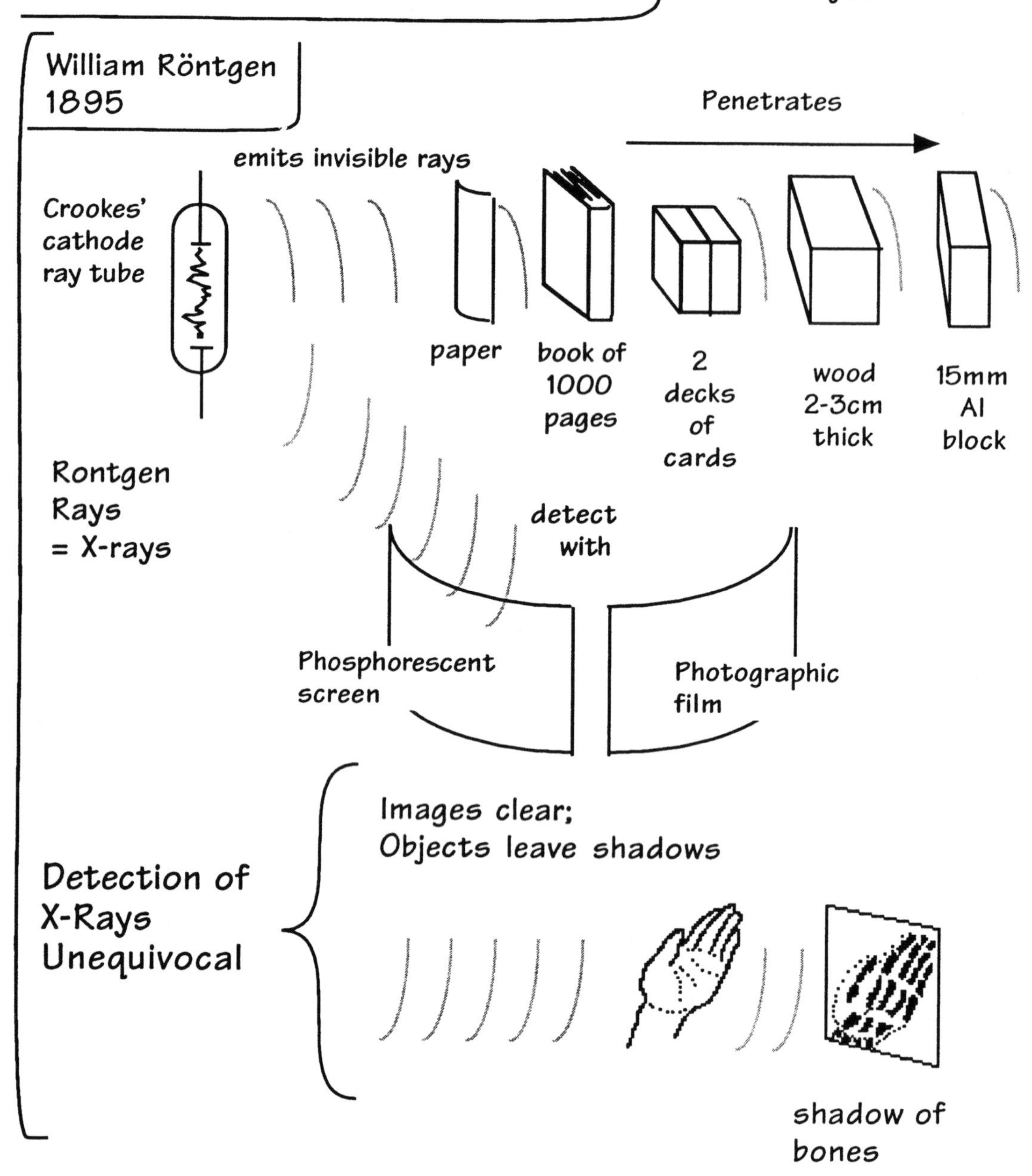
Classic case of failed objective judgment : N-rays
Background: The 1890s was the era of rays.
William Röntgen 1895
Penetrates
emits invisible rays
Crookes' cathode ray tube
paper
book of 1000 pages
2 decks of cards
wood 2-3cm thick
15mm Al block
Rontgen Rays = X-rays
detect with
Phosphorescent screen
Photographic film
Images clear;
Objects leave shadows
Detection of X-Rays Unequivocal
shadow of bones

The case of N-rays is a classic case of failure of objectivity in judgment. We shall see that the experimenters never had an objective test of the existence of these rays. Nonetheless because they were convinced that they were there, they managed to find them and catalog their properties.

As background, we should recall that the 1890s was the era of rays. Physicist were discovering a new world of invisible rays with astonishing properties. Röntgen in 1895 had found invisible rays emitted by a Crookes cathode ray tube. This tube is now used as a TV tube; the screen you view is just one end of the tube. The screen glows when the cathode rays strike the chemical screen at this end of the tube. Röntgen found that invisible X-rays are also emitted there.

These rays had enormous penetrating power and could pass through all manner of material including paper, wood and metal. While the rays were invisible, their detection was beyond doubt. If cast on a phosphorescent screen (like the greenish/gray end of the TV tube), they would make the screen glow. If they struck a photographic film, they would leave an image. The images left on the screen or film were striking. If the rays first passed through part of a human body such as a hand, an image of the hand would appear in which the bones were visible. Röntgen had discovered X-rays.

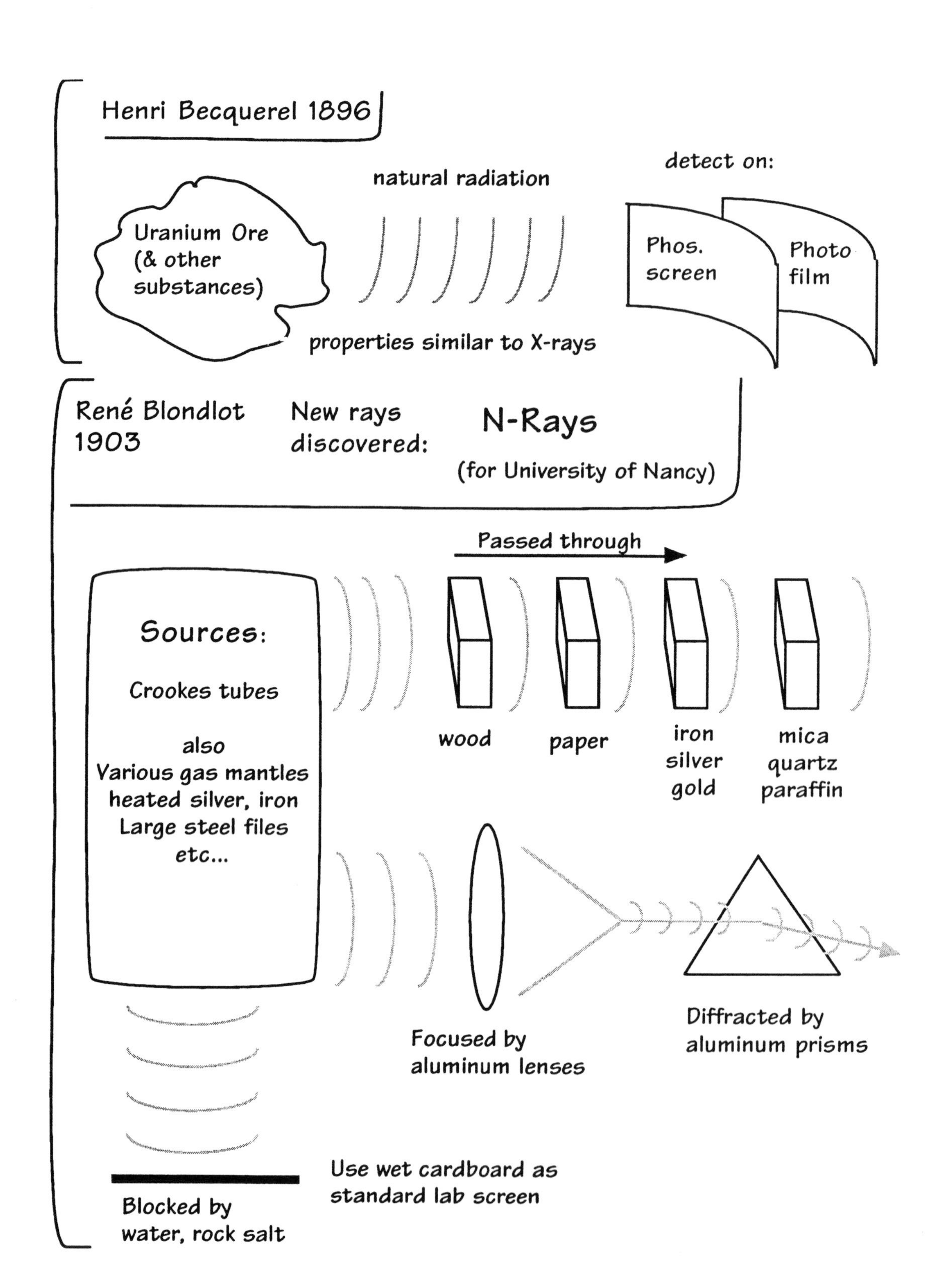

Henri Becquerel 1896
natural radiation
detect on:
Uranium Ore
(& other
substances)
Phos.
screen
Photo
film
properties similar to X-rays
René Blondlot
1903
New rays
discovered:
N-Rays
(for University of Nancy)
Passed through
Sources:
Crookes tubes
also
Various gas mantles
heated silver, iron
Large steel files
etc...
wood
paper
iron
silver
gold
mica
quartz
paraffin
Focused by
aluminum lenses
Diffracted by
aluminum prisms
Use wet cardboard as
standard lab screen
Blocked by
water, rock salt

Röntgen's discoveries were soon followed by others. In 1896, Henri Becquerel found that minerals such as Uranium ore emitted radiation of their own accord. This natural radiation had properties very similar to Röntgen's rays. They had vast penetrating power and their presence could be revealed by phosphorescent screens and photographic film.

So René Blondlot's discovery of yet another type of ray followed this mainstream of discovery. His "N-rays"--named after his home University of Nancy--had the usual penetrating powers. They could pass through wood, papers, metals and minerals. Lenses of aluminum would focus them in the way glass lenses focus light. Aluminum prisms would diffract them as glass prisms diffract light. But the penetrating power of N-rays was curiously limited. They could not pass through water or rock salt. Blondlot used wet cardboard in his laboratory as a screen for N-rays.

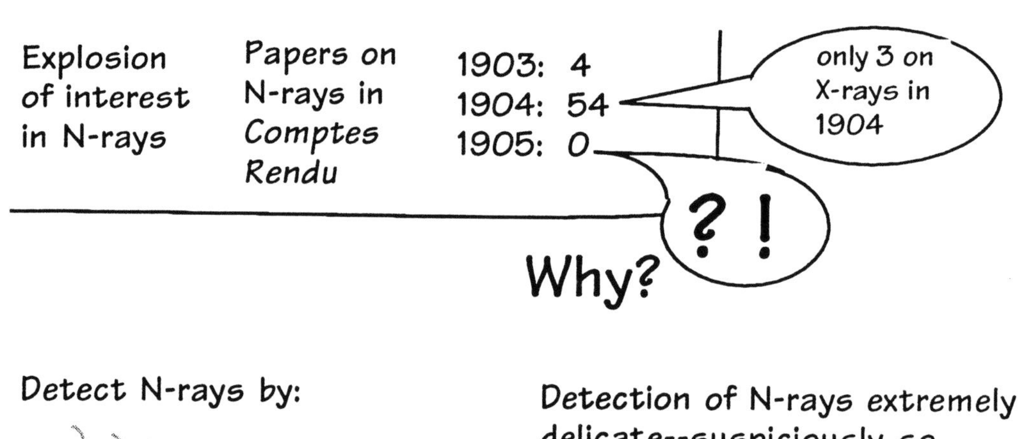

Detect N-rays by:

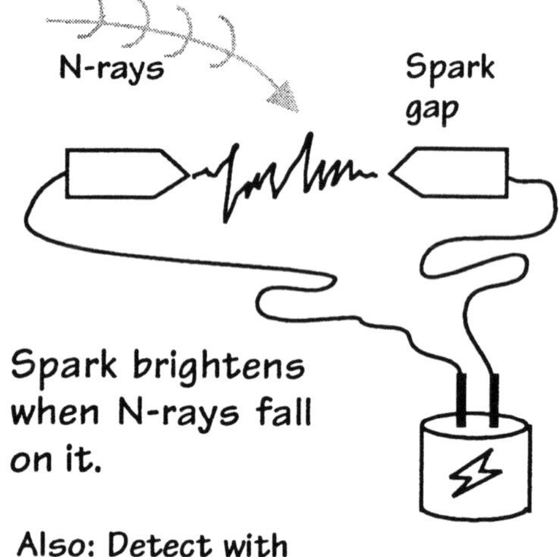

Spark brightens when N-rays fall on it.

Also: Detect with phosphorescent screen

Detection of N-rays extremely delicate--suspiciously so

Blondlot:
- At very boundary of perceptibility
- Need elaborate training & self discipline to see changes in spark gap
- Avoid straining vision and conscious fixing on source
- See without looking
- Maintain quiet & passivity

Other experimenters failed to reproduce Blondlot's experiments: Kelvin, Crookes, Lummer, Rubens, Woods

Suggests N-rays a delusion

Proved by Wood in visit to Blondlot's lab 1904

The discovery of N-rays was welcomed in the physics community. the numbers of papers published on them in *Comptes Rendu*, the leading French journal, soared. 54 were published in 1904, while only 3 were published on the subject of X-rays. But in 1905, no papers were published in *Comptes Rendu* on N-rays.

The reason for this dramatic shift began with the difficulty of detecting N-rays. They could not be used to make nice picture like X-rays. Their detection lay at the outermost edge of our capabilities. N-rays would just perceptibly intensify an electric spark, Blondlot claimed. One could only see the intensification after careful training and self-discipline. They were also supposed to produced a slight glow on a phosphorescent screen. The conditions Blondlot described as conducive to seeing the effect just also happened to be those conducive to self-deception. For when an effect is just on the boundary of the perceptible, whether one sees it or not may well be closely bound up with whether one expects to see it. If we have strong expectations for when the detector will reveal N-rays, might that not be sufficient to convince us that we have seen them when we expected to?

What deepened these worries was the growing list of experimenters who were unable to find any trace of N-rays. This list included some of the leading experimenters of the day. One had to begin to wonder if there were N-rays at all. Catastrophe for Blondlot struck with the visit of an American physicist R. W. Wood, who came to observe Blondlot's experimental technique.

Wood observed many defective experiments.

The most telling of these was:

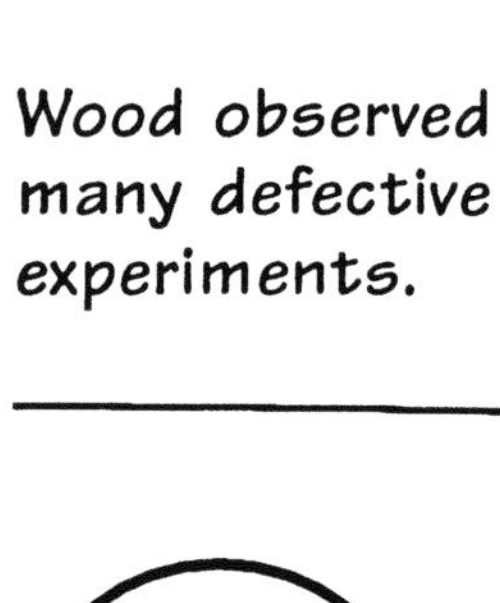

Blondlot and coworkers all found same 4+ peaks

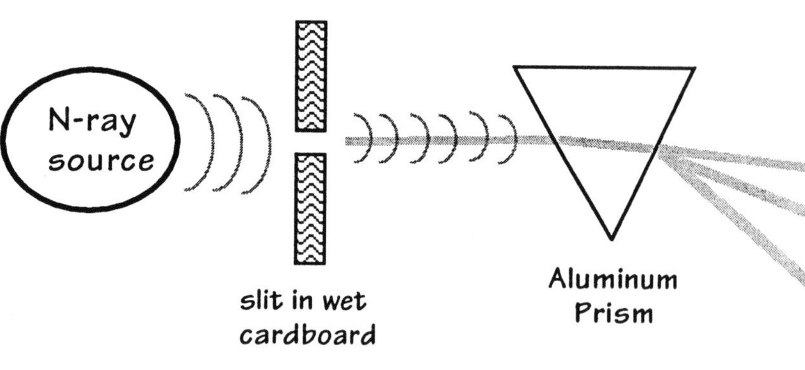

spectrum of N-rays revealed by phosphorescent detector

Wood could see no variation in brightness of detector

THEN

Wood surreptitiously removed the prism... Blondlot & coworkers still got perfect results!!

Blondlot refused all "blind" experiments that could distinguish real results from self delusion.

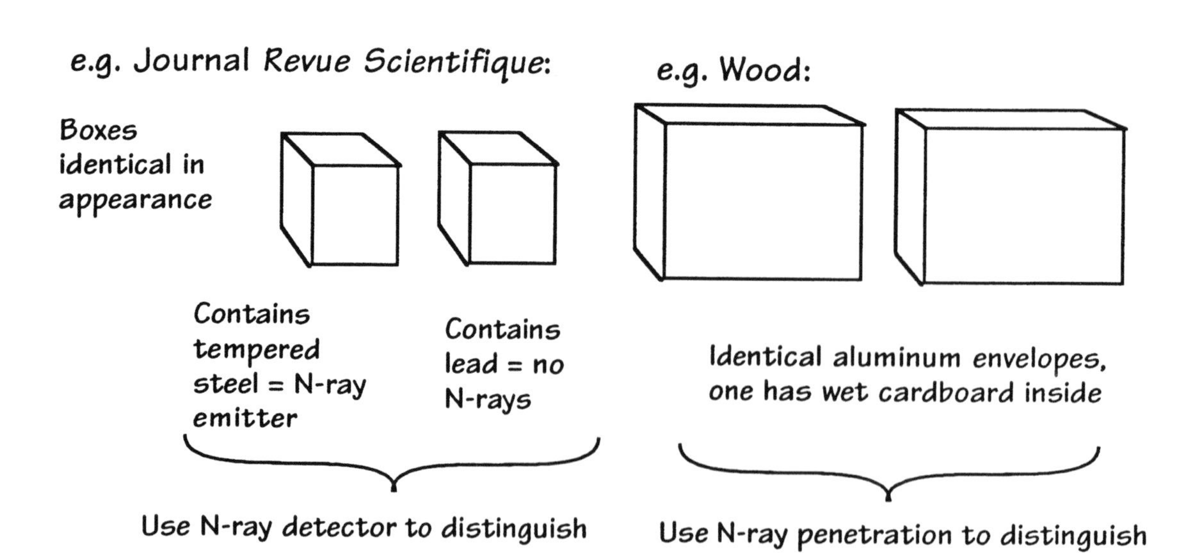

Wood found many defective practices in Blondlot's laboratory and became convinced that there were no N-rays. They were just the product of self-deception by Blondlot and his co-workers. What gave decisive evidence of this unhappy verdict was a clever, if unkind, prank that Wood played on Blondlot. In one experiment, an aluminum prism was used to diffract a beam of N-rays. The diffracted beam was claimed to produce a pattern of N-ray that could be revealed by a phosphorescent detector. Blondlot and his co-workers repeatedly found the pattern; Wood could not. The experiment was conducted in a darkened room. This gave Wood the opportunity to reach out surreptitiously and remove the crucial aluminum prism. Not knowing the prism had been removed, Blondlot and his coworkers still found the pattern--even when by their own account there ought to be none.

Once the phenomenon of N-rays was suspected of being a result of self-deception by the experimenters, new trials were proposed that could decide the issue. They were designed to be "blind" tests, in a similar sense to blind controlled studies. Wood suggested that Blondlot be given two identical aluminum envelopes. Only one would contain wet cardboard and thus only one would be impervious to N-rays. Without being told which was which, Blondlot ought to be able to distinguish them merely by checking which allowed N-rays to pass. Similarly the journal *Revue Scientifique* proposed that Blondlot be given two boxes identical in appearance. Only one would contain an N-ray emitter. Without known which was which, Blondlot ought to able to distinguish them by looking for N-ray emissions. Blondlot refused to accept all such challenges, dismissing them as simplistic.

See Irving M Klotz, "The N-ray Affair" *Scientific American*, 242 (1980).

(3) Comparison of New Results with Well-Established Theory

Test (new results) against (accumulated results of many earlier studies = well established theories.)

Example of application

Jacques Benveniste
Elizabeth Davenas
et al. 1988

"Human basophil degranulation triggered by very dilute antiserum against Ige"
Nature, 333: 816-818 (1988)

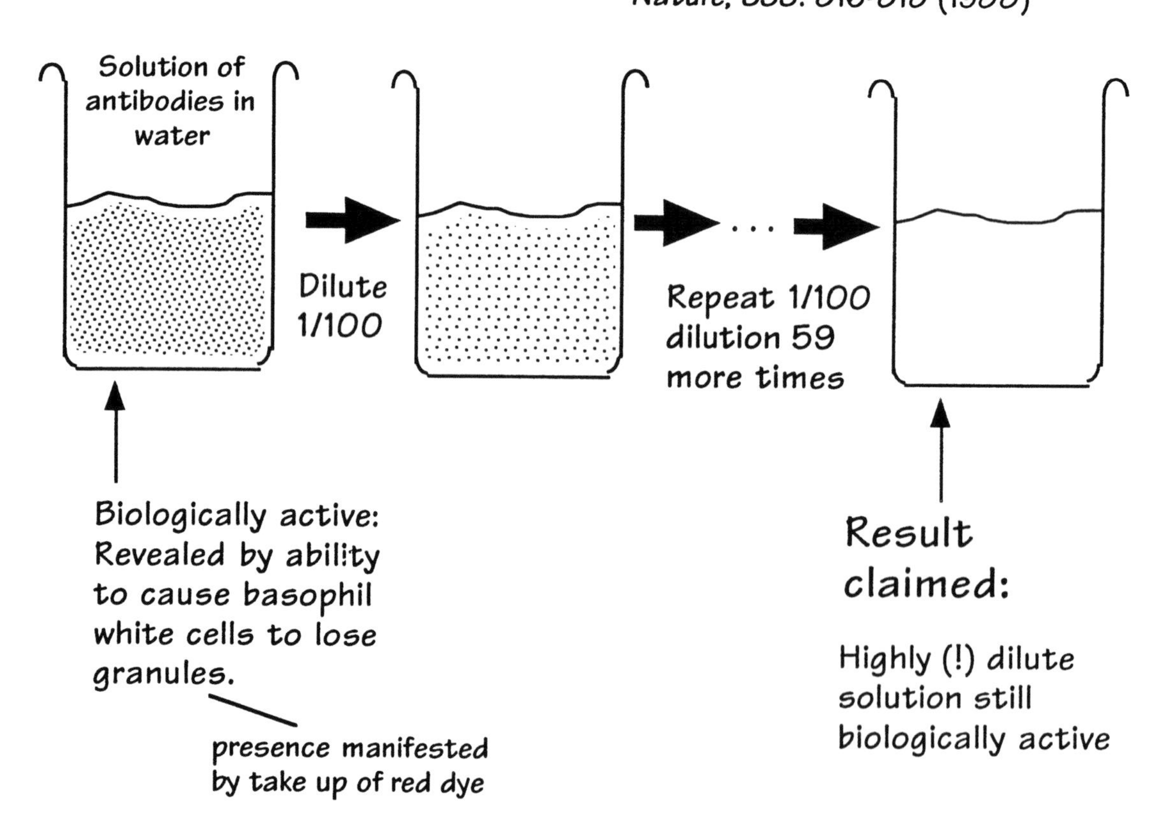

The next device we can use in our inquiry exploits the fact that we already know a very great deal about our world. This knowledge is cataloged in the many, well-established theories that comprise standard science. New results should be checked against what we already know. The supposition is not that our standard science is infallible. But it is more secure than any other reference point. If a new result conflicts with standard science, that ought to arouse the gravest suspicions. We should realize what is at stake in accepting this new result. If it is right, then we must overturn a well-established theory. Such "over-turnings" do occur and they are spectacular. But they are extremely rare. In the vast majority of cases, when a new result conflicts with standard science, that new result will prove to be erroneous in one way or another.

This device is illustrated by recent, controversial work by a French research team, led by Jacques Benveniste, Elizabeth Davenas and others. They investigated solutions of antibodies in water. These antibodies exhibited biological activity; specifically, they could produce a testable change in human white blood cells. The team investigated what happened to this biological activity as the antibody solution was diluted repeatedly with water.

The dilution was extreme. They diluted the solution to 1/100th strength. That means they discarded 99/100th of the original solution and kept only a 1/100th part. They then returned the solution to its original volume with pure water. This was *repeated sixty times* overall--a huge dilution. The result claimed was that the highly diluted solution still retained its biological activity.

The journal *Nature* is a leading scientific journal. Its editor, Maddox, was very skeptical of the result. But he decided to publish it since word of the controversial result had begun to circulate informally. It seemed best for a clear statement of it to be made available. However he took the unusual move of prefacing the article with a note of editorial reservation to alert readers that the article's publication was not to be taken as verification of its results.

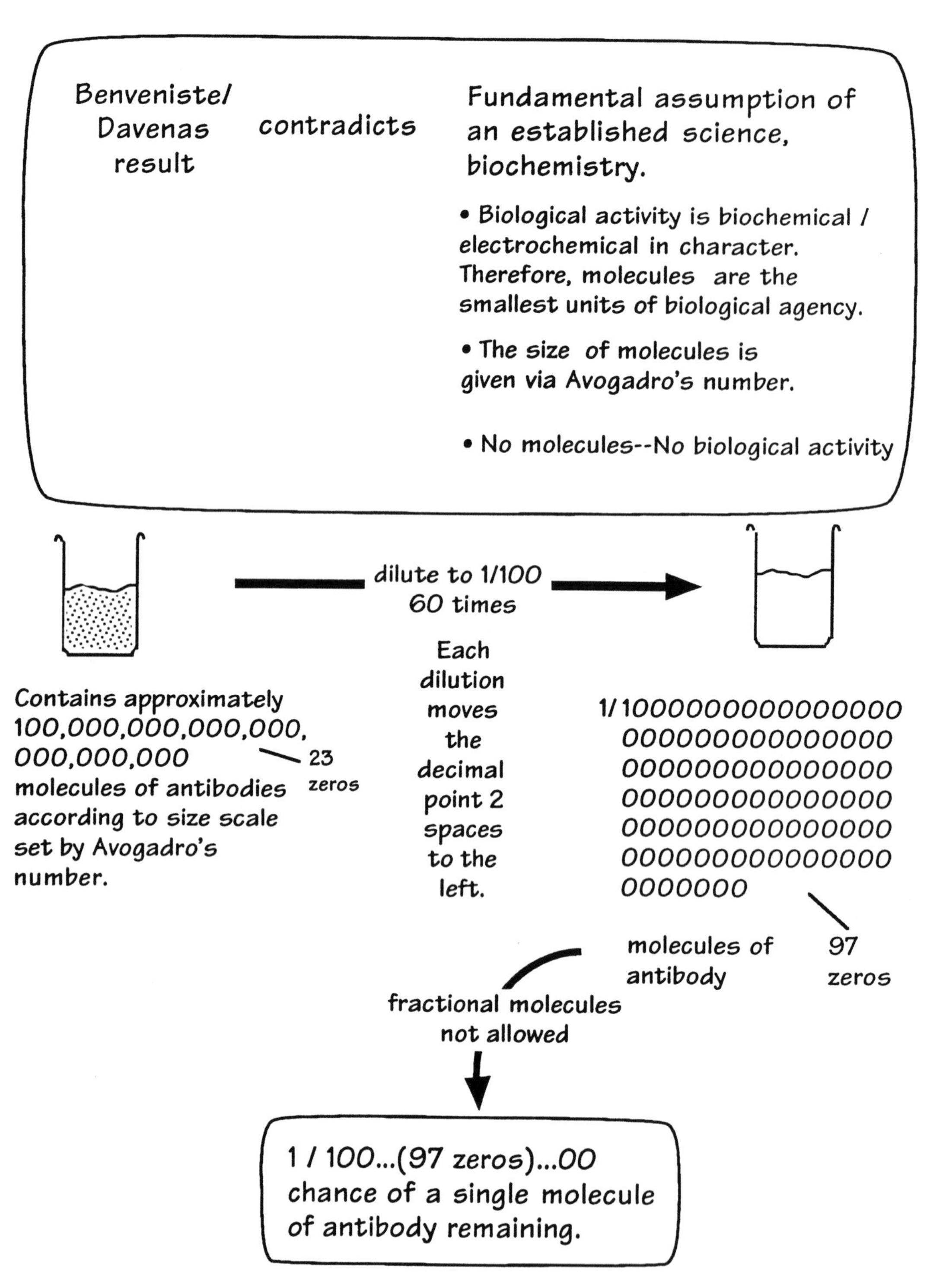
Benveniste/
Davenas
result
contradicts
Fundamental assumption of an established science, biochemistry.
• Biological activity is biochemical / electrochemical in character. Therefore, molecules are the smallest units of biological agency.
• The size of molecules is given via Avogadro's number.
• No molecules--No biological activity
dilute to 1/100
60 times
Each dilution moves the decimal point 2 spaces to the left.
Contains approximately 100,000,000,000,000, 000,000,000 molecules of antibodies according to size scale set by Avogadro's number.
23 zeros
1/1000000000000000
000000000000000
000000000000000
000000000000000
000000000000000
000000000000000
0000000
molecules of antibody
97 zeros
fractional molecules not allowed
1 / 100...(97 zeros)...00 chance of a single molecule of antibody remaining.
Impossible for solution to retain biological activity

Why was the Benveniste/Davenas result greeted with such skepticism? The problem is that the result directly contradicts a basic tenet of a well-established science. In biochemistry it is assumed that all biological activity is ultimately chemical in nature. That is, biological processes--the growth of our bones, the digestion of our food, the contraction of our muscles--are just the molecules of the chemicals of our bodies interacting chemically.

Thus molecules, the smallest chemical units of matter, are also the smallest units of biological activity. If there are no molecules present, then there can be no biological activity. We have established in basic chemistry just how big molecules are. The size estimates are drawn directly from Avogadro's number which gives us a count of how many molecules are present in a substance of given composition and given weight.

From biochemistry, we can estimate how many antibodies are present in the original solution. With each dilution, we can compute how this number drops. After 60 dilutions we find that the antibody solution has been so thoroughly flushed through with diluting water that it is unlikely that even a single of the original antibody molecules remains in the solution. "Unlikely" understates the improbability: fantastically-hugely-enormously unlikely is closer! The calculation sketched opposite shows a 1 in 1000...(97 zeros)...00 chance of even a single molecule of antibody remaining.

Thus biochemistry insists that it is impossible for the highly diluted antibody solutions to have any biological action. This verdict does not depend upon some fancy and fragile result from some obscure corner of biochemistry. It just follows from the most mundane components of biochemistry. We have established how big the biochemical molecules are and that none remain after the repeated flushings of the dilution operations. Without these molecules, there can be no biological activity.

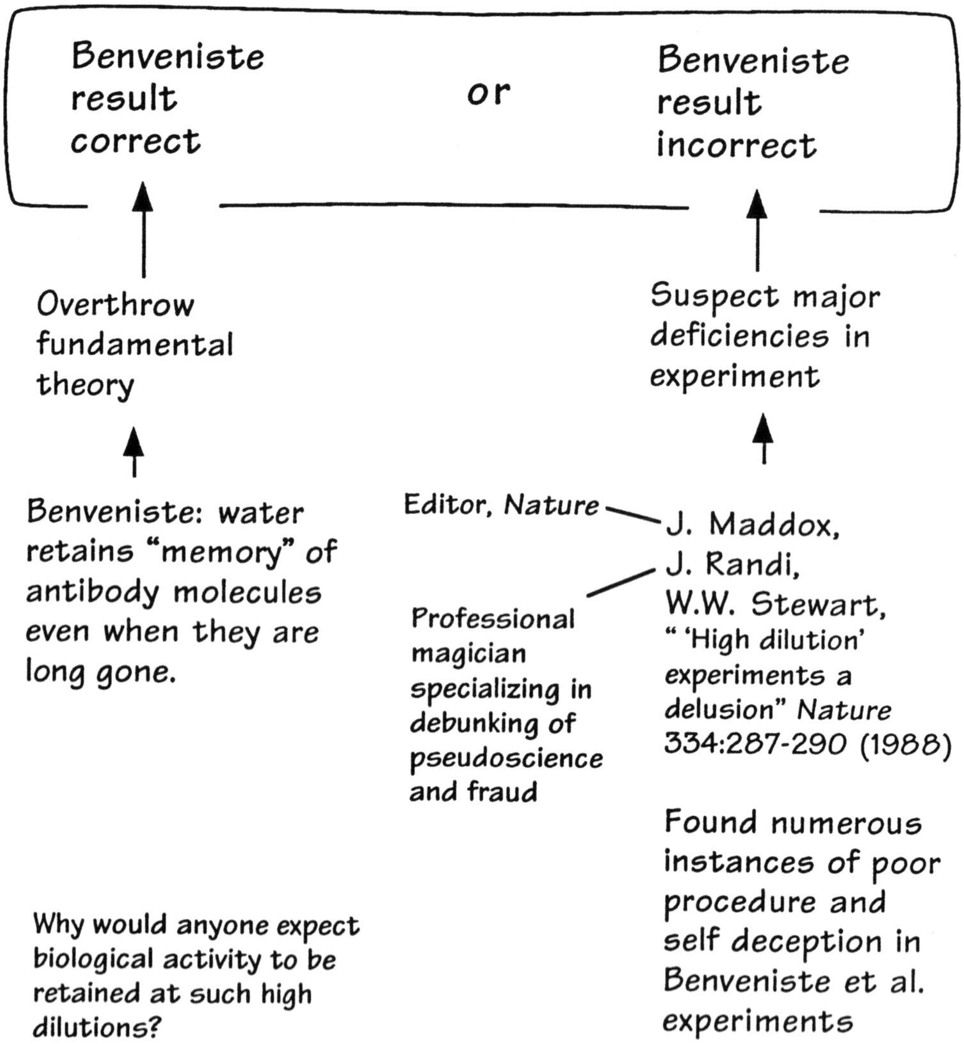

Benveniste result correct — or — Benveniste result incorrect

↑ Overthrow fundamental theory

↑ Benveniste: water retains "memory" of antibody molecules even when they are long gone.

↑ Suspect major deficiencies in experiment

↑ J. Maddox, J. Randi, W.W. Stewart, " 'High dilution' experiments a delusion" *Nature* 334:287-290 (1988)

Editor, *Nature* — J. Maddox

Professional magician specializing in debunking of pseudoscience and fraud — J. Randi

Found numerous instances of poor procedure and self deception in Benveniste et al. experiments

Why would anyone expect biological activity to be retained at such high dilutions?

Homeopathic medicine: based on supposed efficacy of highly diluted agents

Benveniste experiments in part funded by homeopathic medicine manufacturers.

Here is our choice: either the Benveniste result is correct or it is not. The evidence points strongly towards it being incorrect.

If it is correct, then we must discard the basic supposition of an established theory. The simple biological activity observed in the experiment cannot have a chemical basis. This is an outcome that is hard to accept given the huge body of data that supports the chemical basis of biological activity.

How else might we explain biological activity? Benveniste suggested that water retains a "memory" of antibody molecules long after they are gone and this memory allows residual biological activity. This suggestion is very implausible. There seems no credible mechanism for such memory. Worse, if the effect does occur, then we should expect ordinary water to have a huge array of diverse activity, since virtually any sample of water will have been in contact with very many different molecules at the levels of dilution of the experiment. The sample should "remember" them.

If the Benveniste result is incorrect, then we need only suppose that there are deficiencies in the experiments, sufficient to produce a spurious result. Some evidence for this was provided by the journal *Nature*. It assembled a scientific hit team, which included the journal's editor. They visited the Benveniste laboratory and observed the procedures used. They reported numerous instances of poor procedure and self-deception in the researcher's evaluation of the results.

There is one final puzzle. The result claimed by Benveniste seems so fantastic that we must wonder why it was even subject to the time, effort and expense of investigation in the Benveniste laboratory. As to the plausibility of the result, it is the basic postulate of a fringe medical system, homeopathy. This system administers hugely diluted agents as medicines. Because of the high dilutions, their efficacy is strongly doubted by mainstream medicine. If Benveniste could prove his result, that may lend some credibility to the homeopathic system. But why would Benveniste be interested in performing this service? It turned out that his experiments were funded in part by the manufacturers of homeopathic medicines.

(4) Use All Channels of Information

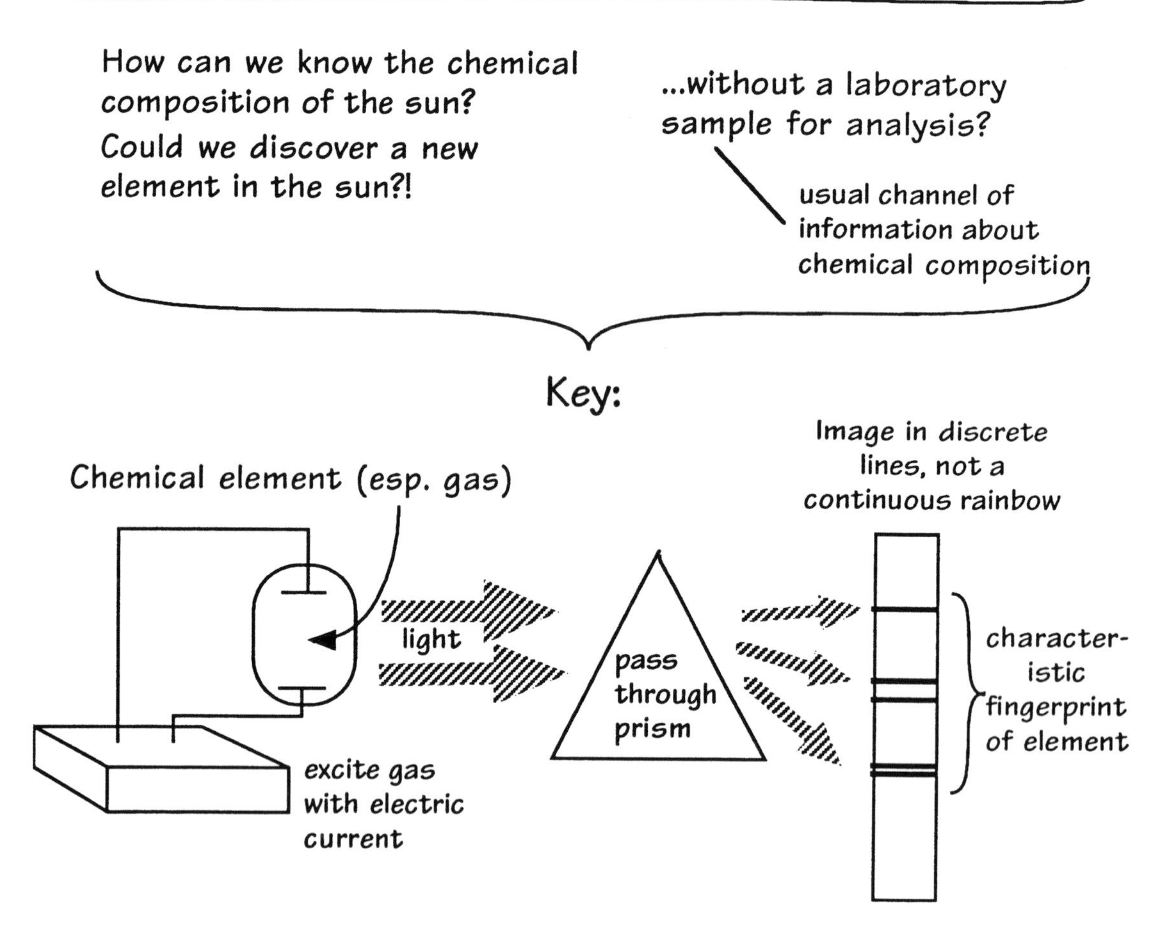

One type of obstacle to knowledge of the world seems insurmountable. Our curiosity often extends to things that are too tiny for us to see, too distant for us to discern, or otherwise beyond our powers of perception. It is here that science has enjoyed huge successes. Repeatedly science has discovered new channels of information that extend the reach of our investigations.

One vivid example illustrates how we may be surrounded by the information we seek without knowing it. How could we know what the sun is made of? Could we discover a new element in the sun before we find it on the earth? To our ancestors in centuries past, this must surely have seemed impossible. Our frail bodies are trapped on the earth; the sun is vastly distant; and it is far too hot to sample even if we knew how to approach it.

What provided the key to these investigations was the study of excited gases. If one takes a gas enclosed in a glass tube and excites it--energizes it--by passing an electric current through it, it will emit light of various colors. One familiar example is the the sodium vapor lamps that are sometimes used for street lighting. These lamps glow a brilliant orange, the color mix emitted by excited sodium vapor.

This light turns out to have a very special structure. It is a mixture of different colors, that is, of light of different wavelengths. Where we might expect a continuous mix of colors, instead we find that only a very few colors (that is, distinct wavelengths) are present. The precise mix of colors turns out to provide a characteristic fingerprint of the emitting element. We read the fingerprint by passing the emitted light through a prism or other device like a diffraction grating that splits the light into its different colors. The resulting pattern--the "atomic emission spectrum"--will not be a continuous rainbow, but a distinctive pattern of lines. This pattern can be used to identify the emitting element.

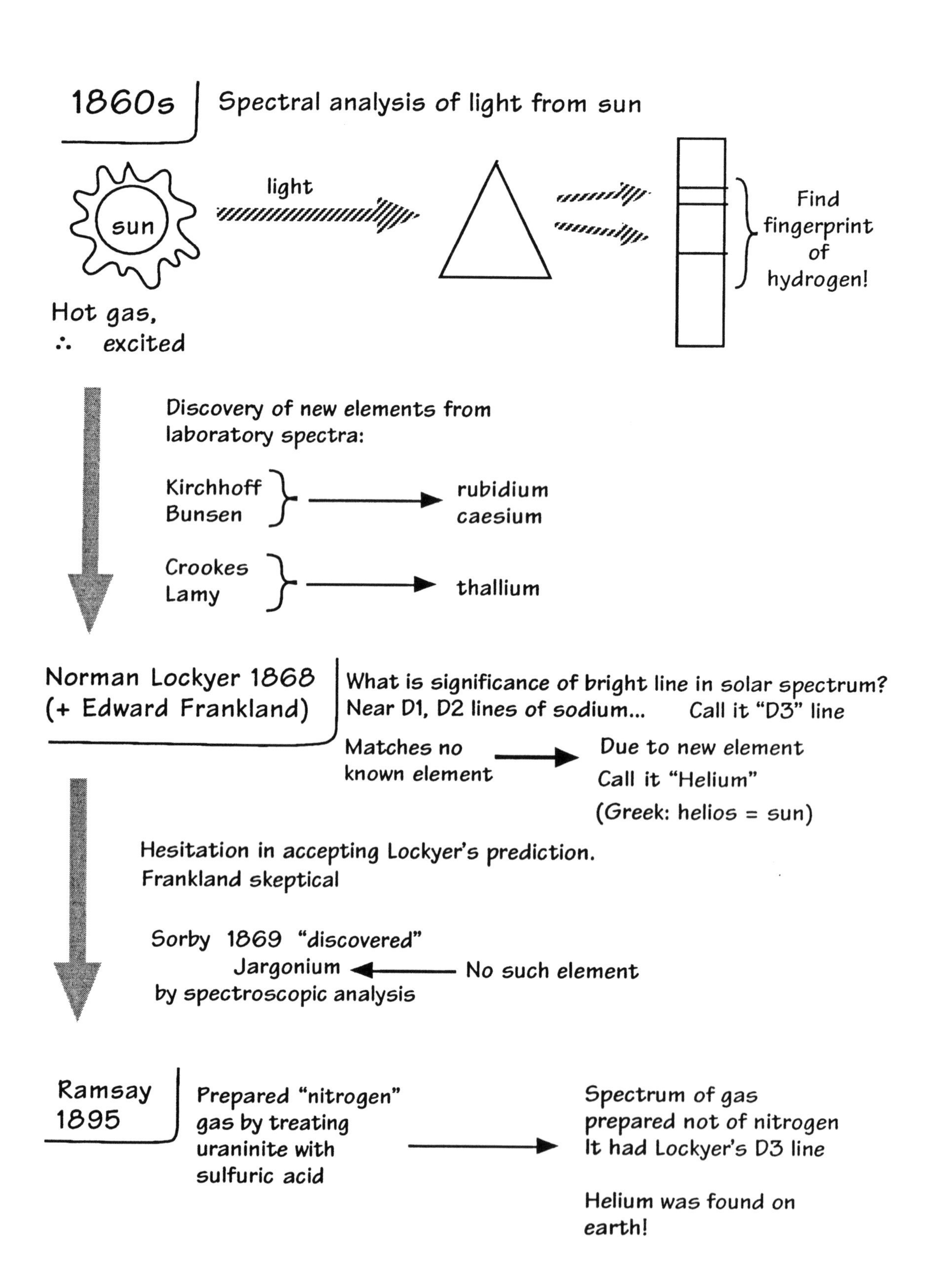
1860s
Spectral analysis of light from sun
sun
light
Find fingerprint of hydrogen!
Hot gas,
∴ excited
Discovery of new elements from laboratory spectra:
Kirchhoff
Bunsen
rubidium
caesium
Crookes
Lamy
thallium
Norman Lockyer 1868
(+ Edward Frankland)
What is significance of bright line in solar spectrum?
Near D1, D2 lines of sodium... Call it "D3" line
Matches no known element
Due to new element
Call it "Helium"
(Greek: helios = sun)
Hesitation in accepting Lockyer's prediction.
Frankland skeptical
Sorby 1869 "discovered" Jargonium by spectroscopic analysis
No such element
Ramsay
1895
Prepared "nitrogen" gas by treating uraninite with sulfuric acid
Spectrum of gas prepared not of nitrogen
It had Lockyer's D3 line
Helium was found on earth!

In the 1860s scientists began to investigate the spectrum of sunlight. They found the emission spectrum of the element hydrogen, thus establishing that hydrogen is present in the sun. Ironically that piece of information about the sun's chemical structure has poured down on us on every sunny day.

But more was to come. Scientists had found that they could discover new elements by pursuing spectra of laboratory samples that matched no known spectra. Could the same be done with the spectrum of sunlight? In 1868, Normal Lockyer and Edward Franklin noted a bright line in the solar spectrum. It was near the "D_1" and "D_2" lines of the sodium spectrum, so they called it the "D_3" line. Lockyer conjectured that it was the line of new, a hitherto unknown element. He called it "Helium" after the Greek for the sun. There was some hesitation in accepting Lockyer's suggestion. Not all such predictions were successful. Henry Sorby had used similar spectroscopic techniques to discover a new element Jargonium. But it turned out to be a mistake; there was no such element.

Vindication for Lockyer's conjecture came in 1895. Ramsay used a standard technique for preparing what was thought to be nitrogen, treating uraninite with sulfuric acid. But the spectrum of the gas proved not to the that of nitrogen; it was a new element. It turned out to have Lockyer's D_3 line. It was the helium he had "seen" decades before in the sun. Helium had been found on the earth.

This use of spectra is but one of the many ways we can extend our reach. Often quite simple devices yield results. Atoms are so very small, that we can never hope to see them. How could we know their size? In 1905, Einstein found a beautifully simple way. Robert Brown had noticed that tiny pollen grains jiggle in an endless dance in water. Einstein realized that these jiggles were due to the endless jostling of the grains by the rapidly moving molecules of water. From the size of these jiggles, Einstein was able to estimate the size of the water molecules.

(5) Use of Evidence

Basic rule:
Believe that for which we have good evidence.
How do we marshal evidence so that it makes the strongest case for our hypotheses and theories?

*** Evidence should overdetermine theory**

Evidence should not just be compatible with the hypothesis; it should test it.

Example:
Fixing position from compass bearings

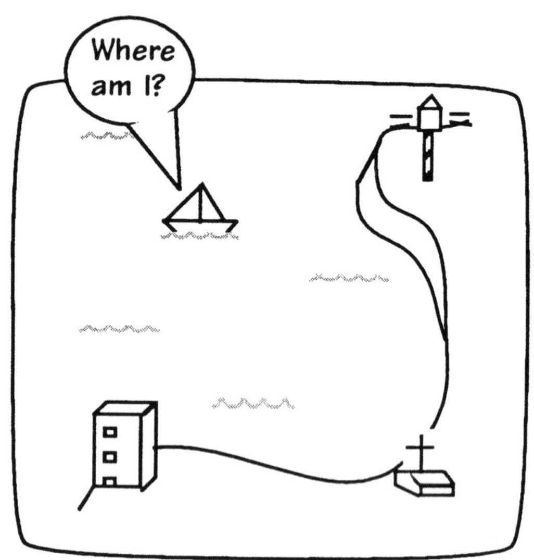

Take compass bearings from known landmarks

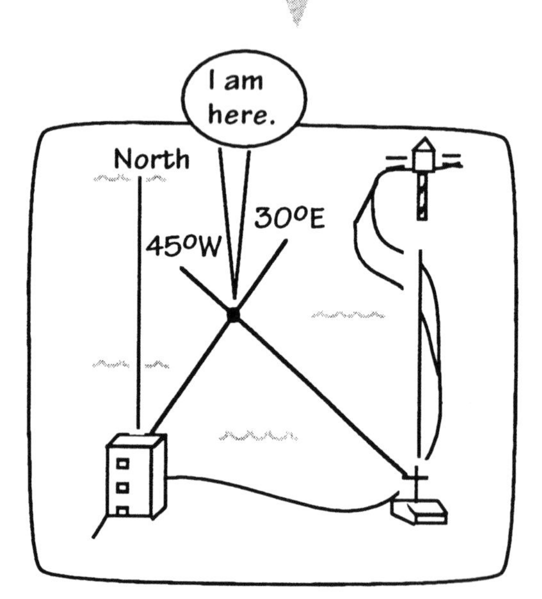

BUT
How accurate is the position fix?
Is the fix correct?
Was a landmark misidentified?

In the end all of the devices we use reduce to one principle: believe that for which we have good evidence. The methods discussed so far are all designed to enable the amassing of evidence that supports (or refutes!) the hypothesis or theory at issue. Evidence can provide weak or strong support. There are some general ways in which one can seek and collect evidence so that it makes a stronger case.

To begin, we hope that the evidence will be compatible with our hypothesis--otherwise the evidence will have refuted it. But mere compatibility does not make a strong case. That strength comes when the evidence overdetermines the hypothesis. Then the evidence tests the hypothesis. This principle is hard to understand when stated in these vague, general terms. An example will make its content clear.

Consider the problem of sailors trying to locate their position at sea. If they can see known landmarks on the coast, they can use compass bearings to decide the direction in which they lie from these landmarks. These directions can be drawn as lines on a map. Where they cross is the boat's position. Opposite, the sailboat is 30° East of the tall building and 45° West of the church.

The precise point the sailors plot on their map is compatible with these bearings. But some doubts must remain. Is the position determined accurately? Have they made some blunder that means it is completely incorrect. Did they mistake a different building for the church? These worries mean that the hypothesis (the plotted position) is not strongly supported by the evidence (the bearings), although clearly it does provide some support. How can the sailors improve their confidence in their hypothesized position?

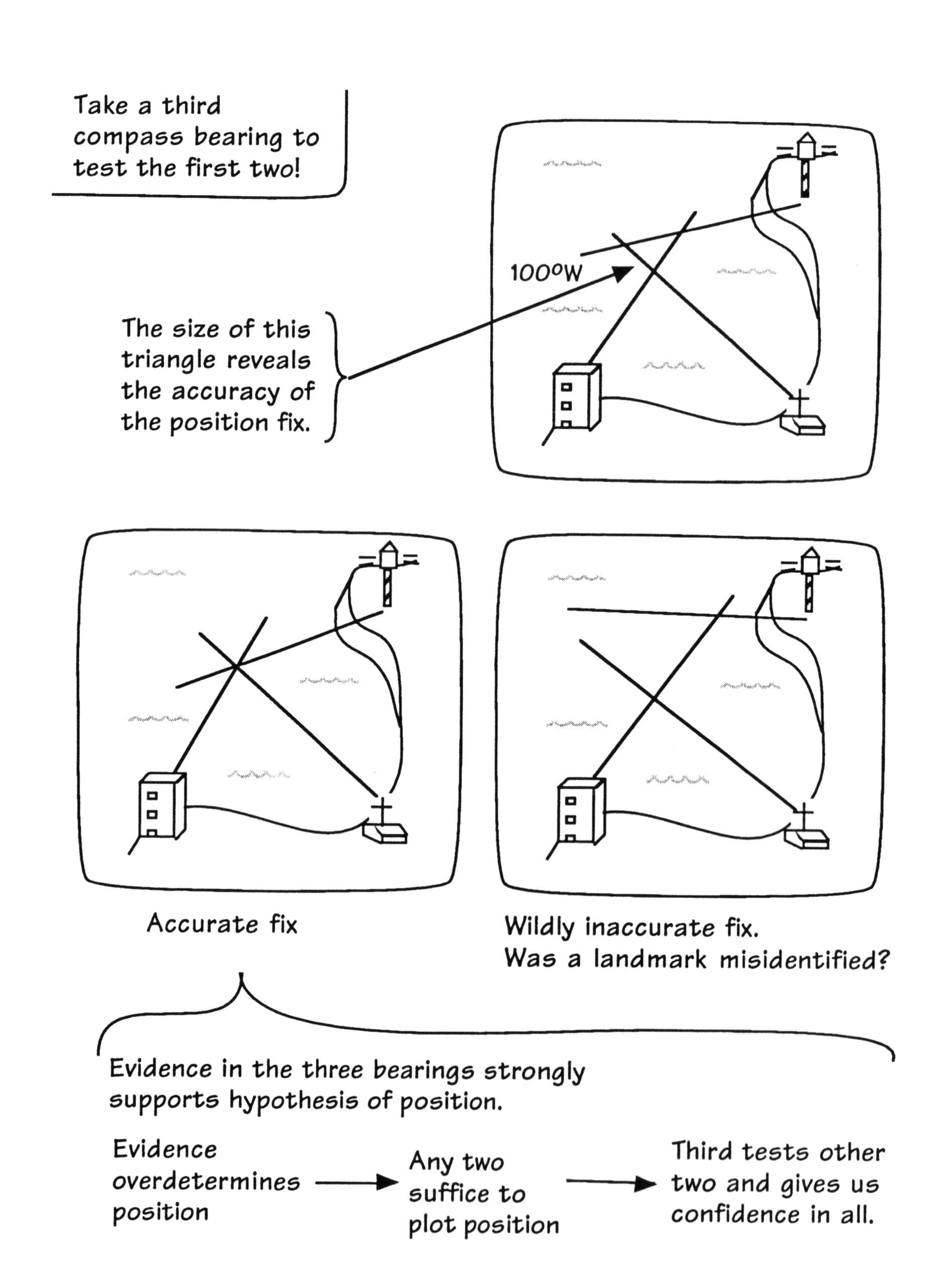

Evidence overdetermines position → Any two suffice to plot position → Third tests other two and gives us confidence in all.

All these worries can be settled by taking a bearing off a third landmark. If everything has been done correctly, the third line added to the map should agree exactly with the position plotted from the first two. This is the case of the accurate fix shown opposite. But if anything has gone wrong, this ought to be revealed in inconsistencies between the different bearings. If the compass readings are inaccurate, then the three lines will not meet at a single point. They will miss one another and form a triangle. The amount by which they miss is measured by the size of the triangle they form. This triangle gives a vivid indication of the accuracy of the position fix.

If there are more serious problems, they too should be revealed. If, for example, one or more of the landmarks have been misidentified, then it is very unlikely that the three bearings will agree on the position, even approximately.

These effects illustrate how a sufficient body of evidence can overdetermine an hypothesis. Any two bearings are needed to plot a position; but three do a much better job. Each of the three bearings tests the other two. If there is an error in even one, that will become clear. If the fix is accurate, as revealed by the smallness of the triangle, we can have very high confidence in our position fix.

✱ Confirm theory with independent evidence

"Independent witnesses" are most convincing.

Locate position with:

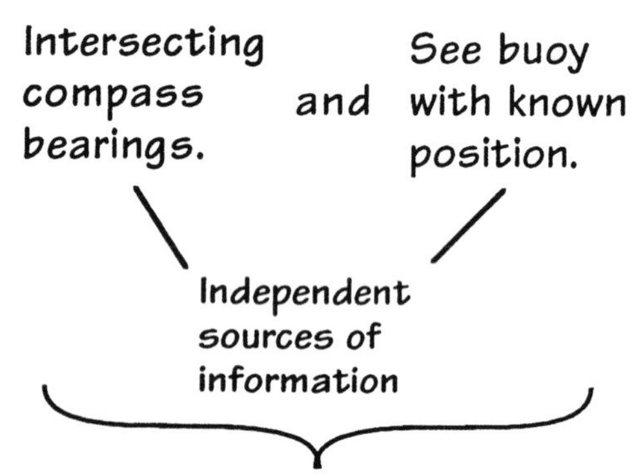

If they agree, you can be very sure of your position fix.

Perrin's (1910s) estimate of size of atoms from many independent phenomena

Brownian motion
Alpha decay
Black body radiation
Electrochemistry
.
.
.

All Yield → Same size for atoms

Strong evidence that correct size has been found and thus for the reality of atoms.

Our search for evidence that overdetermines our theory or hypothesis leads us to prefer evidence from independent sources or, more briefly, independent evidence. This is something that we are familiar with in the legal context. The testimony of one witness can be good evidence. But it is all the stronger to have a second witness with no connection to first and who saw the same thing.

The third bearing in the position fixing example already had this independent character in the sense that it was distinct from the first two. Generally we seek independence in the more extreme sense that the evidence is of a quite different type. Thus our sailors may seek to compare their position fix with the known position of a buoy. If evidence from these independent sources agree--the bearings and the position of the buoy--then the sailors can have great confidence in their position fix.

This use of independent evidence is very common and very powerful in science. In the early 1910s, when the existence of atoms was still under debate, Jean Perrin mounted a powerful case for the existence of atoms using independent evidence. He cataloged all know methods for determining the size of atoms. He found that each of these independent methods agreed. That gave strong evidence that the size estimates were correct and therefore that the atoms measured did exist.

Massive overdetermination by evidence sustains our confidence in well established theories

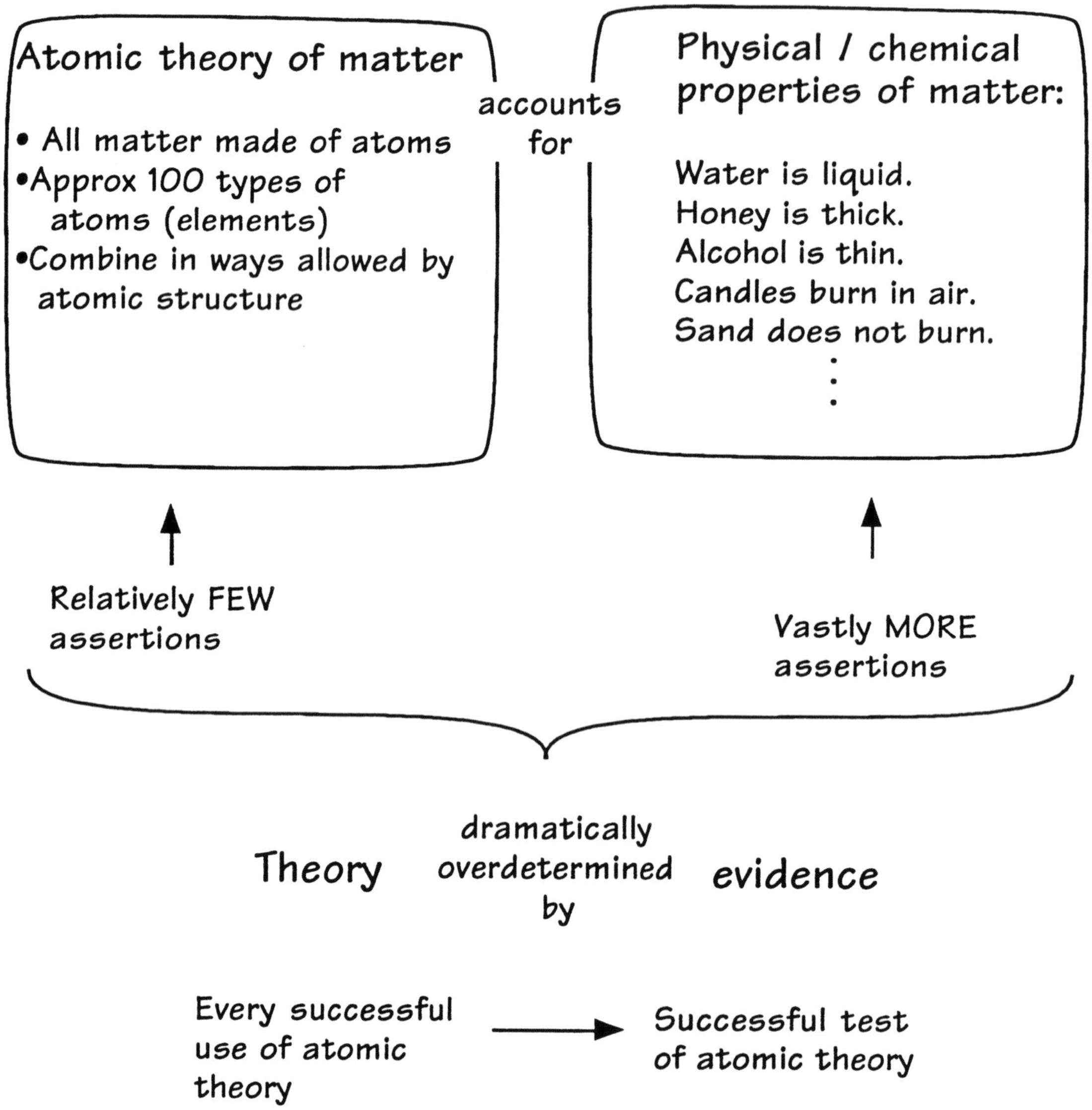

Atomic theory of matter

- All matter made of atoms
- Approx 100 types of atoms (elements)
- Combine in ways allowed by atomic structure

accounts for

Physical / chemical properties of matter:

Water is liquid.
Honey is thick.
Alcohol is thin.
Candles burn in air.
Sand does not burn.
⋮

Relatively FEW assertions

Vastly MORE assertions

Theory — dramatically overdetermined by — evidence

Every successful use of atomic theory → Successful test of atomic theory

These examples of overdetermination and independence of evidence are simple and almost trivial when taken in isolation. However the accumulated effect of many such instances is what leads us to the high degree of confidence that we have in our established theories.

Take the case of the atomic theory of matter. The theory, and the chemistry that it supports, requires a large but still relatively modest number of assumptions. Yet these assumptions are capable of accounting for a huge range of observable physical and chemical properties of matter. The possibility for overdetermination lies in the fact that the number of assertions the theory makes is far less than the number that forms the evidence for it. Whenever the theory is used in a new domain of experience, each successful application amounts to a successful test of the theory, further enhancing the strong case for it.

Replication of an experiment = Confirmation by independent evidence

General expectation: Accept no result of experiment until the experiment has been successfully repeated in another laboratory.

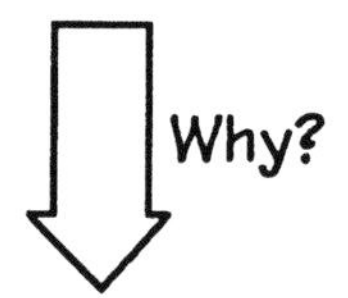

Either the new result properly reflects Nature or the new result merely reflects errors in the experiment undertaken.

Decide which by checking if a completely different group of people in a different lab get the same result.

= independent evidence that the result reflects Nature and not some vagary of the context of the original experiment.

Experiments that were successfully replicated:

* Salk polio vaccine.
* Röntgen's X-rays
* Most major results of science.

Experiments that were not successfully replicated:

* Blondlot's N-rays
* Benveniste's massive dilution experiments
* Pons and Fleischmann's cold fusion experiments.

There is a very important application of our preference for results supported by independent evidence. It lies in a common expectation in science for any new experimental result. It must be possible for another laboratory to repeat the experiment and get the same result. Indeed, if a new result is surprising or implausible, belief in the result will routinely be suspended until replication has been achieved. This was certainly the reaction to Benveniste's dubious results on massive dilution and these doubts proved to be justified.

The demand for replication at first seems odd. If we measure Mount Everest and find it the tallest mountain in the world, what point is there in measuring it again? We just repeat ourselves. The point, of course, is that the first measurement may have been in error. If we are anxious to prove that Everest is the tallest, we might inadvertantly overlook some error that would compromise the result. A good way to check that this has not happened is to get someone else to redo the measurement. We would like the second team to be as independent from the first as possible. When they both produce the same result, we conclude that this result reflects something that they have in common. But we have sought to ensure that the only thing they have in common is Mount Everest and not, for example, the people and instruments used in the first measurement. So the result reflects the mountain and we have two independent items of evidence.

The results of any experiment can be tainted by poor procedures, wishful thinking, simple blunders or a myriad of other factors peculiar to who does the experiment and where it is done. The demand for replication is a simple mechanism for controlling these sorts of errors.

We have seen several examples of experiments that could be replicated and several that could not. The new one listed opposite pertains to an announcement by Stanley Pons and Martin Fleischmann of the University of Utah in 1989 that they had successfully produced a nuclear fusion reaction in a test tube at room temperature. This announcement was astounding. A fusion reaction is the nuclear reaction that powers a hydrogen bomb. Exhaustive efforts over preceding decades have only ever succeeded in producing the reaction at temperatures and pressures comparable to those found within a hydrogen bomb explosion. Skepticism was vindicated by failures of attempts to replicate the experiment.

* Build a case that shows how evidence supports the chosen theory and that no other can reasonably be expected to compete.

Trial for murder

Show evidence points strongly to accused and there is no other likely culprit.

e.g. fingerprints point uniquely to presence of accused

Controlled Studies

Test population shows reduction in rate of polio.

Control population fails to show reduction.

Only difference between the two populations is that the test population was given the Salk vaccine.

Conclude: Salk vaccine reduced rate of polio and there is no other plausible cause.

Ideally, what we would like to show is that our hypothesis or theory is strongly supported by the evidence and that there is no other hypothesis or theory that can compete. For then, accepting the hypothesis or theory is essentially irresistible. Now it may seem an impossible dream to mount a case that is this strong. That is not so. It is difficult and, when it is done, the arguments involved may be very complicated. A few examples at least make its possibility plausible.

In the legal context, let us say in a trial for murder, this is the type of case prosecuting attorneys try to mount. They do try to show that there is good evidence that the accused committed the crime. But they will also expend considerable effort to show that there are no other viable candidates for the culprit. Various forensic tools, such a fingerprint analysis and DNA analysis, can be invaluable here, for they are able to point with a high degree of certainty to a particular individual and no other.

This type of exclusive case is also the goal of controlled studies. Consider the trials of the Salk polio vaccine. The investigators will be able to make such a case if their test population shows reduced rates of polio compared with the control population. If the study was set up properly, then the only systematic difference between the two populations would be that the test population was given the vaccine and the control population was not. The cause of the different rates would have to be a systematic difference in the populations. So we would have to conclude that the vaccine was responsible for the reduced rate of polio and that there could be no other plausible cause.

Techniques for finding Nature's secrets.

1. Controlled Studies

2. Recast Claims in Objective Terms

3. Compare New Results against Well Established Theory

4. Use All Channels of Information

5. Use of Evidence: Overdetermination, Independence and Uniqueness

Modern science provides us with an image of the natural world of great depth and detail. It assures us of that the cosmos came to be out of a fiery conflagration some 100 some billion years ago; that our home is merely a speck of rock orbiting a ball of luminous hydrogen and helium; that this ball glows through continuing explosions just like our terrestrial hydrogen bombs; that our bodies are just complex electrical and chemical systems; that their construction is directed by plans encoded in vast DNA molecules enclosed in every cell; and so on in levels of detail so great that no one person can hope to master any more than the tiniest corner of it.

Immense as it is, there is no mystery in the origin of this image. The climbing of a mountain results from many small steps; the erection of a huge building from the accumulation of many small parts. The edifice of science was built from many small and mundane pieces, each tested and retested to ensure that it can sustain what will follow. There are no recipes assured to provide success in finding the next piece. In principle we may need to invent new approaches and methods for each. In practice, however, it turns out that many of the approaches that have worked, continue to work. The five approaches discussed in this chapter have earned their place in the list of tried and trusted techniques.

Assignment 3: How We Know

Time and again science delivers results about nature whose discovery seems beyond ordinary human powers. The purpose of this assignment is to give you a sense that such discoveries can be made by quite ordinary methods if one can only find a sufficiently ingenious way of applying them. For this assignment, submit

1. A clear statement of a result in science that cannot be know by simple observation.
   Example: The sun contains a new element Helium.

2. A brief statement of why this cannot be know by simple observation.
   Example: The sun is too far away and too hot for us to be able to recover a sample of it for laboratory analysis.

3. A brief account of how ordinary methods were sufficient after all to establish the result.
   Example: We did not need to travel to the sun to obtain a sample. The light from the sun provided the clues. It contains spectral lines typical of terrestrial elements, but coinciding with none of the lines of the known elements. Thus Lockyer conjectured that the lines were due to an element then unknown on earth. His conjecture was confirmed when Ramsay found the element Helium on earth and it was verified that it had the same spectral lines.

4. A citation to the source you used for information.
   Example: A. J. Meadows, *Science and Controversy: A Biography of Sir Normal Lockyer, Founder Editor of Nature* (MIT Press, 1972), pp.50-60, 194-198.

In 3., the point is not just to say who is credited with the discovery and when it happened. The point is to find the method and reasoning used to arrive at the result.

You will need to do a bit of digging for this assignment. Almost any scientific source should provide material, although you may find that sources that lean somewhat towards historical aspects are more helpful. One suggestion is the text: B. Rensberger, *How the World Works*. Quill, 1986. Here are a few possibilities, however, *you are urged strongly to find your own examples.*

- The Universe began in a fiery conflagration 100 billion years ago.
- The stars are at many *different* distances from us.
- There is no air in outer space.
- The Earth is roughly spherical.
- The moon is made of rock and dust. (How was it known before the Apollo landings?)
- All matter is made of atoms too tiny to see.
- Water is not an element.
- Fossils are not of recent origin but can be as much as 3 billion years old.
- Blood circulates through the body.
- Maggots and vermin are not spontaneously generated by decaying organic matter.
- The plague is communicated by fleas.
- Malaria results from mosquito bites.
- AIDS is caused by a virus.
- Isaac Newton (died 1727) suffered some level of Mercury poisoning.
- An individual's traits are encoded in the DNA molecules.
- There is no biggest prime number.

Chapter 4

The Conservation of Energy:

The First Law of Thermodynamics

Why are we studying thermodynamics?

(1) Illustration of a deep theory

We will see how a few basic principles and concepts can govern an enormous range of phenomena in unexpected ways. } How science works!

(2) Provides a very novel perspective on familiar processes

The theory has useful and direct applications. It is useful knowledge for a good citizen.

What is thermodynamics?

Literally: Study of motion of heat

More generally: Study of energy & its transformations

Basic Principle: "No free lunch", "You cannot get something for nothing"

make precise & investigate consequences

In the preceding chapters, most of our discussion of science has been on a general level. We cannot achieve a deeper understanding of science without a deeper and more systematic exposure to one example of a real science. Thermodynamics turns out to be an ideal science to study. We can proceed very rapidly from core principles and postulates to seeing how the theory is structured and how these principles and concepts allow us to make sense of an enormous range of our experiences of the world. In addition, thermodynamics turns out to be of enormous importance in technology. The wise and efficient exploitation of energy is of the greatest significance to our lives. A knowledge of thermodynamics provides citizens an unparalleled insight into the great energy industries (transportation, power generation, etc.) and leads to perspectives not otherwise attainable.

In the five chapters to follow, we will develop basic notions of the science of thermodynamics and some of its applications. The science of thermodynamics is, read literally, the study of the dynamics of heat, that is, the motion of heat, the things that cause that motion and its effects. The science arose historically in studying the dynamics of heat. However it rapidly became the general science of energy and its transformations.

In everyday life one has a deep suspicion of a scheme that offers something for nothing, a free lunch. Everyday wisdom assures us that there are no free lunches. A scheme that seems too good to be true probably is not true. It is the same in the study of energy. In that context it is possible to make very precise the senses in which you cannot get something for nothing. These precise statements are the laws of thermodynamics; and the science of thermodynamics is the study of their consequences, which turn out to be rich and far-reaching.

The laws of thermodynamics

1st law: Energy conservation — more familiar ideas

2nd law: Restrictions on energy transformations — rarely covered outside thermodynamics class

Origins of thermodynamics

Society needs

MOTIVE POWER → moves things to useful ends
=ability to move things.

-move plow to till fields
-turn millstone to grind grain
-haul barges and carts
-turn the great wheels of industry

Is a system generating motive power? —test→ Can it be used to raise weights?

Our introduction will concentrate on the two most important laws of thermodynamics. The first law, to be covered in Chapters 4 and 5, states that energy cannot be created or destroyed; it is always conserved. This notion and its power will be familiar to many of you. The conservation of energy is often covered in elementary science classes. The second law is more subtle. It places limits on our freedom to transform energy into different types. In particular it limits our freedom to transform heat energy into the more valuable form of work energy (such as electrical energy). The law allows us to set up a general theory of heat engines. The theory is especially interesting since most engines commonly used--from automobile engines to electrical power generating plants--are heat engines.

To begin our study of thermodynamics, we recall one of the most basic needs of human society, the ability to move things. Much of our efforts in human history has been given over to setting things into motion--in agriculture, in transportation and in industry. A system that allows us to bring things into motion is one that provides what I shall call "motive power"--the ability to move things. This is a somewhat old-fashioned term and no longer in common use in the sciences. It will prove very useful since it cuts to the core of our concern in thermodynamics.

You may not always be sure if a system is genuinely generating motive power. A simple test will settle your doubts. Can the system be used in some way to raise weights? If yes, then it is generating motive power; if no, then it is not. An automobile engine generates motive power. That is obvious since its function is to bring the automobile into motion. It also can be used to lift weights. We can use the car to raise weights up a hill side; more directly we could connect a rope to the engine so that it would wind up the rope and raise a weight.

Traditional sources of Motive Power:

Muscles	-	Humans, beasts of burden, oxen, horses...
Wind	-	Sailboats, Windmills
Water	-	Water mills
Fire	-	Steam engines (18th century +)
.		.
.		.

All these sources are expensive or erratic or both.

Long tradition seeks **Perpetual motion machine**

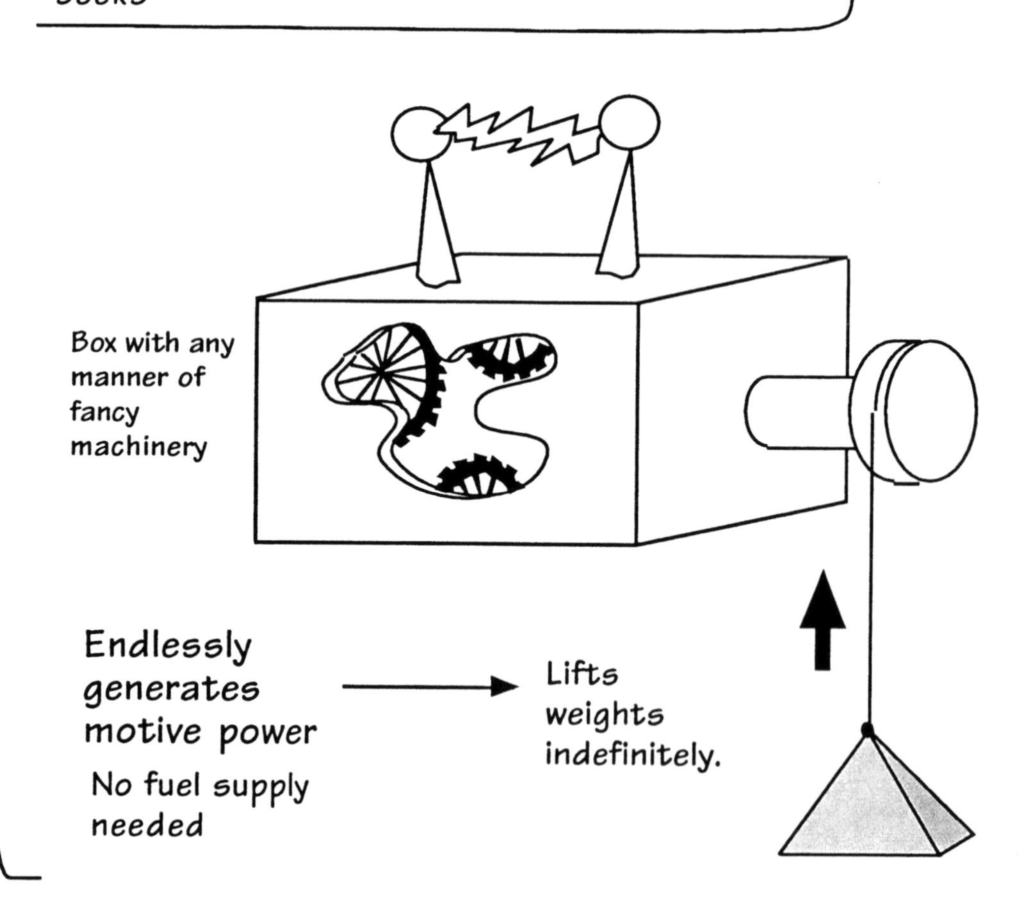

The development of human civilization has proceeded hand in hand with the development of stronger sources of motive power. The earliest sources were muscle power, originally human muscles. Since we humans are weak, all our activities are restricted. One human can only run so far and till so much soil in a day. Any big project requires large numbers of people simply to have enough motive power. Pyramids were built by thousands of workers and galleys were rowed by many slaves.

Use of animal muscles greatly relaxed these restrictions. Horses and oxen generate far more motive power. Better yet are sources such as wind and moving water. They enable sailboats to navigate the globe. Windmills and water mills provided power for much of the machinery of the early industrial revolution. The decisive development came with the steam engine, the use of fire as a source of motive power. Beginning in the eighteenth century, the steam engine was developed and perfected until it became the preferred power source in virtually all areas: it drove us across the land in locomotives, across the seas in ships and turned the great wheels of industry. Our search for motive power continues today. We now exploit sources as diverse as a nuclear power plants to move a submarine and the pressure of geothermal steam--steam from water boiled deep within the earth--to generate electricity.

All these sources of power have a cost. They may be erratic. Horses and oxen are tame, but they do need considerable care and must rest. Wind and water mills are limited by the wind and water available. Steam engines require a continuous supply of an expensive fuel, such as wood or coal. Thus it has long been a dream of inventors to find some machine that would generate motive power without restriction; in particular, no fuel would be required. The machine they sought is the "perpetual motion machine". It would be a device in the most general sense; it did not matter if it used simple mechanical cogs, pipes with fluids, heated gases, electrical components or whatever. All that was required was that it be some contraption that produced motive power without limit and without the need for a fuel. Such a device would revolutionize society to the same extent as the introduction of the steam engine--if only one could be designed.

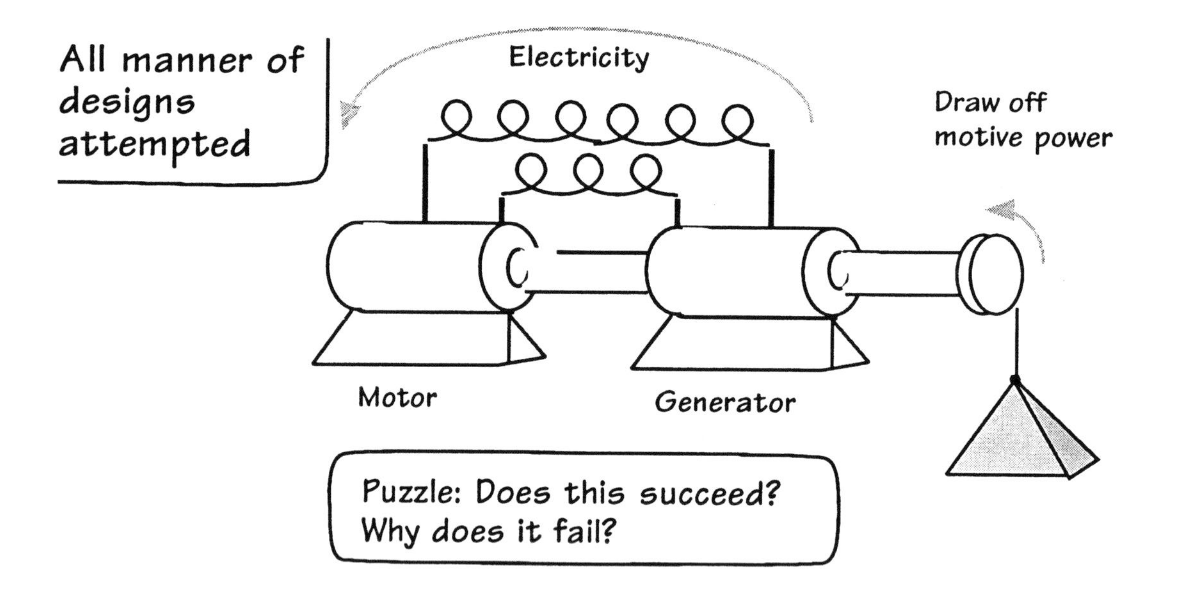

Perpetually over-balanced wheel

One of oldest traditions in perpetual motion:

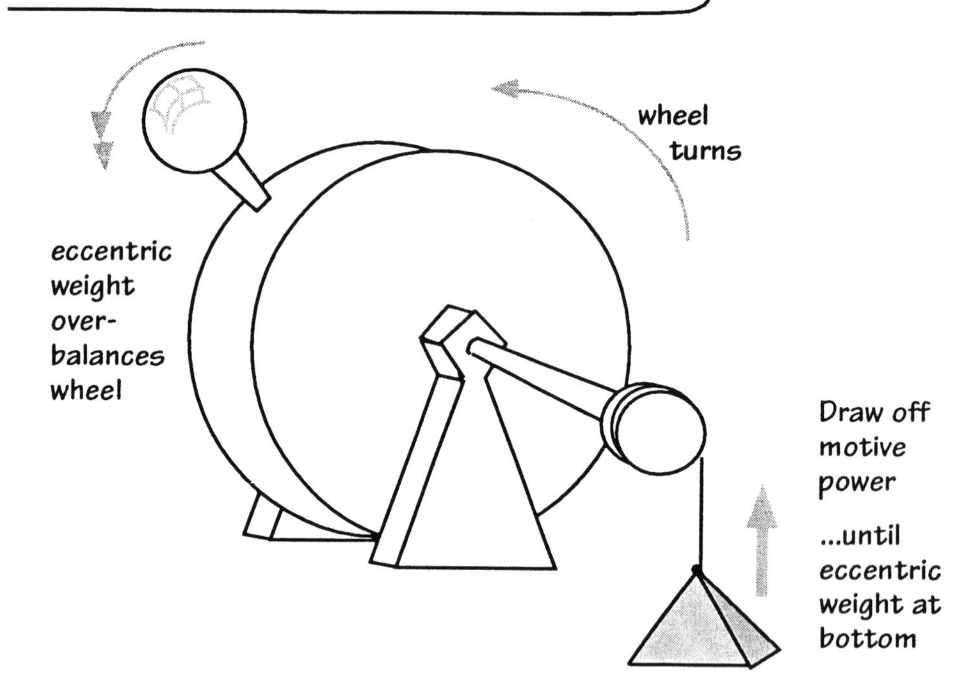

Is it possible to design a wheel that is always overbalanced and always turns of its own accord?

The lure of perpetual motion machines has been strong. Innumerable designs have been tried. Some are extremely simple. One design occurs to anyone who tinkers with electric motors. If you supply electricity to an electric motor, it will turn a shaft. From that turning, one can draw off motive power. A generator is the reverse. It has a shaft which requires motive power if it is to turn. When it turns, the generator produces electricity. So, one asks, what happens if we connect the motor's shaft to the generator's shaft? The motor will run the generator. But where do we get the electricity to run to motor? From the generator, of course! We expect this coupled device to run indefinitely and--now the crucial piece--that we can draw motive power off the common rotating shaft for other uses. Of course the scheme is too good to be true. Can you see where it fails?

One of the oldest traditions in perpetual motion machine design is concerned with an overbalanced wheel. The simplest example of such a wheel is an ordinary wheel with a weight on one side. It will turn until the weight has moved to its lowest position and the wheel becomes balanced. In the course of the motion, motive power can be drawn off. We could use the motion, for example, to turn a pulley that raises a weight.

Now this overbalanced wheel is not a perpetual motion machine. Would it be possible to redesign the wheel so that it becomes perpetually overbalanced? That is, somehow the motion of the wheel is to reposition the weights automatically so that the wheel never attains an even balance. The wheel would turn indefinitely and supply endless motive power.

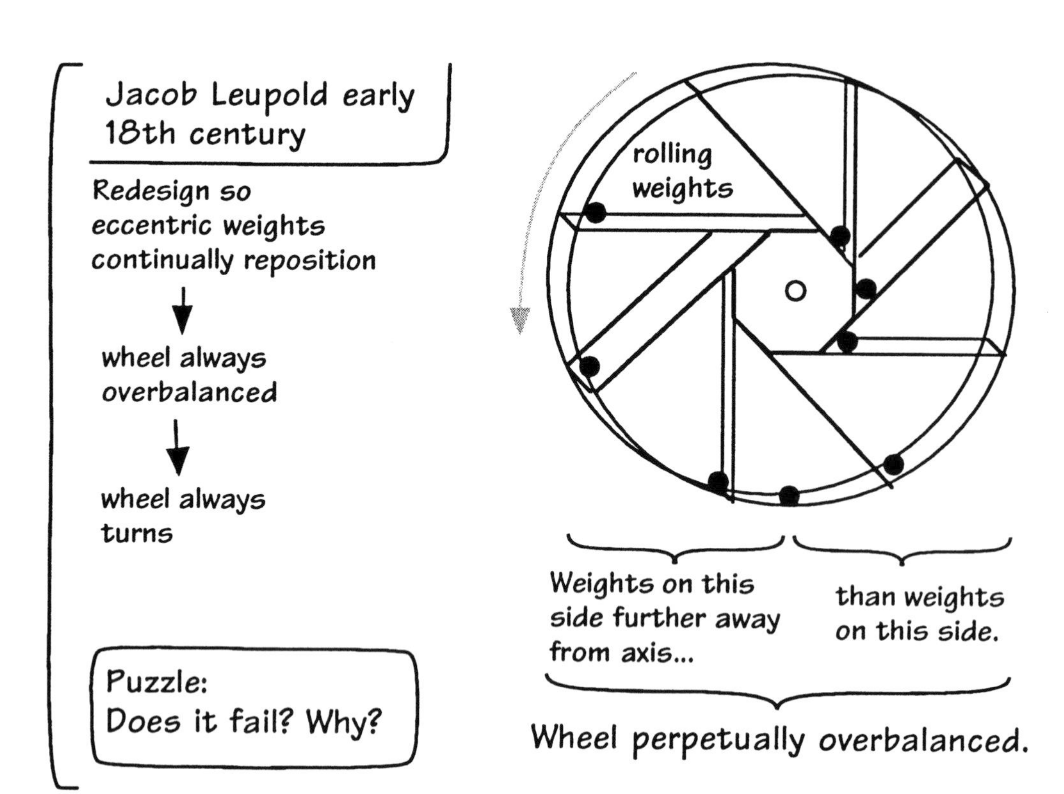
Jacob Leupold early 18th century
Redesign so eccentric weights continually reposition
wheel always overbalanced
wheel always turns
Puzzle:
Does it fail? Why?
rolling weights
Weights on this side further away from axis...
than weights on this side.
Wheel perpetually overbalanced.

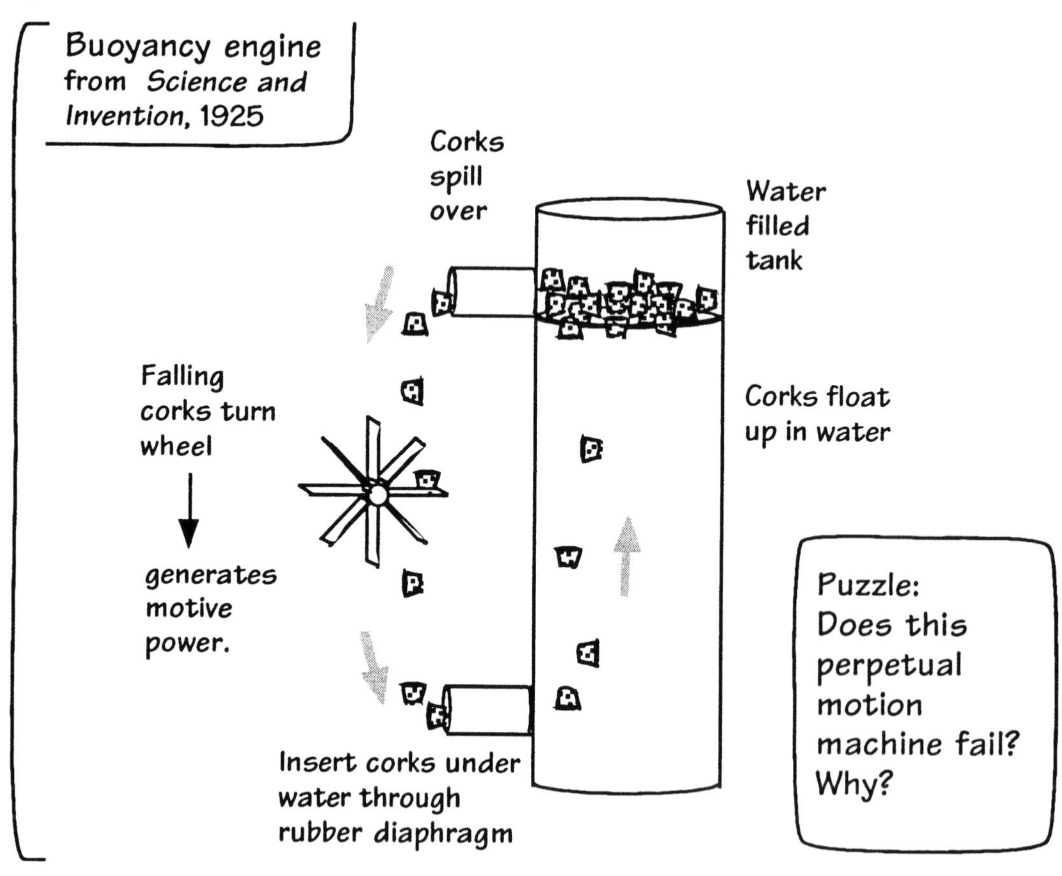
Buoyancy engine from *Science and Invention*, 1925
Corks spill over
Water filled tank
Falling corks turn wheel
generates motive power.
Corks float up in water
Insert corks under water through rubber diaphragm
Puzzle:
Does this perpetual motion machine fail? Why?

Jacob Leupold made one of many attempts to design a perpetually overbalanced wheel. His design was simple. As the wheel turned, the rolling weights would reposition themselves continuously, rolling back and forth in the chambers of the wheel as shown opposite. When they were on the right hand side of the wheel, they would locate themselves close to the wheel's axis. When the were on the left hand side, they would be close to the rim. This imbalance of weights would keep the wheel turning and enable motive power to be drawn from its motion.

Notice that a perpetual motion machine is NOT just a system that perpetually moves. A heavy flywheel set spinning on frictionless bearings will continue to move indefinitely. But it does not count as a perpetual motion machine in the present sense, since it cannot supply limitless motive power. One could couple a spinning flywheel to some device that would draw off motive power by lifting weights, for example. But that would slow down the flywheel and eventually stop its motion completely. Thus a flywheel spinning on frictionless bearings is not a perpetual motion machine since it cannot supply limitless motive power. Leupold's design was intended to overcome this problem by ensuring that the wheel was permanently out of balance. This permanent lack of balance rather than merely the wheel's motion is intended to be the permanent source of motive power.

There have been many ingenious proposals for perpetual motion machines. In the buoyancy engine shown opposite, falling corks turn a wheel from which we draw motive power. Once the corks have fallen, they are inserted through a rubber diaphragm into a tank of water. The corks float to the top of the tank. There they spill over and fall onto the wheel, completing the cycle from which we are supposedly able to draw limitless motive power.

Do these perpetual motion machines work as intended? Is there a flaw in each design? Can you see what they are?

Centuries of failed attempts to redesign perpetual motion machine using all manner of systems (mechanical, hydraulic,...)

Perpetual motion machines are impossible

PROVE special case with Newtonian mechanics

Purely mechanical perpetual motion machine is impossible: (Uses only wheels, cogs, pulleys, springs, etc.)

Academie des Sciences(France) 1775:

Resolves to review no more proposals for perpetual motion machines.

"The construction of a perpetual motion machine is absolutely impossible"

Work of Mayer, Helmholtz, Joule... 1840's

Notion of conservation of forces in nature

"force" not used in modern sense

Clausius, Thomson (Kelvin) 1850's

1st law of Thermodynamics: Conservation of Energy

"energy" first used by Thomas Young 1807 for *vis viva* = energy of motion

If nothing else, there are immense financial rewards to be expected by the successful inventor of a perpetual motion machine. So the quest for a perpetual motion machine has been long and vigorous. But the outcome has invariably been the same. No design works. They all fail, no matter what type of design is employed or what type of process is used--mechanical, thermal, hydraulic, pneumatic, magnetic or any combination of them. Centuries of failure has made the conclusion inescapable: perpetual motion machines are impossible. In certain limited areas, this impossibility could even be proved. If the machine was to be purely mechanical operating only with cogs, wheels, pulleys, springs and the like, then the impossibility followed from Newton's mechanics. The inevitable pattern of failure was so prominent that the French *Academie des Sciences* decided that it would waste no more time reviewing proposals for such machines. They are "absolutely impossible," it announced.

How are we to understand the failure of all perpetual motion machines? The failure is an indication of the operation of a "you cannot get something for nothing" principle at the deepest scientific level. For whatever the principle, it must apply to any branch of science rich enough to allow us to tinker with the design of perpetual motion machines. In the course of the nineteenth century that principle began to take form. In the 1840s, there was growing talk of a "conservation of forces" in nature. A gain of force in one area must always be compensated, they postulated, by a loss of force elsewhere. In so far as we equate "force" with generation of motive power, this amounts to a prohibition on perpetual motion machines. (Notice that in the 1840s the word "force" was not used in the modern precise sense of physics.)

In the 1850s these speculations matured into a science. Rudolph Clausius and William Thomson (later Lord Kelvin) developed the science of thermodynamics. Its first law dealt with the notion of energy. Energy, it asserted, is always conserved. In all systems, a gain of energy in one place can only result from a loss of energy somewhere else. Since the generation of motive power was a form of the production of energy, the first law amounted to a prohibition on perpetual motion machines. If a machine was to produce energy, it must be supplied with energy; in effect, all such machines must be fueled, where that fuel is the source of the energy they produce.

First Law of Thermodynamics
Conservation of energy

In all processes, energy is conserved

Total energy in all forms at start = Total energy in all forms at end

Loss/gain energy in one place ALWAYS EXACTLY compensated by gain/loss energy in another.

All systems contain energy:

- heat energy of fire
- kinetic energy of moving body
- electrical energy of lightning
- chemical energy of food

All processes are a conversion of energy from one form to another.

Wood fire burns:

chemical energy in wood → heat energy in fire

If one considers a particular system, one must also count energy that passes in or out of the system.

e.g. Human for a day

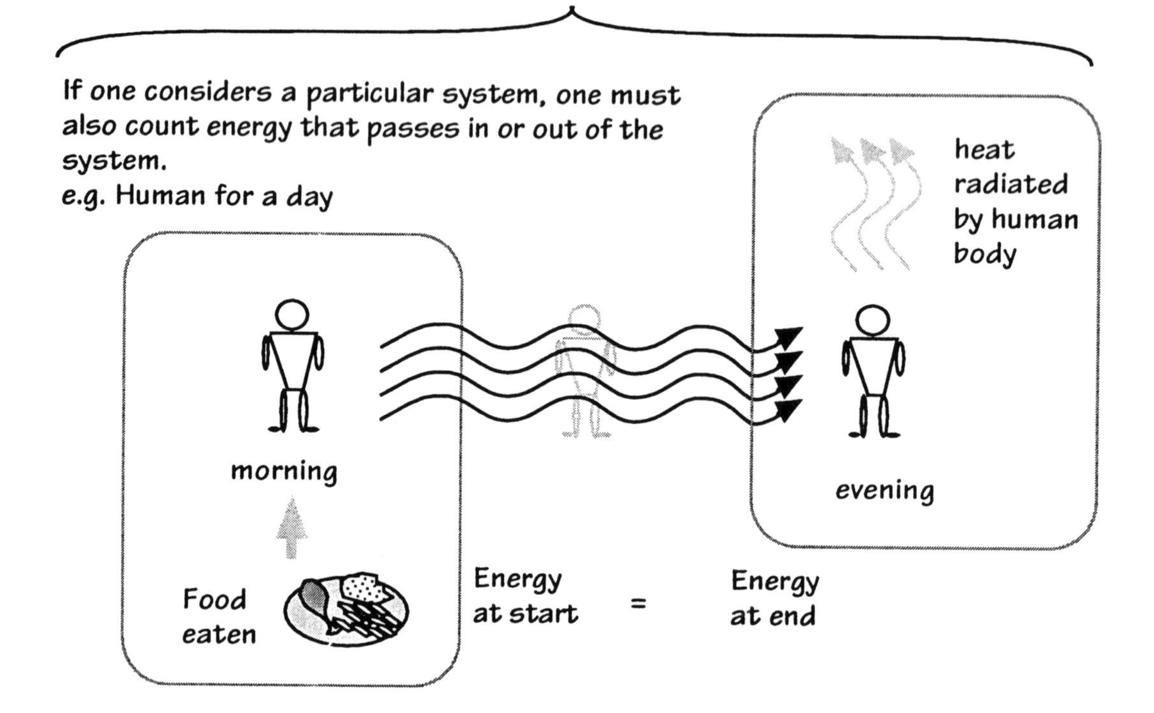

We have now come to the core of thermodynamics, so let us go over this more slowly. There can be no perpetual motion machines since they violate the first law of thermodynamics, the law of conservation of energy. This law presupposes a particular picture of nature. All systems are conceived as containing a certain amount of energy and that energy will have a particular form. For example, a fire contains heat energy. A moving body has energy of motion, called "kinetic" energy. Lightning has electrical energy. Food contains chemical energy. Now all physical processes or changes involve the conversion of energy from one form to another. When wood burns, the chemical energy of the wood is converted to heat energy in the fire. When we humans pursue our daily lives, the chemical energy of our food is converted into energy of motion as our muscles move our bodies, heat energy in our muscles as our metabolism keeps us warm, and so on.

The first law of thermodynamics gives us a fundamental law of accounting. It says that if we count up the total amount of energy at the start of a process, it will be exactly the same as the total energy at the end. To use the technical term, we say energy is "conserved."

What this means is that if we gain energy in one place, we must have lost some in another. Otherwise the total amount of energy in all places involved could not remain the same.

The first law tells us what the outcome of an accounting sum will be. The total energy we start with with exactly equals the total energy we finish up with. Doing this sum can be quite complicated. Imagine that the process we pursue is the life of a human over a day. To add up all the energy we start with, we have to consider every source of energy: they include the energy already in the human's body in the morning, as well as all the energy in the food the human eats. To add up the energy we finish with, we have to consider many components: they include the energy remaining in the human in the evening, as well as all the energy given off in the form of heat, energy of motion and so on. If we track down and add up all of these correctly, the law assures us, we will find that the total amount of energy we started with equals the amount we have at the end.

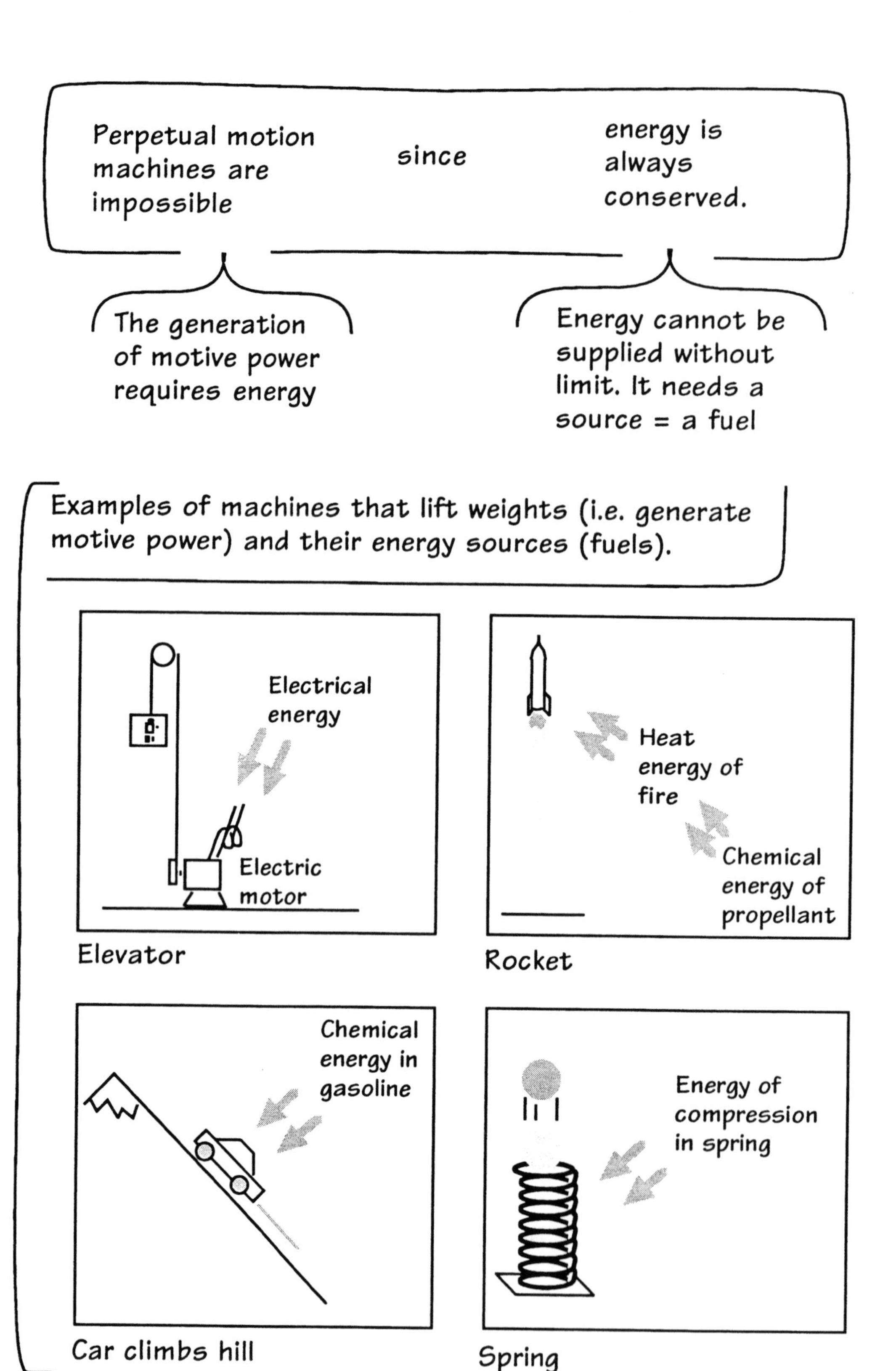
Perpetual motion machines are impossible
since
energy is always conserved.
The generation of motive power requires energy
Energy cannot be supplied without limit. It needs a source = a fuel
Examples of machines that lift weights (i.e. generate motive power) and their energy sources (fuels).
Electrical energy
Electric motor
Elevator
Heat energy of fire
Chemical energy of propellant
Rocket
Chemical energy in gasoline
Car climbs hill
Energy of compression in spring
Spring

We can now see in more detail why the law of conservation of energy prohibits perpetual motion machines. If a possible effect of the generation of motive power is to raise bodies, then the generation of motive power amounts to the supplying of energy. For when we raise a body we increase its energy of height, the energy stored in it because of its height. So a machine that generates motive power is losing energy, which reappears as the energy of height of the bodies raised. If it is to continue generating motive power indefinitely as in a perpetual motion machine, this loss of energy must be made good. That is, it must be supplied with energy. But that supplying of energy is the supplying of a fuel in the most general sense. Therefore the machine cannot be a perpetual motion machine since such a machine is assumed not to require a fuel.

We can see what this need for an energy fuel looks like by taking a few familiar examples of machines that raise weights. Elevators in tall buildings raise the weight of people. They operate with electric motors that wind up cables. The energy that fuels the electric motors is the electrical energy supplied from the power company. A rocket can be used to raise a weight, the "payload," to great heights. The immediate source of the energy used to raise the payload is the heat of the fire in the rocket motors. This energy is supplied in turn by the chemical energy stored in the propellant fuel. A car raises its own weight when it climbs a hill. The energy to raise itself comes from the gasoline fuel supplied to the engine. A compressed spring can hurl a weight skyward. The energy to raise the weight is stored in the tension of the spring when it is compressed. This energy is supplied by whatever force compresses the spring. That force may be your own muscle power.

The Different Forms of Energy and their Conversions

- Different forms of energy tend to be measured in different units
- All the units are interconvertible

In principle and in practice, any of the units can measure any form of energy.

Analogy

We can measure lengths in any of many systems of units:

inches, feet, yards, miles, ...

Use inches for small objects. Use miles for large distances.

But we do not have to use inches for small objects and miles for large distances. We can always convert a measurement in one unit to the other.

Review

There are 63,360 inches in a mile:

(a) How many inches are there in 1.9 miles?

Answer:
1.9 x 63,360 = 120,384

(b) How many miles are there in 158,400 inches?

Answer:
158400/63360 = 2.5

Now that we have our notion of energy and a law that governs it, how are we going to use it? In practical terms, the bulk of our efforts will be spent in considering how forms of energy convert into each other. In pursuing these conversions, we will notice repeatedly that energy in different forms tends to be measured in different units. Heat energy is measured in calories or Btu; electrical energy in kilowatthours; energy of motion in Joules; and so on. The use of different units ought not of itself to be a barrier. If a quantity of energy is described in one unit, we can always reexpress it in another. A good deal of our efforts will be spent on such conversions.

These conversions ought not to trouble us. The idea is actually a familiar one. We are used to lengths being measured in many different units. Our tendency is to use different units for certain ranges of lengths. We use inches for small objects, miles for long distances and yards for intermediate lengths. But we could always use inches for long distances, if we really wanted to, and miles for objects, if we really had to. We can say that a race track is 3600 feet long or 0.68 miles. Both are equally correct, but one may just be more familiar or more convenient. To see they are the same, we need to see how to convert distances measured in one unit, feet, to another, miles.

This type of conversion should be familiar. Two sample review questions are shown opposite. If you have any hesitation about how one does these conversions, you should review this material. The energy conversions and other related calculations that follow are the same as these inch/mile conversions. The only difficulty that you may find is that it is a little harder to visualize a conversion from Joules to Btu than it is from inches to miles.

We now turn to a review of the different types of energy and the units used to measure them.

Energy of height

= energy stored in a weight when it is raised

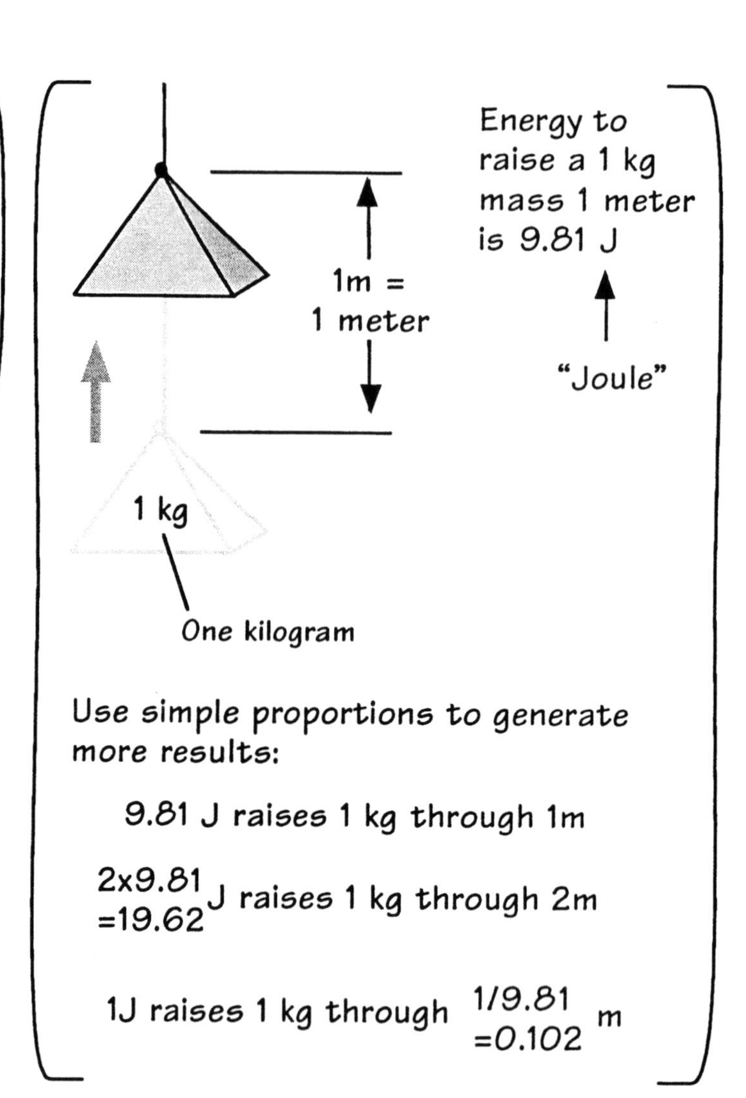

Elsewhere you may find this energy called "potential energy." Beware! The term "potential energy" can be vague and may describe any form of energy that is latent or hidden.

Conversion:

1000 J = 1 kJ

Joule Kilojoule

1 J = 1/1000 kJ = .001 kJ

The energy of height of a body is the energy stored in the body when it is raised to some height. When a brick layer works to lift bricks up to the top of a building, the bricklayer has worked to increase the energy of height of the brick. That energy could be released if the brick were to fall.

Sometimes this energy of height is called "potential energy." We shall avoid the term "potential energy" here. In some contexts the term can have a quite precise meaning. If the context is restricted to the elevation of weights, for example, then the term will just mean energy of height. However in other contexts it can have a far vaguer meaning. There it may just refer to any form of energy that is hidden or latent. Thus the range of energies that may be described in this looser way is huge and vague.

The unit of energy commonly used to measure energy of height is the "Joule". How much energy is there in a Joule? 9.81 Joules (write "9.81 J") is the energy needed to raise a one kilogram ("1 kg") mass by one meter ("1 m"). Once you know this basic fact, many others can be generated by simple proportions. For example, 2x9.81=19.62 J will raise the same kilogram mass by two meters. Or just one Joule will raise the kilogram by 1/9.81 = 0.102 meters.

Throughout we will tend towards the metric units of kilograms and meters. This system is becoming an international standard in science. Also many of the calculations concerning energy are simplified by this system. One kilogram is a little more than two pounds. One meter is a little more than a yard.

A unit of energy derived from the Joule is the kiloJoule. It is 1,000 Joules. The prefix "kilo" designates a thousand and it used quite widely. A kiloton, for example, is a thousand tons.

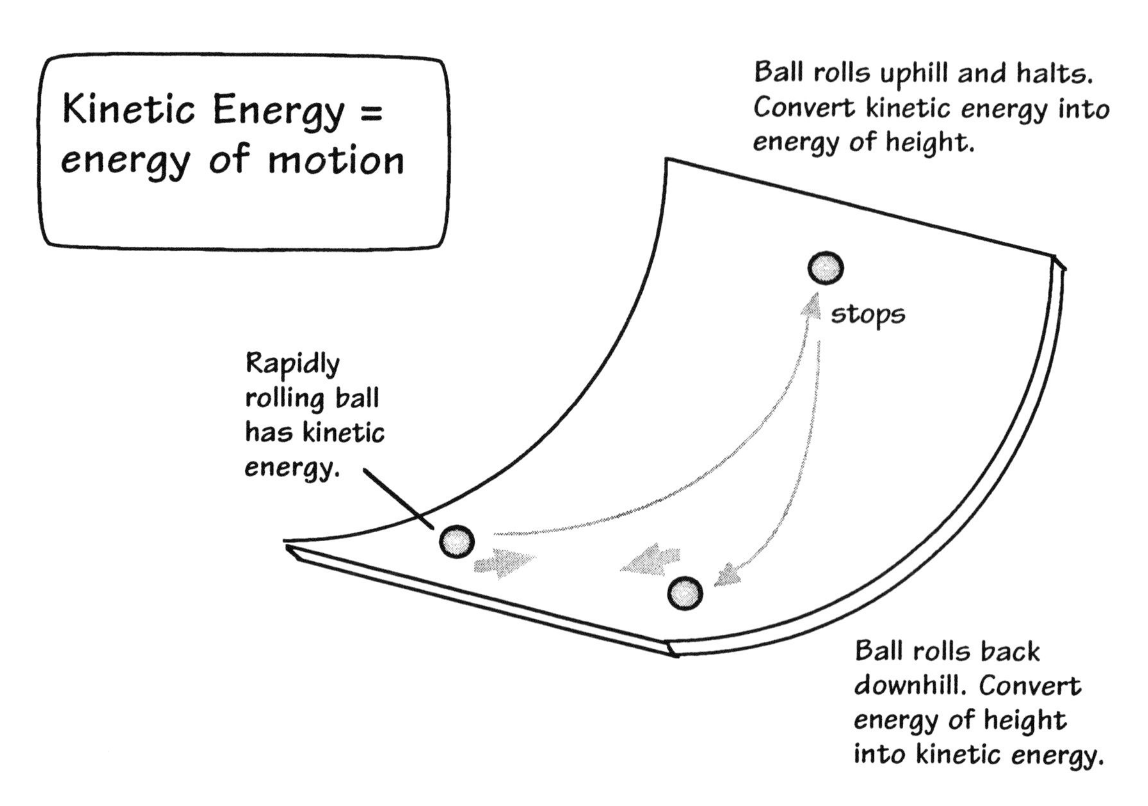
Kinetic Energy =
energy of motion
Ball rolls uphill and halts.
Convert kinetic energy into
energy of height.
stops
Rapidly
rolling ball
has kinetic
energy.
Ball rolls back
downhill. Convert
energy of height
into kinetic energy.

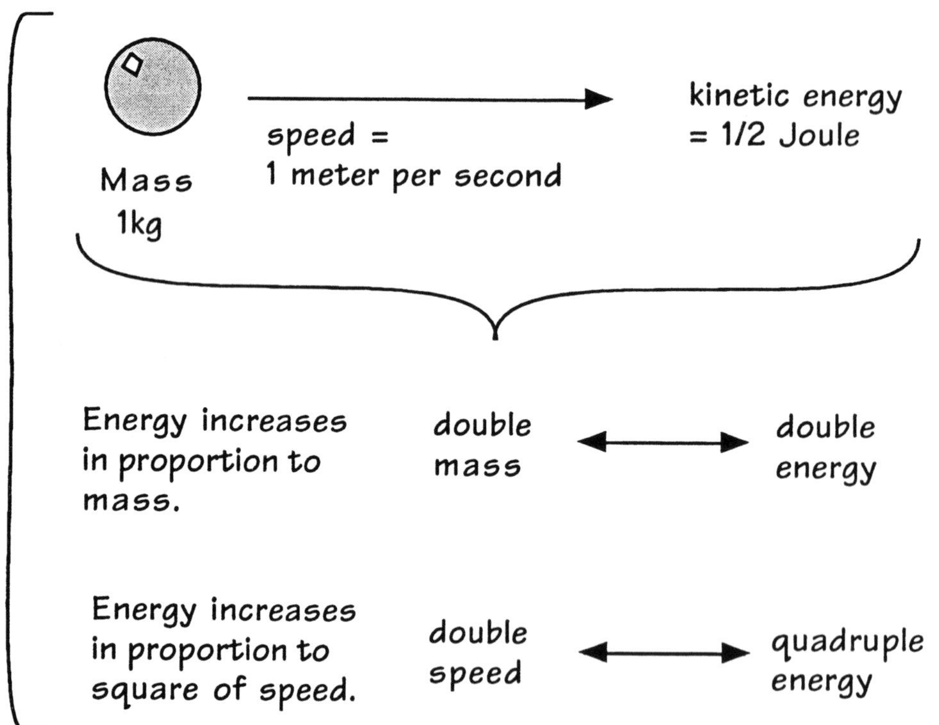
Mass
1kg
speed =
1 meter per second
kinetic energy
= 1/2 Joule
Energy increases
in proportion to
mass.
double
mass
double
energy
Energy increases
in proportion to
square of speed.
double
speed
quadruple
energy

Kinetic energy is the energy a body has because it is moving. A ball rolling up and down an slope illustrates the conversion between kinetic and energy of height. At the bottom of the slope, the rapidly moving ball has a lot of kinetic energy. It loses that kinetic energy as it climbs the slope. At the top of its climb, where it momentarily stops, it has lost all its kinetic energy, which has been converted into energy of height. The ball now rolls back down the slope. At the bottom, where it has recovered its speed, the energy of height of the ball has been converted back into kinetic energy.

Kinetic energy is commonly measured in Joules. For example, a one kilogram mass moving at a speed of one meter per second will have a kinetic energy of 1/2 Joule. That kinetic energy will increase in proportion to the mass of the body. At one meter per second, a 2 kg mass has kinetic energy of 1 J; a 10 kg mass of 5 J; and so on.

Beware, however, kinetic energy does not scale proportionally when we increase the speed; kinetic energy increases in proportion to the square of the speed. Therefore doubling the speed quadruples the kinetic energy. So a 1 kg mass moving at 2 meters per second has kinetic energy of 4x1/2 J = 2 J.

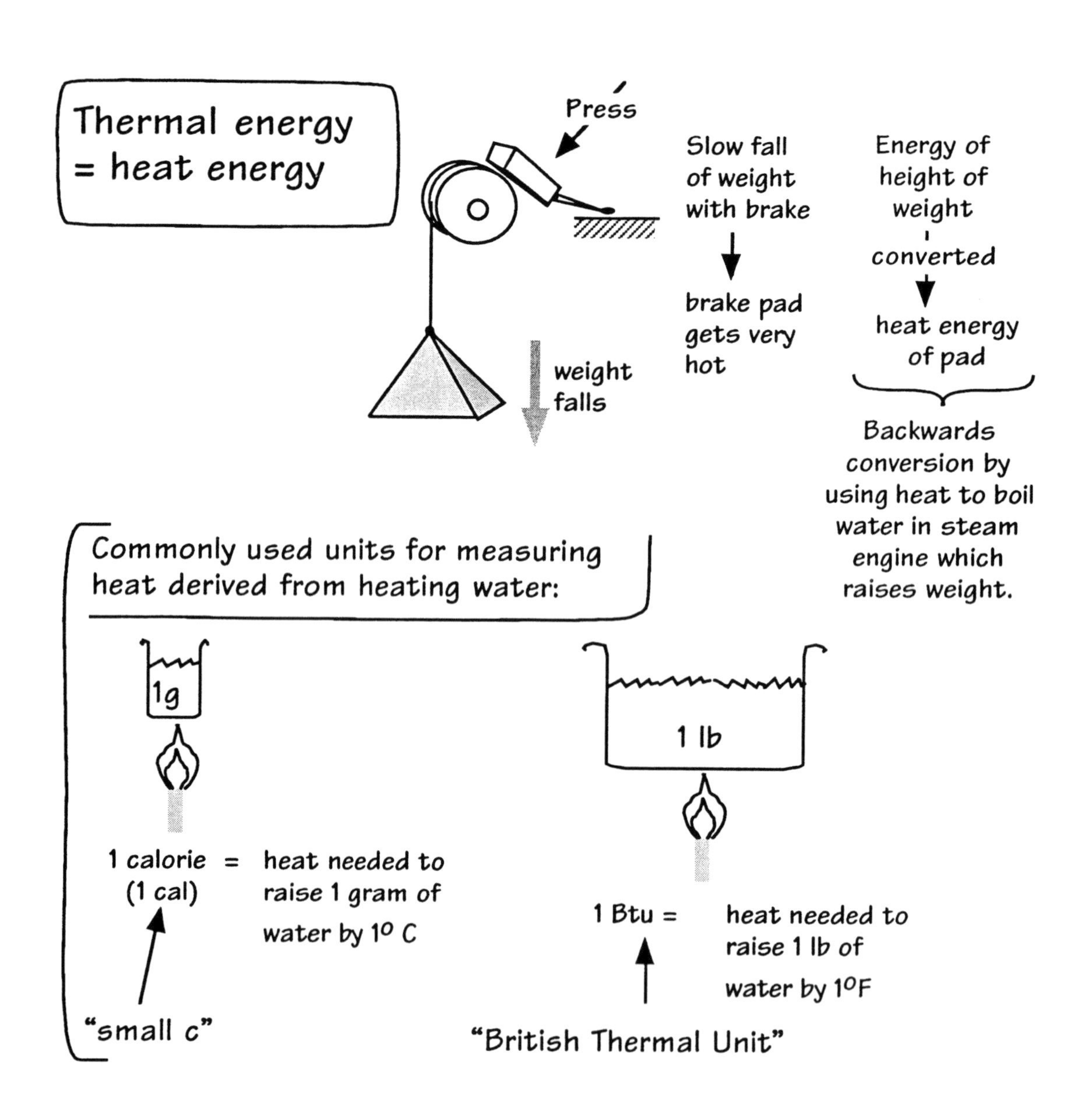

Conversions:

1000 cal = 1 kcal
4.186 J = 1 cal
4.186 kJ = 1 kcal

3.968 Btu = 1 kcal
0.948 Btu = 1 kJ

"kcal" = kilocalorie = 1,000 calories
"kilo" means 1,000

Thermal or heat energy is the energy a body has because it is hot. Heat energy can be readily converted to and from other forms. For example one could use a falling weight to turn a wheel. If the motion of the wheel is kept slow by the applying of a brake pad, then the brake pad will get hot. This process is the conversion of the energy of height of the weight into the heat energy of the pad. It is inevitable that the pad gets hot. For the law of conservation of energy requires that the energy of height lost by the falling weight reappear somewhere else. The same effect arises when a car descents on a steep road from a mountain top. If the brakes are used to slow the descent of the car, they will become hot. If the descent is too rapid, the brake pads can overheat and damage the brake system.

One can convert heat energy back into energy of height--but not so easily. One way is to use the heat energy to generate steam in a steam engine. The power of the steam engine could then be used to raise the weight.

The units that are used to measure heat energy are derived from the heating of water. A calorie ("cal") is the amount of heat needed to heat one gram of water by $1^{o}C$. (Note that this calorie is spelled with a small c. We will see shortly why this matters!) A British thermal unit ("Btu") is the amount of heat needed to heat one pound of water by $1^{o}F$.

The conversions between the various units of heat we have seen are shown in the box opposite. Notice the use of the prefix "kilo" to define the unit "kilocalorie", which is 1,000 calories.

Chemical energy = energy stored in chemical structure of a material

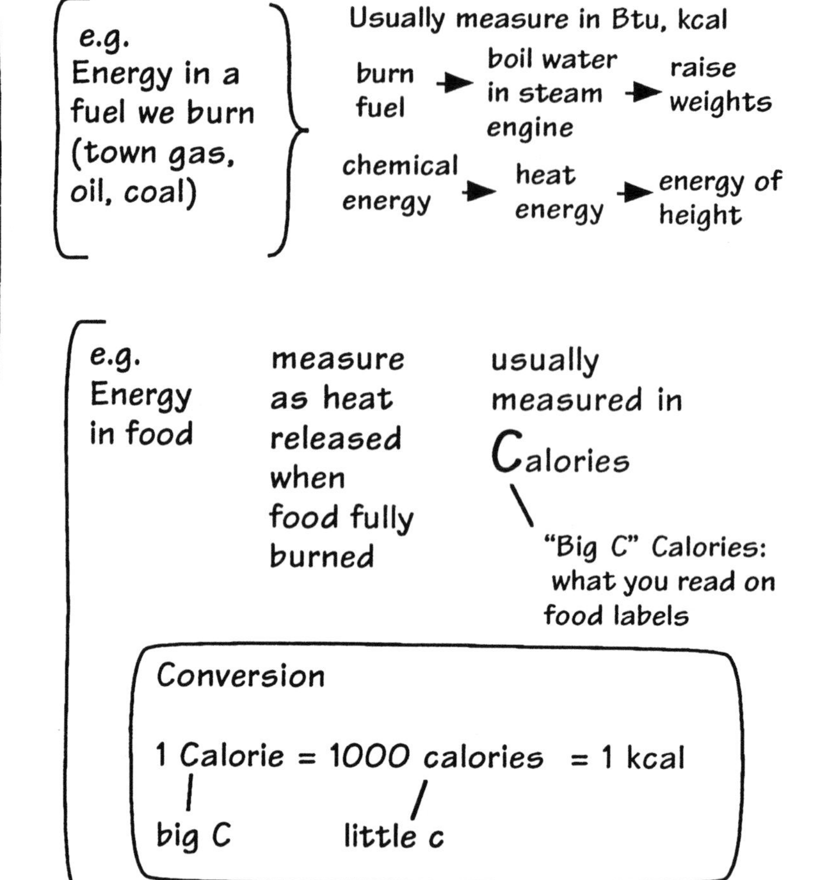

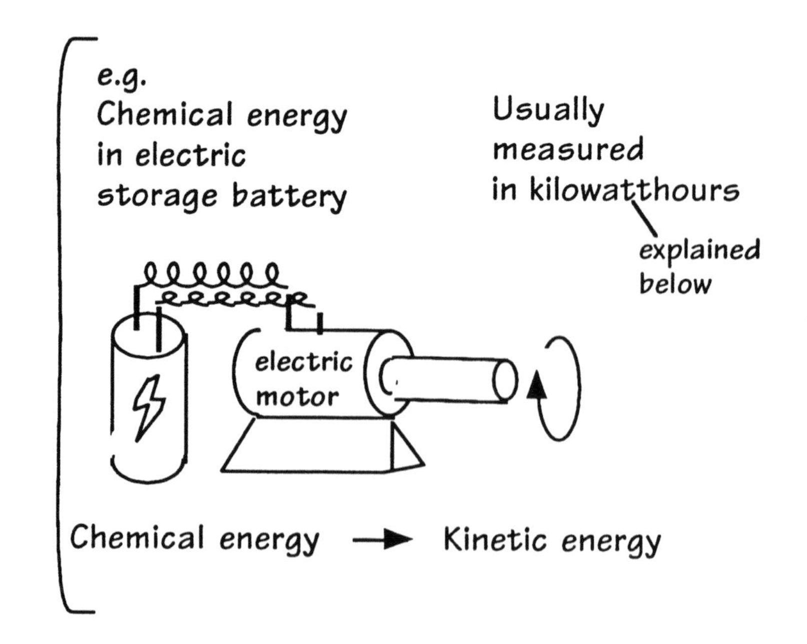

Chemical energy is the energy a body has that is stored in the chemical structure of the body. The most familiar example of chemical energy is the energy stored in fuels that we burn: town gas, oil, coal, wood, etc. Because this chemical energy is most commonly converted into heat energy, it is customary to use a unit of heat energy to measure the chemical energy stored in a fuel. Thus the town gas we buy is bought in units of Btu; we buy it for its energy content. Look on your gas bill. You will see that you are billed for the number of Btus of gas energy you consumed.

The burning of a fuel can initiate a sequence of important energy conversions. For example, we may burn a fuel to generate steam in a steam engine. The pressure of the steam may be used to raise weights. This illustrates the conversion of chemical energy in the fuel to heat energy in the steam to the energy of height of the raised weight.

We eat food for the chemical energy it contains; that is what "fuels" our bodies. We measure the chemical energy of a food by burning it; this is similar to what our bodies do with the food--"burn" it slowly in our cells. So again the unit used to measure the energy content of a food is a unit of heat. The unit usually used in the U.S. is the Calorie. Unfortunately there is a potentially misleading complication. The food energy Calorie is spelled with a "big C". This big C calorie is actually equal to 1,000 of the little c calories we saw before; that is, a big C Calorie is the same as a kilocalorie.

An electric storage battery also holds energy in chemical form. When we connect wires to the battery, we allow electric current to flow from the battery. This current carries energy, so that there is a loss of chemical energy from the battery. Thus when a battery drives an electric motor, we have the conversion of chemical energy in the battery into the electric energy of the current in the wires. The electric motor then converts this energy into the energy of motion of the motor's shaft.

Electrical Energy

Chemical energy in battery

↓

Electrical energy flowing in wires

1 watt = Flow rate of 1 joule of electrical energy per second

A watt is a unit of energy flow, not of energy.

If there is an energy flow of 1 watt, then 1 Joule of electrical energy flows past this point in each second.

Flow of 1 kilowatt of electrical energy

Electrical capacitor stores electrical energy

In one hour, **one kilowatthour** ("1 kWhr") of electrical energy accumulates

Conversions

1000watt = 1 kW (kilowatt)

1 kWhr = 3600 kJ

energy of light, energy of radiation.

Electrical energy is the energy associated with electricity. It is the form of energy that drives most of our household appliances--in particular, all those that need to be plugged in to a power socket. The most vivid manifestation of electrical energy is lightning. When one sees a lightning bolt, one has a sense of seeing pure energy. Nice as that sounds, it is an illusion. There is a huge amount of energy in a lightning bolt. But what we see is light emitted by excited atoms in the air as the electricity of the bolt passes through the air.

A complication with electrical energy is that its most familiar unit, the watt, actually measures energy flow, not energy. If a wire carries one watt of electrical power, this means that one Joule of electrical energy is passing through the wire in each second. (It may be helpful here to think of water volume measured in gallons as opposed to water flow rate measured in gallons per minute. If a pipe carries a flow of one gallon per minute, then one gallon of water passes through each point of the pipe in a minute. The watt as a unit of energy flow corresponds to the water flow unit, gallons per minute.) A very common unit of power is the kilowatt ("kW") which is 1,000 watts.

A familiar source of electrical energy is a battery. It holds chemical energy which can be converted into electrical energy when wires are attached. Electrical energy can be stored directly in an electrical component called a capacitor. If electrical energy flows at a rate of one kilowatt into the capacitor for one hour, we will have stored one kilowatthour of electrical energy in the capacitor.

This unit, the kilowatthour, is the unit commonly used to measure electrical energy. It is equal to 3600 kJ. You can also understand the unit as follows: a kilowatthour of electrical energy is enough energy to run a 1 kilowatt electrical heater (a small room heater) for one hour. When you buy electrical energy from the power company, you buy it in kilowatthours. This is the unit you will find on your bill from the power company. A kilowatthour of electricity costs a few cents. The actual number varies according to the power company and special conditions. Five cents per kilowatthour is typical.

Electrical energy can easily be converted into other forms. For example, if we connect a light bulb to the power line, we convert the electrical energy into light energy, a form of energy of radiation.

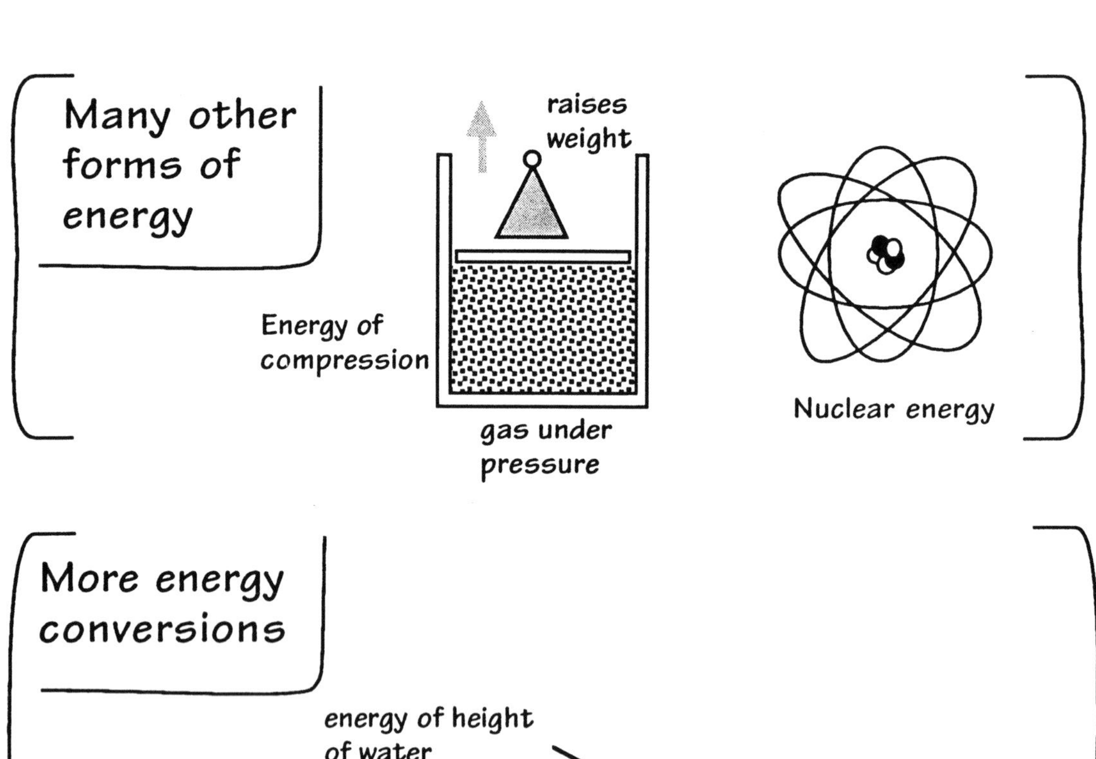
Many other forms of energy
raises weight
Energy of compression
gas under pressure
Nuclear energy

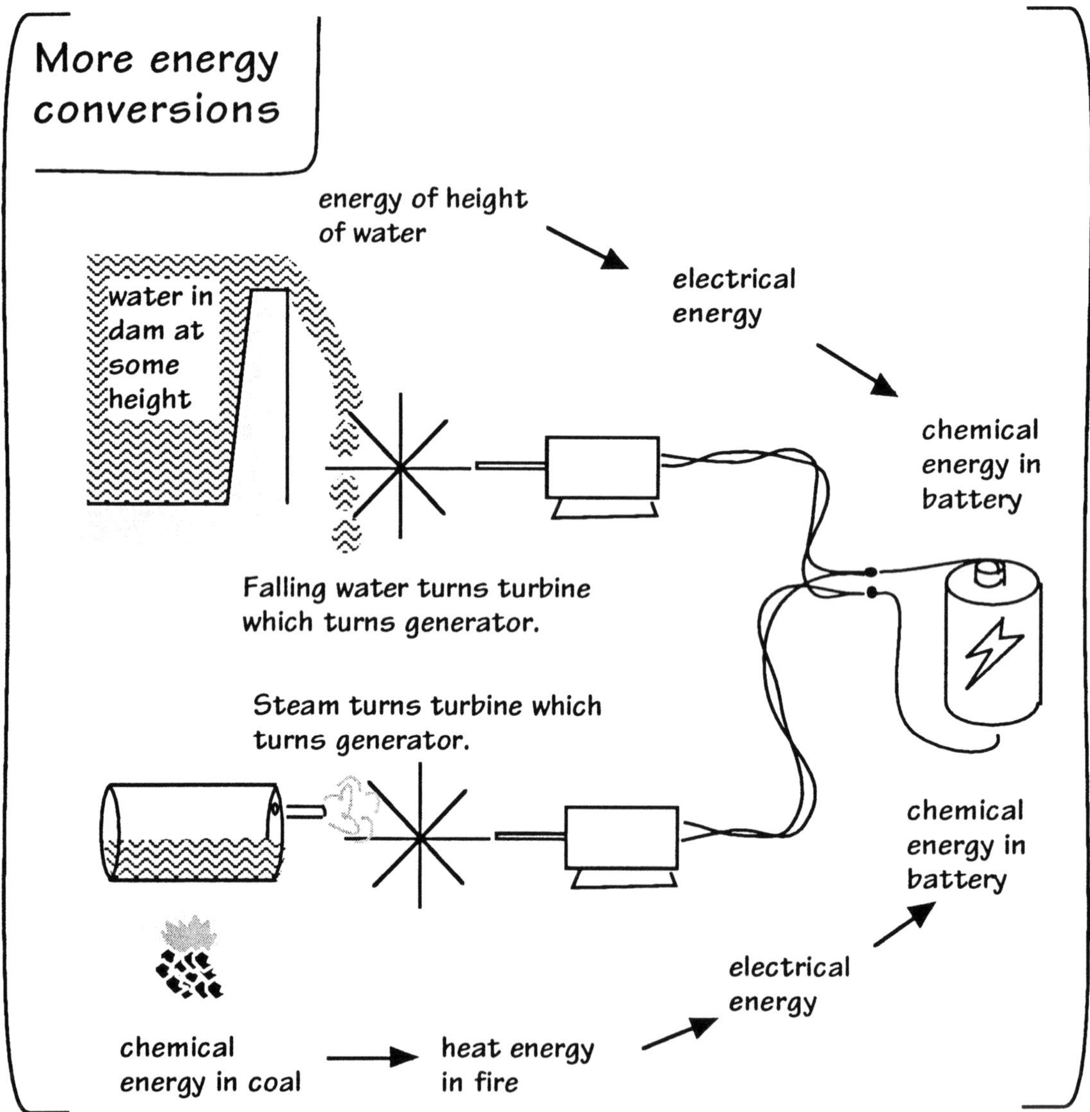
More energy conversions
energy of height of water
electrical energy
chemical energy in battery
water in dam at some height
Falling water turns turbine which turns generator.
Steam turns turbine which turns generator.
chemical energy in battery
electrical energy
chemical energy in coal
heat energy in fire

There are many other forms of energy. Gases under high pressure, for example, carry energy of compression. This energy can be used to raise weights. Nuclear energy is the energy that binds the nucleus of atoms together. We now know how to convert substantial amounts of this energy into other forms. When it is done explosively, we have an atomic or nuclear bomb. When it is done in a more controlled way, we have a nuclear power plant, where the nuclear energy is converted into electrical energy.

In general all these forms of energy can be converted into one another. For example, we have seen how the chemical energy in a battery can be converted into electrical energy. This process is reversed when we use electrical energy from a power outlet to recharge a battery. This process of recharging comes at the end of a long chain of conversions. It may have begun with falling water in a hydroelectric power station, turning a generator that produces electrical power. This process is the conversion of the energy of height of the water into electrical energy. Or we may have burned coal in a coal fired power station to produce high pressure steam which turns the generator. In this process, the chemical energy of the coal is converted into heat energy in the fire and then eventually into electrical energy in the generator. In either case, this electrical energy is then converted into chemical energy in the battery that it recharges.

Not all energy conversions are so straightforward. For example, consider the conversion of substantial amounts of nuclear energy into other forms. That this was possible in principle was known for decades and it did happen spontaneously in small amounts in natural radioactivity. But it required the sustained efforts of a large group of scientists and engineers of the Manhattan Project in the Second World War to do it.

What we observe:

"The brake pads is pressed onto the spinning wheel. The wheel slows and the brake pad gets hot."

vs.

What we say in the theory:

"When the brake pads slows the wheel, the wheel's kinetic energy is converted into heat energy in the pad."

What justifies the extra pieces in the theoretical account? Energy cannot be seen, touched smelled or felt.

Thermodynamics is a mature science. Its theoretical entities and their properties are fixed by observation and experience.

The argument to be sketched in more detail in the following:

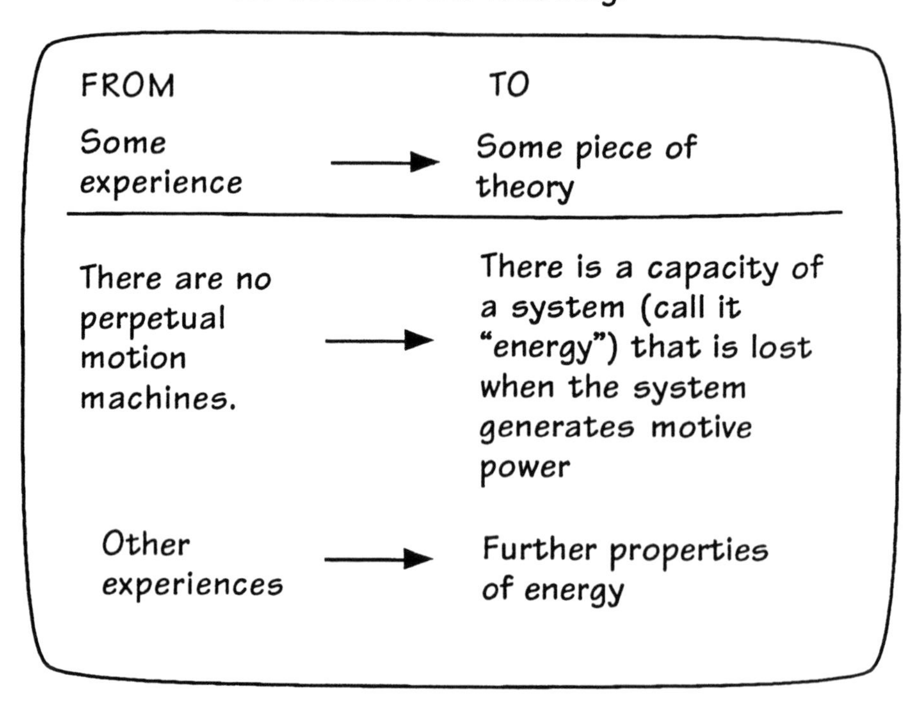

In learning about thermodynamics, we have, in effect, learned to describe nature differently. We no longer just say that a brake pad gets hot. We now say that it gains heat energy. The difference is not merely a different way of talking. The new description involves an entity "energy" that we do not directly observe in nature. Yet we suppose that it exists in all things in a variety of forms and that it moves from form to form when any change occurs.

What gives us such assurance that energy exists with all the properties assumed for it? Scientists are not the only ones who talk about unseen entities that pervade all things. In the occult literature, one finds talk of all sorts of mysterious spirits and forces that pervade all things. Scientists routinely dismiss such talk as deluded. What is different about energy? Why is it any less dubious? The difference is that the existence and properties of energy are supported strongly by an impressive body of evidence of the type we discussed in the last chapter.

One part of the case for energy and its properties is that thermodynamics makes many predictions about what will happen in nature: how much electrical energy is needed to heat this amount of water to boiling; how high the energy of the rocket fuel can fling a rocket; and so on. Each of these predictions, when successful, lends support to the theory. The accumulated weight of all such support is now enormous, after well over a century of continuous use of the theory in many diverse domains.

There is a more direct way to see how observation and experience lead us to the notion of energy and its properties. It is possible to map out a series of arguments in which we start with some experience and from it infer directly to the existence and properties of energy. These arguments are laid out in the steps 1 to 4 below. Through them, we will see that the existence and properties of energy is essentially completely fixed by experience. In Chapter 3 we saw that one strives to build a case for a theory in which the evidence is seen to pick out just that theory and no other--it forces acceptance of just that one theory. The arguments that follow have this character. If we accept the evidential experiences that are their assumptions, we have no choice but to accept the theory that follows from them.

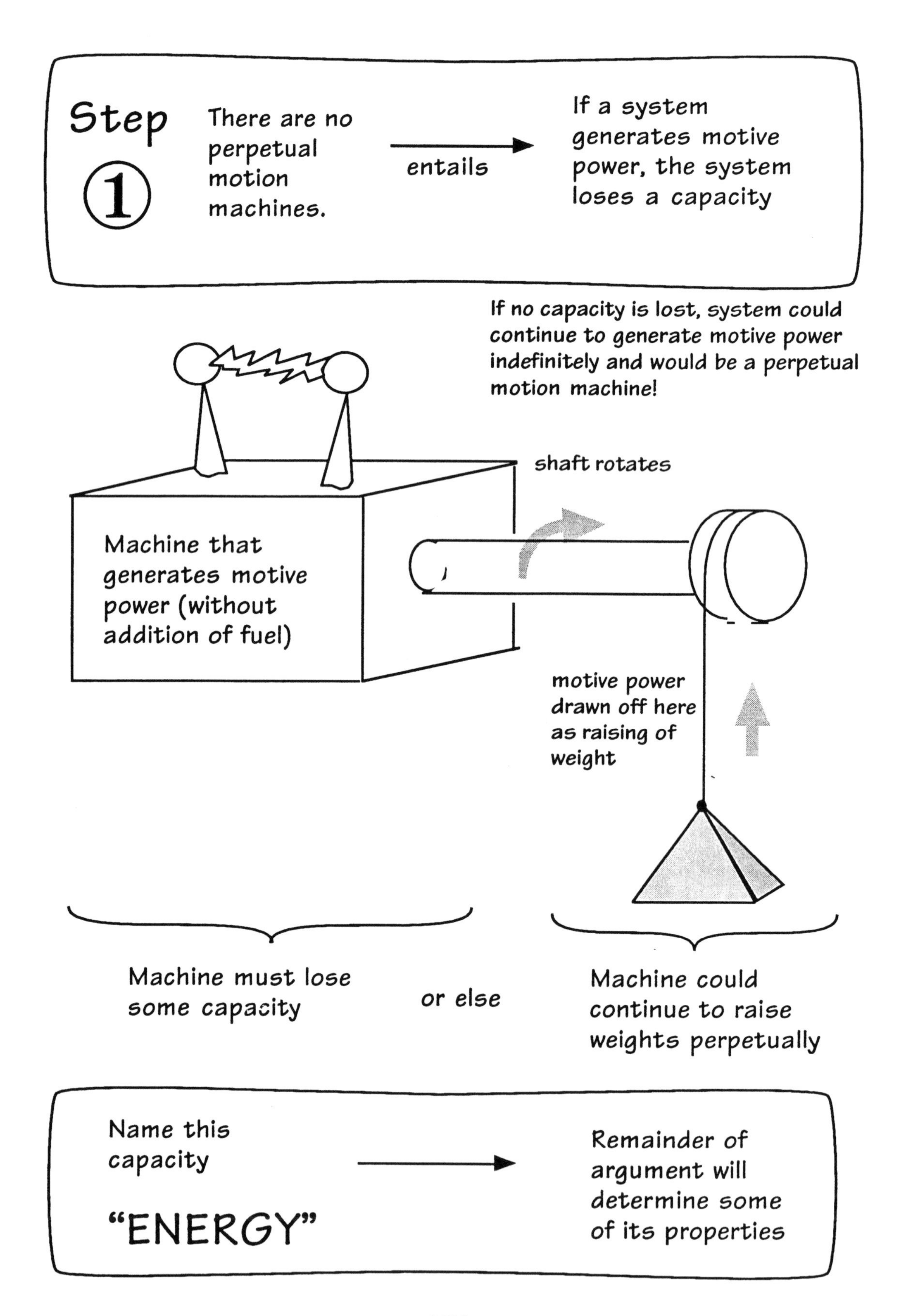
Step
1
There are no perpetual motion machines.
entails
If a system generates motive power, the system loses a capacity
If no capacity is lost, system could continue to generate motive power indefinitely and would be a perpetual motion machine!
shaft rotates
Machine that generates motive power (without addition of fuel)
motive power drawn off here as raising of weight
Machine must lose some capacity
or else
Machine could continue to raise weights perpetually
Name this capacity
"ENERGY"
Remainder of argument will determine some of its properties

Our first step is to infer to the existence of a thing that we will decide to call "energy". We begin with our experience that there are no perpetual motion machines. It follows from this that any system that produces motive power must be changed or depleted in some way. To see why we have to conclude this, imagine otherwise. That is, imagine that we have a machine that is not depleted in any way as it produces motive power, for example, it does not require any addition of fuel. This machine could remain in the state in which it started, or, at least, return to that state in the course of its operation. Now, the ability to generate motive power comes from the state of the machine. So, if the machine can maintain its state, then it will never lose the ability to generate motive power. That is, it will be a perpetual motion machine.

Therefore any system that generates motive power must be depleted in some way. Consider this depletion to be a loss of a capacity in some general sense. We need a name for this loss of capacity. We will chose to call the capacity lost "energy."

Note that we do not yet know much about the capacity lost that we have decided to call "energy." We will need to investigate it further to find out its properties. All we know about it is that it is lost by a system when that system produces motive power.

Step ② When a machine generates motive power, there is a definite physical change inside the machine. → Loss of capacity (energy) corresponds to a specific physical change, which characterizes the type of energy.

Label type of energy according to the type of physical change: "chemical energy" "electrical energy" "heat energy" etc...

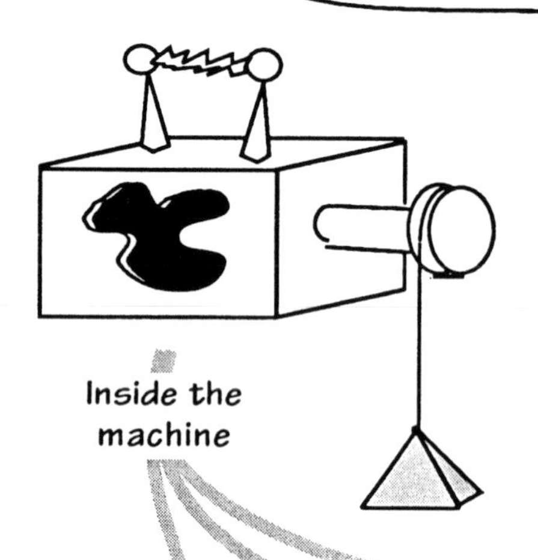

Inside the machine

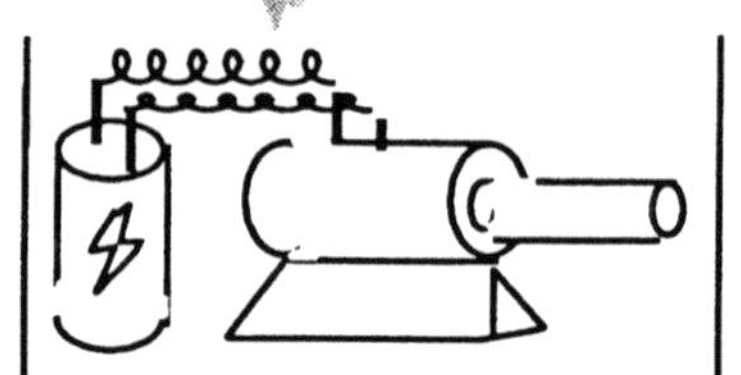

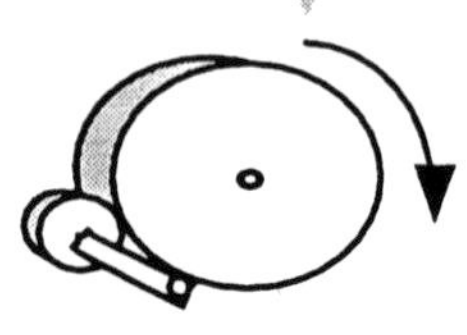

Compressed spring raises weight	Battery & electric motor raises weight	Rapidly spinning flywheel raises weight	type of machine
↕	↕	↕	
Spring expands	Battery runs flat	Fly wheel slows	physical change
↕	↕	↕	
Energy of compression lost	Chemical energy lost from battery	Flywheel loses energy of motion	form of energy

The first thing we can conclude about what we have defined as energy is that it comes in many different types. We know this since our experience is that any machine that produces motive power changes in definite ways as it loses the capacity to produce motive power.

Assume the inner working of the machine involves a compressed spring that unwinds and generates the motive power. This loss of capacity corresponds to the loss of compression of the spring. In this case, we say the engine loses the energy of compression stored in the spring.

Assume the engine employs a battery to power an electric motor that then yields motive power. The loss of capacity corresponds to the flattening of the battery, that is, the exhausting of the chemical fuels in the battery. In this case, the engine loses chemical energy from the battery.

Assume the engine employs a rapidly spinning wheel--usually called a "flywheel." We recover motive power by bleeding off motion from the flywheel. The loss of capacity corresponds to the slowing of the flywheel's motion. The engine loses energy of motion from the flywheel.

Thus, in general, there are as many types of energy as there are types of system. We can identify the type of energy lost in generating motive power by examining the type of change that accompanies the inevitable loss of capacity of an engine producing motive power.

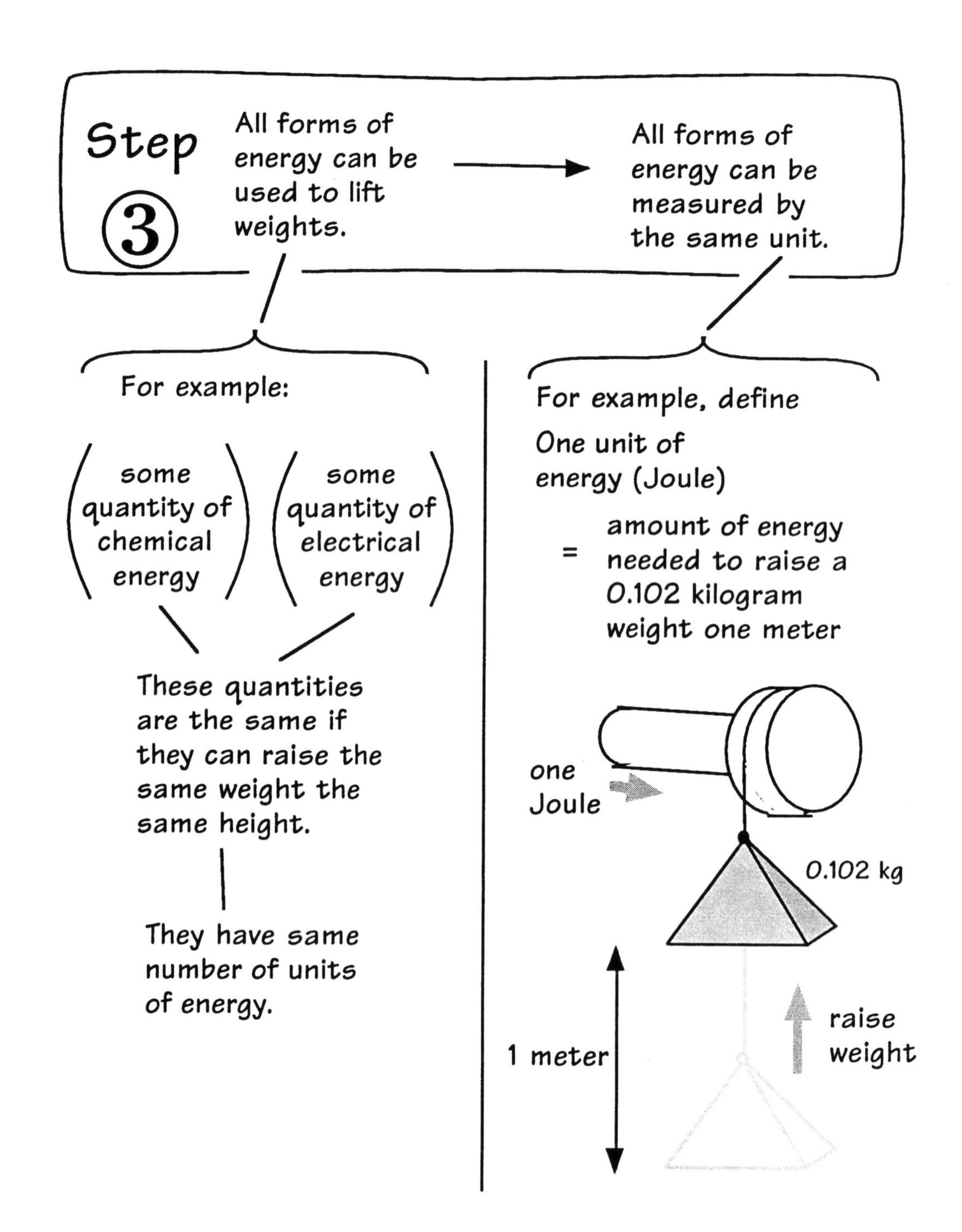
Step
3
All forms of energy can be used to lift weights.
All forms of energy can be measured by the same unit.
For example:
some quantity of chemical energy
some quantity of electrical energy
These quantities are the same if they can raise the same weight the same height.
They have same number of units of energy.
For example, define
One unit of energy (Joule)
= amount of energy needed to raise a 0.102 kilogram weight one meter
one Joule
0.102 kg
1 meter
raise weight

Now that we know that there are different types of energy, we may well wonder how we are to compare them. Does it make sense to say that we have more or less of one type of energy compared to another? It turns out that there is a simple numerical measure that can be used to measure and compare all the forms of energy. The existence of that measure depends on the fact that all forms of energy may be used to generate motive power. More compactly, we may say that all forms may be converted into motive power. The amount of motive power generated in this conversion measures how much energy was present. This allows us to compare whether two quantities of energy are the same. Some quantity of chemical energy would be the same as some quantity of electrical energy if both could be used to raise the same weight the same height.

We need some arbitrary definition of a unit if we are to assign numbers to quantities of energy. The internationally preferred unit of energy is the Joule. A Joule of energy is the amount of energy needed to raise a weight of 0.102 kilograms (about a quarter of a pound) a distance of one meter (about a yard). So we have a Joule of any form of energy--chemical, electrical, etc.--if that energy is consumed exactly in raising the 0.102 kg weight one meter. If the quantity of energy can raise the weight ten meters, then we have ten Joules of that energy type.

Step

For any energy form, if there is a machine that converts it into motive power, we can find a machine that effects the backwards conversion. → All forms of energy can be converted into each other

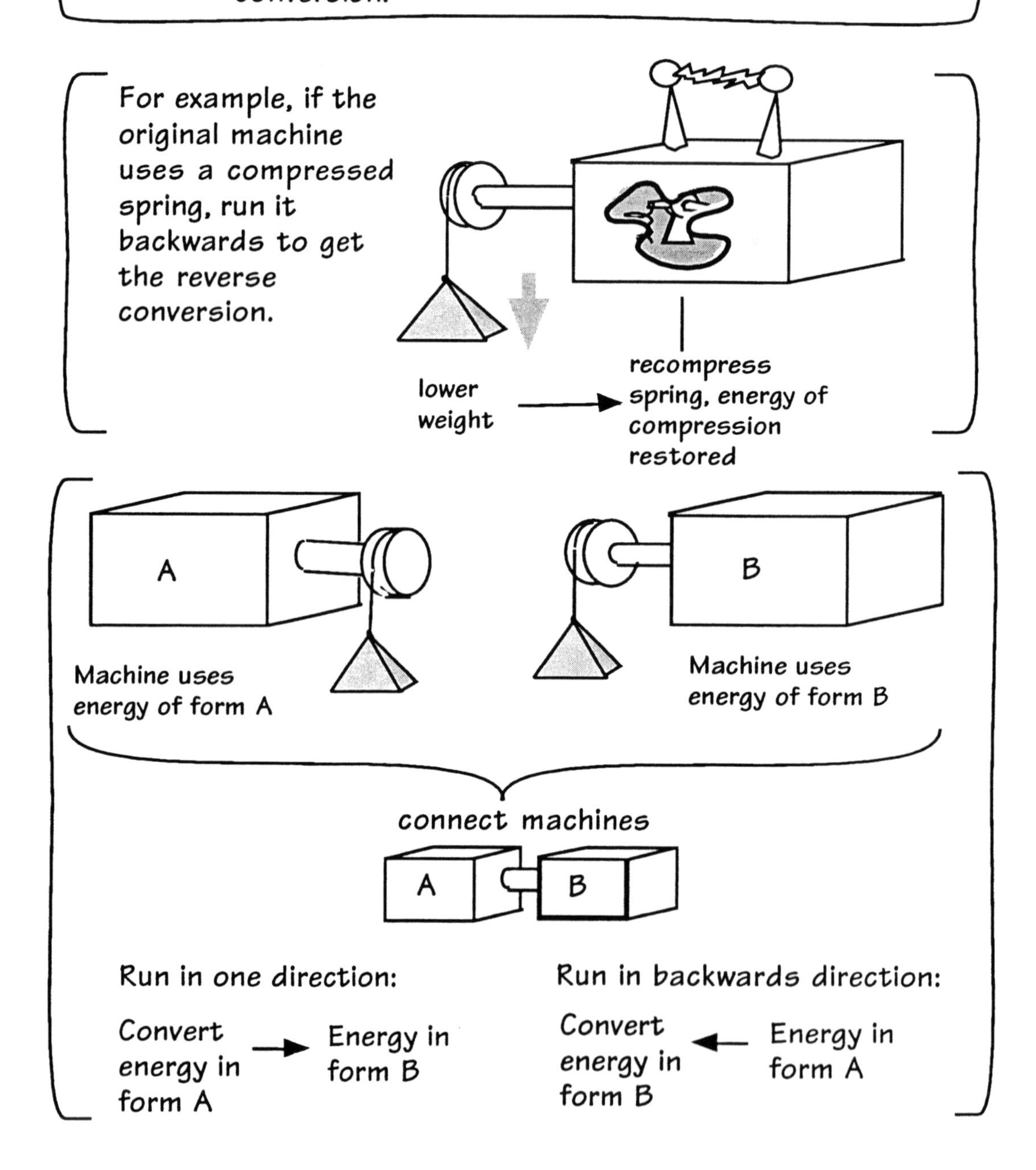

All forms of energy may be compared in magnitude and measured on the same scale. That does not yet mean that all forms of energy may be converted into one another. In principle we may have many forms that remain distinct. This turns out not to be the case, however; any energy form can be converted into any other. That this is so depends on another experience. All forms of energy can be converted into motive power through an engine. That is the way we originally defined energy--the loss of capacity of a system when it generates motive power. It turns out that for every such conversion, the backwards conversion is possible. There is always a machine that can take motive power and return that energy form, that is, restore the lost capacity.

For example, if the energy used by a machine is energy of compression in a spring, we need only run the machine backwards to recompress the spring. The motive power used to enforce the backwards operation is converted into energy of compression in the spring. Similarly, an electric motor run backwards becomes a generator whose electricity can recharge a battery. This restores the chemical energy lost from the battery.

More generally if we have machines that use any energy forms A and B and can also be run backwards, then we can couple them as shown opposite. If we run the machines in one direction, energy form A will be depleted and transferred as motive power to the second engine. In the second engine, running backwards, this motive power will be converted to energy form B. Overall, we have converted energy in form A into energy in form B. Running the coupled machines in the other direction produces a conversion of energy form B into energy form A.

Synopsis of results:

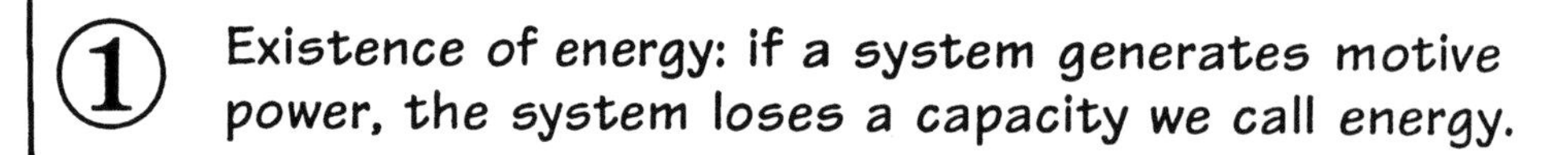

1. Existence of energy: if a system generates motive power, the system loses a capacity we call energy.

2. The loss of capacity (energy) corresponds to a specific physical change which characterizes the types of energy.

3. All forms of energy can be measured by the same unit.

4. All forms of energy can be converted into each other.

There are no perpetual motion machines.

↓

No process can create energy.

No process can destroy energy.

↘ ↙

1st Law of Thermodynamics

Energy is conserved.

These four arguments have given us the major properties of energy: that it exists (at least in systems capable of generating motive power); that it comes in different types; that it can be measured; and that it can be converted between the different forms.

We have not yet arrived at the core property that energy is conserved, the first law of thermodynamics. That final step is sketched briefly here.

First we know already that no process can create energy. That follows immediately from the non-existence of a perpetual motion machine. If some system could create energy, we would then couple that to another that converts the energy created into motive power--and we would have a perpetual motion machine!

Second our experience is that no process can destroy energy. This follows automatically for any process that can be run backwards. If such a process could destroy energy, then, when run backwards, it would create energy, which would then enable construction of a perpetual motion machine.

If we combine the two we have the result that no process creates or destroys energy; that is, the energy at the start of the process is the a same as at the end--energy is conserved!

ANSWERS TO PUZZLES

Motor-generator perpetual motion machine

(Electric current produced by generator) never greater than (electric current needed by motor.)

Leupold's overbalanced wheel

If the wheel turns in the direction indicated, then weights A and B would not be in the positions indicated earlier.

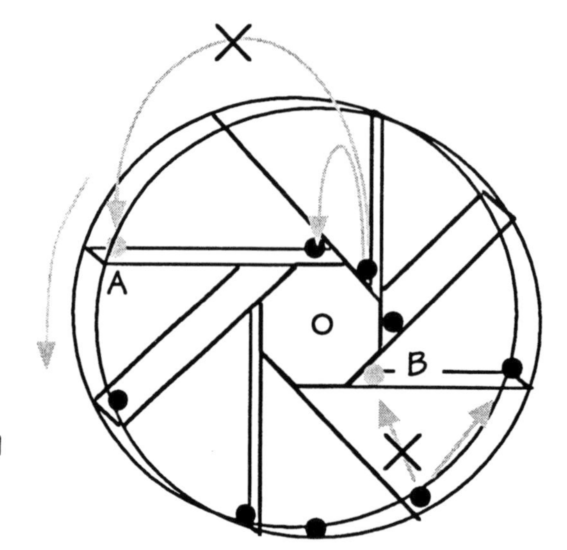

If weights A and B are in the positions to which they naturally roll, the wheel will not be overbalanced.

Buoyancy engine

water
pressure

The effect of inserting a cork is to raise a volume of water equal in volume to the cork.

This volume will weigh more than the cork.

Therefore the motive power needed to insert the cork exceeds the motive power recovered when the cork falls.

Motive power is required to push corks into the water to overcome the pressure of the water.

Why do the perpetual motion machines sketched earlier fail?

In the case of the motor-generator, the generator cannot produce enough electricity to run the motor. A perfectly efficient generator would produce exactly enough electricity to run the motor, as long as the motor was also perfectly efficient. But there would be no excess motive power at all, so the machine would fail to be a perpetual motion machine producing limitless motive power. Real motors and generators are not perfectly efficient, so there would not be enough electricity to run the motor and the coupled devices would not even turn.

That Leupold's wheel appeared perpetually overbalanced depended on an incorrect drawing of the positions of the rolling weights. With the wheel turning as indicated, neither weights A nor B would be in the positions indicated earlier. Weight A would not yet have rolled out to the rim and weight B would not yet have rolled in toward the axis. The net effect would be that the weights on the left and right hand side would balance one another. There would be no perpetual overbalancing to drive the wheel.

The effect ignored in the buoyancy engine is the pressure of the water. Inserting corks to a point underwater requires considerable motive power. The pressure of water at the bottom of a tank can be significant. The question is: is the motive power consumed in forcing the corks into the water greater than the motive power recovered by the falling of the corks? We can convince ourselves that it is, so that the operation of the buoyancy engine would not yield a net gain of motive power; overall, it would require motive power to keep it operating.

To see that more motive power is required, imagine the effect of inserting a cork. A volume of water equal to that of the cork must be pushed out of the way to make room for the cork. This volume will appear at the top of the water tank in the form of a slight rise in the water level. The motive power needed to push the cork in must be sufficient to raise this amount of water. But, since a cork floats on water, we know that it weighs less than an equal volume of water. Therefore the amount of motive power needed to raise the water is greater than the motive power recovered when the cork falls. For, in raising the water, we raise a greater weight than the weight that falls when the cork falls.

Assignment 4: Conservation of Energy

1. Is energy real? It is something that is supposed to exist in all things and be converted into different forms when things change. But it is not something that we can see, hear, touch, smell or taste. Why should we take its existence and properties seriously?

2. You have 1,234 kJ of energy. Re-express this amount of energy in the following units:
(a) Joules J
(b) Kilocalories kcal
(c) calories cal
(d) Calories Cal
(e) British thermal units Btu
(f) Kilowatt hours kWhr

You will find the following converion factors useful:
1000 J = 1 kJ 4.186 kJ = 1kcal 1000 cal = 1 kcal 1000 cal = Cal 0.948 Btu = 1 kJ 1 kWhr = 3600 kJ

3. The table attached considers many possible conversions between different forms of energy. Fill out the table by indicating processes that take energy FROM the designated form on the left TO the form designated by the labels across the top. The process need not be a direct transfer with no intermediate. It must just start and stop with the appropriate forms.

Be careful only to write in real conversion processes in the table. Do not write in processes that merely trigger other conversions. For example, imagine that we switch on a flashlight by bouncing a ball off the switch. The ball does cause the light to come since it trips the switch. But the kinetic energy of the moving ball is **not** converted into the light energy emitted by the bulb. That light energy comes from chemical energy stored in the batteries; the chemical energy of the battery is converted into light energy. This is the real energy conversion involved. The action of the ball is merely to trigger this conversion. Do not include such triggering processes in the table.

4. Bring a device that illustrates one of the conversions of 3. to class for "show and tell".

I will collect the assignments at the start of class and THEN discuss them. Please keep a copy of your assignment or some notes so that we can discuss what you did.

Assignment 4: Conservation of Energy

Name:________________________

TO

F R O M

	Energy of height	Kinetic Energy	Heat Energy	Chemical Energy	Electrical Energy
Energy of height	XXXXXXXX			Falling weight drives generator that charges battery	
Kinetic Energy		XXXXXXXX			
Heat Energy			XXXXXXXX	Steam engine runs generator that charges battery	
Chemical Energy				XXXXXXXX	
Electrical Energy	Electric motor raising weight				XXXXXXXX

Device brought for show and tell: ______________________________

Assignment 4: Conservation of Energy

Name:________________________

TO

FROM

	Energy of height	Kinetic Energy	Heat Energy	Chemical Energy	Electrical Energy
Energy of height	XXXXXXXX			Falling weight drives generator that charges battery	
Kinetic Energy		XXXXXXXX			
Heat Energy			XXXXXXXX	Steam engine runs generator that charges battery	
Chemical Energy				XXXXXXXX	
Electrical Energy	Electric motor raising weight				XXXXXXXX

Device brought for show and tell: ______________________________

Chapter 5

Applications of the Law of Conservation of Energy

Question : What practical use is a knowledge of energy conservation and energy conversion?

Answer : Physical processes of interest to us involve the transformation of energy among different forms.

In many cases:, the course of a process is dominated by energy considerations.

Compare:

How much energy does the process need?	How much energy can be supplied to it?

This chapter : Illustrate how this understanding arises.

As much as possible, we will calculate from information you have readily at hand.

Therefore the calculations are only rough approximations.

In the last chapter we saw how to develop the notion of energy and its transformations. We have seen how we can take everyday processes and redescribe them in the abstract language of a theory. A car driving up a hill is redescribed in terms of the transfer of energy from its gasoline fuel to the potential energy of the car. Is there any practical value in this type of redescription? Or are we merely laboring to learn a new language that makes the familiar seem unfamiliar?

An understanding of the notion of energy and its transformations gives us the power to explain many things that matter to us. All physical processes involve the transformation of energy among different forms. In some, the supply of energy to the process may be the decisive factor in determining whether the process can proceed. In such energy dominated processes we can usually understand a great deal about why the process proceeds as it does once we understand the energy conversions involved. In the analysis, we compare two things: How much energy does the process need to proceed? How much energy can our sources supply? The scope and character of the process is often dictated simply by the need to ensure that the energy supply available matches the energy need of the process.

The calculations that follow illustrate this type of analysis. They will be rather rough and ready. The goal is not to provide an exact analysis, for that would be technically burdensome. Rather the goal is to show how some very simple "back of an envelope" type calculations can yield a great deal of insight. In keeping with this goal, the calculations will rely as much as possible on data that you have readily available to you: What are the typical fuel consumptions of cars? How many Calories does the average person consume daily in food? *etc.* When needed, extra data will be supplied.

We will spend most of the chapter analyzing why we use gasoline to power cars.

Why do automobiles use gasoline powered engines almost exclusively?

auto-mobile = self-moving vehicle, i.e. they carry their fuel with them

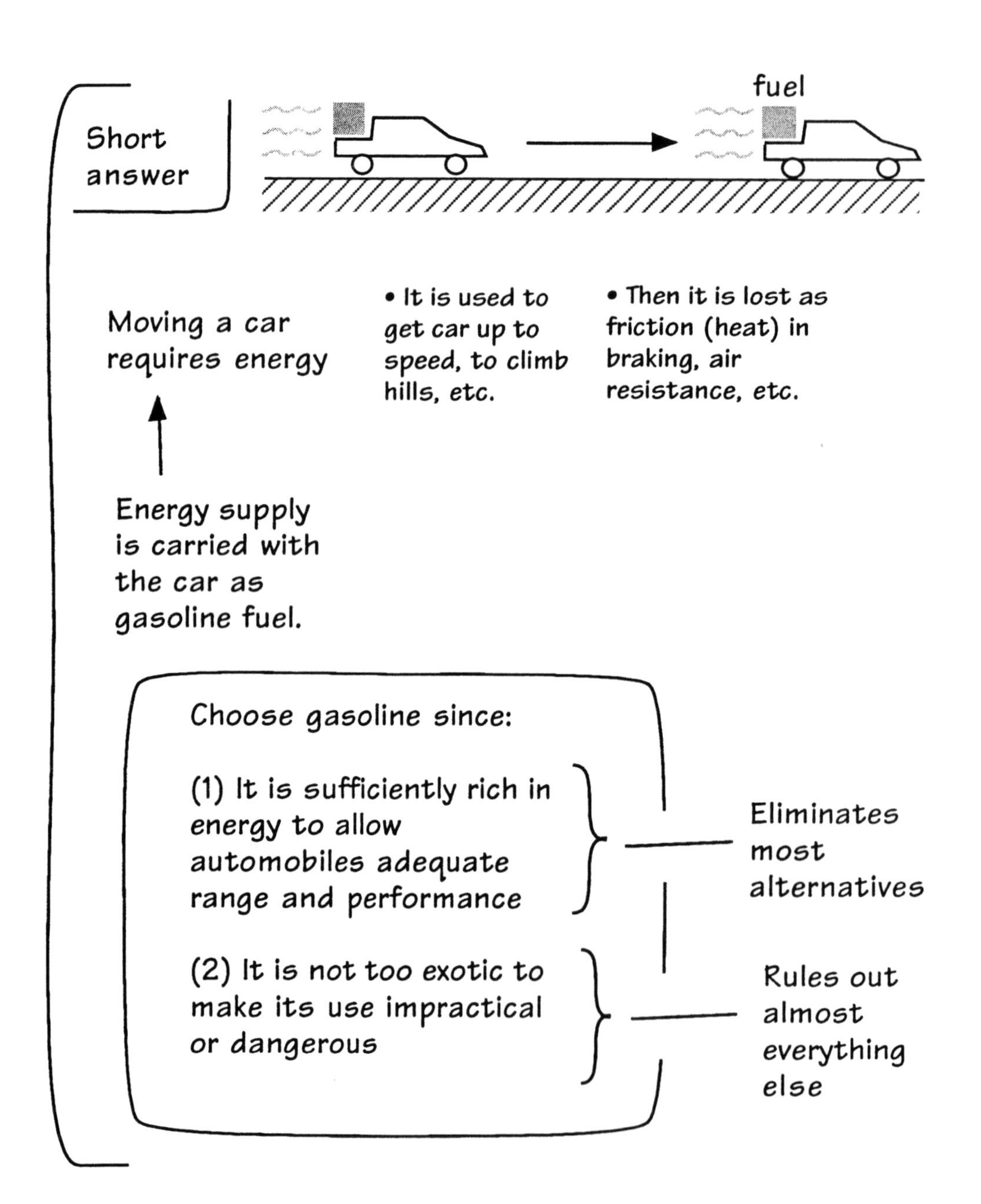

Of all common items of modern technology, the automobile attracts the most sustained and vociferous attacks. We complain that it runs by burning a valuable and non-renewable resource, gasoline made from petroleum. In burning this fuel, automobiles become one of the major sources or air pollution in our big cities. We know of alternatives. Electric cars are most commonly named. In operation they generate essentially no pollution. But they are hardly used at all. Legislators have become so frustrated by this lack of use that recent legislation in California has had the effect of forcing car manufacturers to start selling electric cars.

Why do we persist in our use of gasoline powered automobiles? The short answer is that their use is dictated by energy considerations. To move a car requires energy, a great deal of energy. That energy is carried with the automobile in a fuel. This is what makes the vehicle an automobile--the fuel is carried with it, so that it can move itself. It turns out that gasoline is virtually unrivaled as a fuel that can be carried with the vehicle. It is extremely rich in energy, so that gasoline powered vehicles can readily carry enough energy in their fuel to enable travel over great distances and at adequate speeds. Few other sources of energy can match this. If electric batteries are used to store similar amounts of energy, they turn out to be prohibitively heavy. There are other fuels which carry more energy than gasoline. Hydrogen fuel is an example. But they are sufficiently exotic that their use is not so attractive.

The continued use of gasoline powered automobiles rests on the need for a sufficiently cheap and practical fuel--a movable energy storehouse--and gasoline is almost unique as that storehouse. Its most immediate competition are chemically similar fuels such as LPG (liquefied petroleum gas) and diesel fuel. Virtually all the considerations that make gasoline viable as a fuel apply to these as well. As result they are used. LPG is sometimes used to fuel buses and diesel fuel is the most popular choice of fuel for large trucks and similar larger vehicles.

Long answer

Begins here:

Energy content of gasoline

132,000 kJ in one gallon

This is the amount released as heat energy when one gallon is fully burned.

Car engines are typically 25% efficient at converting this heat energy into energy of motion of the car, so that:

The remaining 75% lost as heat in exhaust, radiator etc.

one gallon gasoline supplies 33,000 kJ usable energy to motion of car

33,000 = 132,000 x .25

Energy needed for 1 mile of CITY driving = 1650 kJ

Assume gasoline consumption = 20 miles per gallon

1 mile driving requires usable energy of 1/20 gallon gasoline

= 1/20 x 33,000 = 1650 kJ

Energy needed for 1 mile of FREEWAY driving = 660 kJ

Assume gasoline consumption = 50 miles per gallon

1 mile driving requires usable energy of 1/50 gallon gasoline

= 1/50 x 33,000 = 660 kJ

What other sources are there for this energy?

Application of law of conservation of energy:

Energy dissipated in driving car = Energy supplied in fuel

To carry out the analysis, we need first to determine the energy needs of our automobiles. The common experience of car drivers gives us most of the information we need. Most drivers know roughly how many miles their cars will run on a gallon of fuel. If we know how much energy is in a gallon of gasoline, we can then determine how much energy is needed to move the car.

The energy content of one gallon of gasoline is 132,000 kJ. That is the amount of energy released as heat when one gallon of gasoline is fully burned. This figure is higher than the one we want. Gasoline engines are not able to capture all of this energy and use it to drive the car. As a rough average, car engines can capture about 25% of this energy and use it to move the car. That is, there is about 33,000 kJ of *usable* energy in each gallon of gasoline.

The gasoline consumption of cars vary with the conditions. We consider two representative cases:

In ordinary city driving, a typical car may average about 20 miles of driving to a gallon of gasoline. This means that 1/20th of a gallon of gasoline is needed to drive the car one mile. Therefore the energy needed to drive the car one mile in the city is the usable energy of 1/20th of a gallon of gasoline; that is, 1/20 x 33,000 = 1,650 kJ.

In freeway driving, a car will have much better fuel consumption; say about 50 miles of driving to a gallon of gasoline. This means that 1/50th of a gallon of gasoline is needed to drive the car one mile. Therefore the energy needed to drive the car one mile in the city is the usable energy of 1/50th of a gallon of gasoline; that is, 1/50 x 33,000 = 660 kJ.

We now ask what other sources may be used to supply this energy. We will look at several possibilities, including an electric battery. Notice that we are applying the law of conservation of energy in asking this question. We know that one mile of city driving requires 1,650 kJ of energy. The law tells us that this energy must be supplied from somewhere else, for energy cannot be created or destroyed. Any alternate automobile must have a comparable source of energy.

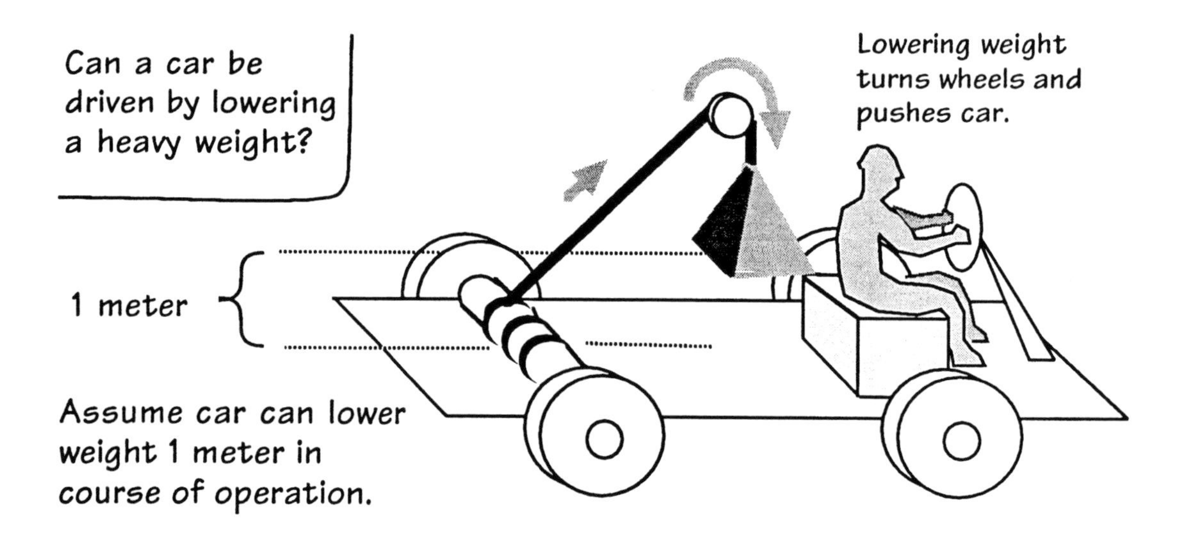

How heavy must the weight be to supply energy for 1 mile of city driving?

Energy sought:

= 1650 kJ
= 1,650,000 J

Energy available:

1 kg lowered 1 meter releases 9.81 J

weight needed for one mile $= \dfrac{1,650,000}{9.81}$ = 168,000 kg

= 168,000 / 907 tons (907 kg = 1 ton)

= 185 tons

Impossibly heavy!!

Energy supplied by 185 ton weight

-Can only take car 1 mile !

-DOES NOT ALLOW FOR ENERGY NEEDED TO MOVE THE 185 TON WEIGHT!

-This will be substantial since an ordinary car weighs about 1 ton

Puzzle: This design is a complete failure for another energy related reason. What is it ?

To get us started, consider a very simple source of energy, the lowering of a weight. Would that be a viable energy source for an automobile? We can convince ourselves that it would not from some simple energy calculations.

We will assume that a weight turns the wheels of the car as the weight is lowered. How heavy must the weight be if the car is to perform like an ordinary automobile? Assume that the weight can be lowered by 1 meter in the normal course of operation. If the lowered weight is to provide even enough energy for one mile of city driving, we will see that the design is unworkable. For the one mile of driving, we need 1,650 kJ=1,650,000 J of energy. Now each kilogram in the mass releases 9.81J of energy when it is lowered one meter. Therefore we will need a weight of 1,650,000/9.81 = 168,000 kg. Since there are 907 kg in a ton, this is equivalent to a weight of 168,000/907 = 185 tons!

This result is disastrous. The vehicle would be enormously heavy. An ordinary automobile weighs around one ton. Worse, our calculation that the car needs 1,650 kJ per mile depended on the tacit assumption that the weight of the car was approximately that of an ordinary automobile. If car with the lowering weight of 185 tons were built, it would almost 200 times more. The energy supplied by the lowering weight would come nowhere close to supplying the energy needs of the car. As a rough guess, 200 times more energy would be needed, if the weight lowering car is to match the performance of an ordinary automobile in city driving.

An automobile powered by lowering weights is clearly not one that can be taken seriously. Our certainty of its failure illustrates the power of energy calculations. The issue was decided through a simple comparison of the energy needed with the energy that can be supplied.

Can we run an electric car from storage batteries?

How much energy does an electric battery store?

Typical lead acid battery storage batteries: 1 pound of battery stores 10 watthours = 36 kJ of electrical energy

Sophisticated battery design can improve this...but how much?

How much energy does an electric car need?

Gasoline engine in city driving → 1 mile requires 1650 kJ

Much energy lost as friction in braking

Electric car may have fewer energy losses in braking. → Run motors in reverse to brake and recharge batteries.

Can this be done efficiently?

Therefore, energy needs of electric car in city closer to energy needs of gasoline car on freeway (no braking) 660 kJ / mile

Major energy loss in air friction--not present in low speed driving

Therefore, guess energy needs of electric car in city: 1 mile requires 600 kJ of energy

if performance is to match gasoline car and assuming weights are comparable

An electric automobile is probably the most commonly discussed alternative to a gasoline powered automobile. Such a vehicle is supplied with energy stored in electric storage batteries. Once again we can calculate its energy needs and the energy that the battery can supply.

The most commonly used battery is the lead-acid battery--the type that operates the starter motor in virtually every modern automobile. These batteries store only a modest amount of energy: each pound of battery carries about 10Wh= 36 kJ of energy. More sophisticated designs of batteries can improve on this figure. But, in spite of decades of research, these improved battery designs remain experimental. In any case, we cannot expect enormous improvements in the energy these batteries can carry. Increasing the energy stored ten-fold remains unlikely even with the most exotic of material.

In estimating the energy needs of an electric car, we should allow for the possibility of a wonderful economy in their operation. When gasoline powered cars brake, they are slowed by a pad rubbing on a disk. The energy of motion of the car is lost as heat. To set the car back in motion, more fuel must be burned to replace the energy lost. In principle, in an electric car, we can brake by throwing the electric motors into reverse. This slows the car by converting its energy of motion back into electrical energy, which can then be returned to the battery for later reuse. Electric car designs need not incorporate this feature. And, if they do, it will not be possible to recapture all the energy of motion. Under heavy braking, a lot of energy is delivered rapidly; but batteries can only be recharged slowly.

Thus a well designed electric car will probably need less than the 1,650 kJ per mile that a gasoline powered car needs. Its energy needs may be better approximated by looking at the energy needs of a gasoline car when there is less braking. Here we may think of freeway driving conditions, where the energy needs are 660 kJ for each mile, but there is considerable energy loss through air friction due to high speed.

Combining these considerations, our first guess at the energy needs of a well designed electric car would be around 600 kJ for each mile. As before, we assume that the weight of the car and the performance sought is roughly comparable to those of ordinary gasoline automobiles.

What distance can we expect an electric car to run on a full charge?

Consider a heavy battery = 500lb

Once we make the battery heavier than this we need more energy to move the car because of the extra weight. The figure of 600 kJ for each mile will be too small.

How much energy does this fully charged battery hold?
(500 lb of battery. 36 kJ in each pound)

Total energy = 500 x 36 = 18,000 kJ

How far will the car run on a full charge?
(600 kJ for one mile)

Distance = 18,000 / 600 = 30 miles

Expect: weight of batteries will be the decisive factor in electric car design.

Good performance → High energy demands → Large batteries → Heavy weights to carry → Poor performance

Let us now combine our information on the energy needs of an electric car with the energy available in the battery to assess its performance. We anticipate that the weight of the battery will be the decisive factor. So we begin by assuming the car is equipped with a heavy battery--500 lb. If we assume any higher weight, then our estimate of 600 kJ of energy needed for each mile will be too small, for this figure was computed for vehicles of weight comparable to ordinary automobiles.

We first compute how much energy we have available. If the battery weighs 500 lb and each pound stores 36 kJ when fully charged, then the battery overall stores energy $500 \times 36 = 18{,}000$ kJ. If this energy is used to drive the car, we see immediately that it will have a very restricted range. If 600 kJ are needed for each mile driven, then on a full charge, an electric car will have a range of $18{,}000/600 = 30$ miles. This is barely acceptable.

This short calculation suggests that battery weight will be the decisive factor in the design of electric cars. The desire for good performance turns out to be self-defeating. For good performance--brisk acceleration--requires large amounts of energy. But the supply of large amounts of energy requires large batteries. Large batteries are heavy. If heavy weights are to be carried, we will end up with poor performance.

Our practical experience with electric cars affirms these expectation. In electric cars produced for the consumer market, the ranges on a single battery charge are small, perhaps of the order of 30 to 100 miles. These cars are burdened with heavier batteries, for example, 1000 lb and more in weight. Electric cars are well suited to applications that demand neither high performance nor large range and in which their silence of operation is an advantage--golf carts, for example.

It should be mentioned that electric cars are not automatically an environmental cure-all. While they do not pollute when they operate, there may be considerable pollution associated with the original generation of the electric energy stored in their batteries. Most commonly this energy will come from an ordinary electric power station that is operated by burning coal, a non-renewable resource and producer of greenhouse gases.

Can we eliminate the battery and run the electric car from electricity generated by solar cells?

Sunlight falls on a bank of solar cells on the roof of the car. The cells produce electricity which runs the car.

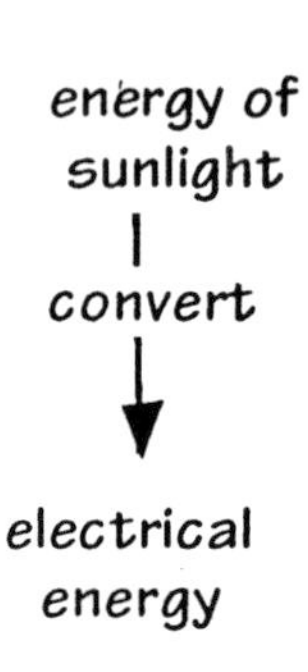

Compute the performance of the car in the best case possible:

Assume a large area of cells:
2 meters x 3 meters = 6 square meters

Best time for sun at mid-northern latitudes: noon, midsummer, no clouds

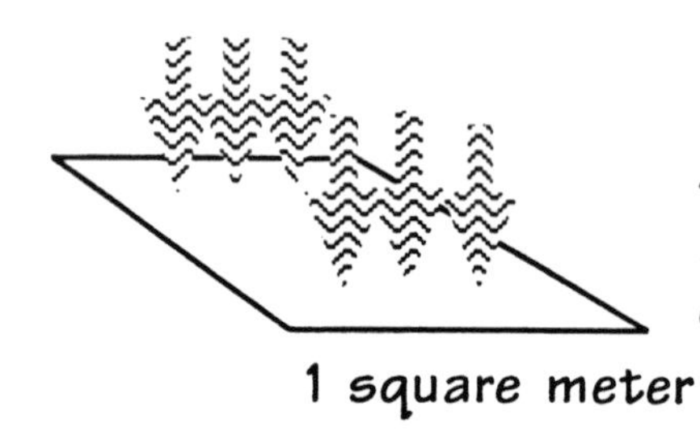

Approximately 4,000 kJ of solar energy in one hour

1 square meter

Total energy falling in one hour on cells = 4,000 x 6 = 24,000 kJ

Electrical energy produced in one hour (high efficiency cells convert 10% of sunlight energy to electrical energy) = 24,000 x 0.10 = 2,400 kJ

Energy needed for one hour of city driving (Need 1650 kJ for each mile and drive at 15 miles per hour) = 1650 x 15 = 24,750 kJ

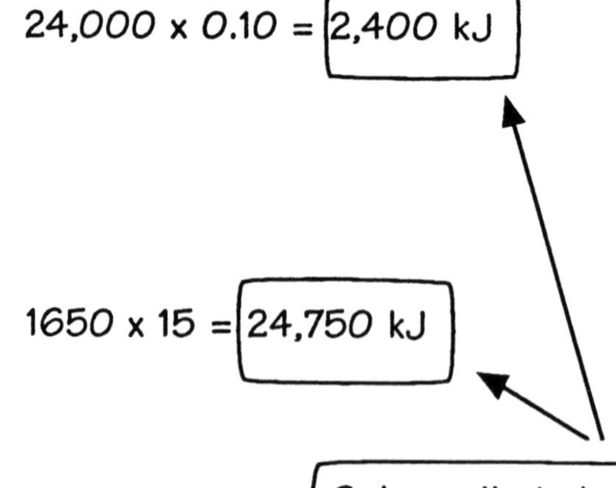

Solar cells in best case provide 10% of the energy needed.

If the weight of the battery defeats ordinary electric cars, should we consider an electric car without a battery, one that runs on electricity from sunlight? The top of the car would be covered with solar cells. When sunlight strikes these cells, they convert part of the energy of the sunlight into electrical energy, which can be used to run the car.

We will estimate the performance of such a vehicle in the best case possible. We will assume a large collector--2 meters by 3 meters = 6 square meters. It would cover the top of an ordinary car. The best time for sun is noon on a cloudless midsummer's day. Under these ideal conditions in the mid Northern latitudes, each square meter of land receives about 4,000 kJ of sunlight energy. Therefore, under these ideal conditions, the total energy in sunlight falling on the car's solar cells is 4,000 x 6 = 24,000 kJ in one hour. Unfortunately solar cells are not very efficient at converting this energy into electricity. Even high efficiency cells can still only convert about 10% of it into electricity. So the cells can supply 2,400 kJ of electrical energy in each hour to run the car. We can soon see that this is too little. For city driving, our car requires 1,650 kJ for each mile. If the average speed of driving is 15 miles per hour, this amounts to a requirement of 1,650 x 15 = 24,750 kJ. Our solar cells can provide barely 10% of this requirement.

We conclude that a solar powered electric car will be dramatically underpowered. Of course the calculation assumes that our car is comparable in weight to an ordinary automobile and that similar performance is required. Thus we may rescue our solar powered design if we make the car very lightweight or if we do not require much performance. Cars designed as entrants for solar vehicle challenge races tend to be very light--a few hundred pounds--and streamlined extensively to minimize air resistance. They are then capable of good performance and high speeds. However these concessions will help us less if the day is cloudy so that there is less sunlight and they will not help at all at night!

A flywheel powered car?

Store energy in heavy spinning flywheel

Draw on this energy to move car

Flywheel:

Mass = 200 kg (about 440 lb)

At maximum speed, it spins at 10,000 revolutions per minute.

Radius = 0.3m (about one foot)

We cannot make the flywheel much heavier or we need more energy just to move flywheel !

Total energy stored = 4,935 kJ

Range in city driving (1650 kJ for each mile) = 4,935 / 1650 = 3 miles

Range in freeway driving (660 kJ for each mile) = 4,935 / 660 = 7.5 miles

Not a serious competitor to the gasoline powered engine!

For experts only

$$\text{Energy of flywheel} = \pi^2 \times \text{mass} \times \left(\frac{\text{turns}}{\text{sec}}\right)^2 \times (\text{radius})^2$$

$$= \pi^2 \times 200 \times \left(\frac{10{,}000}{60}\right)^2 \times (0.3)^2 = 4{,}935{,}000 \text{ J}$$

$$= 4{,}935 \text{ kJ}$$

Another design of automobile is the flywheel powered vehicle. Such a vehicle carries a massive flywheel that is spinning at high speeds. The vehicle is powered by drawing energy from the spinning flywheel.

To estimate the performance of such a vehicle, consider a flywheel with the following properties. It weighs about 200 kg (approx. 440 lb). It is 0.3 meter in radius (about one foot). At its top speed, it spins at 10,000 revolutions per minute. We cannot make the flywheel much heavier without compromising our estimate of energy needed to drive the car. Also if we make the radius or top speed of the flywheel much greater, we risk it being torn apart by centrifugal forces.

A simple calculation shows that the energy stored in the flywheel at maximum speed is about 4,935 kJ. (The calculation--for experts only!--is shown opposite.) We can see immediately that this flywheel car will not be a serious competitor to the gasoline powered automobile. City driving consumes 1,650 kJ per mile. Therefore the flywheel can carry the car a distance of 4,935/1650 = 3 miles in the city. Similarly, freeway driving consumes 660 kJ per mile. Therefore the flywheel can carry the car a distance of 4,935/660 = 7.5 miles on the freeway.

While these ranges compare poorly with a gasoline powered automobile, a flywheel powered vehicle may still have its place if we are prepared to compromise. As before, we may forgo the performance of a gasoline powered automobile. Or we may find an application in which only a small distance must be traveled between opportunities to spin up the flywheel to high speed.

Current research in flywheel design allows construction of lighter flywheels that carry three times the energy calculated above. They achieve this by spinning a much lighter wheel of 50 lb weight made of strong graphite composite material. The high energy storage capacity results from extremely rapid rates of rotation--up to 100,000 revolutions per minute (that is about 1,700 revolutions in each second!). The wheels spin in a vacuum chamber on magnetic bearings, a somewhat exotic and expensive technology.

Key to success of gasoline as a fuel :

Gasoline is energy rich. Each pound carries a lot of energy.

- Half of fuel is air... "carried" for free!
- All mass of gasoline is consumed as fuel.

Compare with an electric battery which has dead weight in its support structure.

Use other fuels that can also burn in internal combustion engine?

	Energy content		
Gasoline	20,000	Btu / lb	
Hydrogen	61,000	Btu / lb	← rocket fuel
Propane	21,600	Btu / lb	← natural gas fueled cars
Ethanol	12,780	Btu / lb	← "Gasohol" when blended with gasoline
Methanol	9,550	Btu / lb	
Lead-acid battery	34	Btu/lb	

We can see why gasoline powered automobiles have enjoyed such popularity. Gasoline has enormous advantages over other fuels. It carries a very large amount of energy in little weight. One pound of gasoline stores 20,000 Btu of energy. One pound of a lead-acid storage battery carries only 34 Btu of energy. Even if only 25% of this 20,000 Btu can be used by the automobile, the advantage over storage batteries is still overwhelming.

It is not hard to see why gasoline enjoys such an advantage. Part of the reason is that the fuel for a gasoline engine is actually gasoline and oxygen. The oxygen is supplied by the air, so that half of the fuel is not even carried by the automobile. A major part of the weight of a storage batter, however, is given over to components like the case and conducting plates that do not partake directly in electricity generation. Also, a gasoline powered engine seeks to burn its gasoline fuel completely in order to wrestle as much energy as possible from it. In a rechargeable battery, the chemicals that store the energy cannot be allowed to degrade too far while the battery discharges. If they do degrade too far, it may become impossible to reverse the process and recharge the battery.

All these advantages derive from the fact that gasoline fuel is used by burning it in air. Therefore we would expect that its closest competition would be other fuels that also burn in air. Hydrogen can be burned as a fuel in automobile type engines. For the same weight, it carries over three times as much energy. But it must be stored under high pressure. (Under atmospheric pressure, it liquefies at -253°C) Also hydrogen is far more dangerous to use. If it leaks, it is far more likely to form an explosive mixture with air. A mixture of anywhere from 4% to 75% hydrogen in air is explosive. The corresponding range for gasoline is far narrower. For butane, a volatile component of gasoline, it is roughly 2% to 10%. Less exotic and more serious competitors to gasoline include propane, a major component of liquefied natural gas, which has comparable energy content. Alcohols--methanol and ethanol, for example--may also be used. They carry less energy for the same weight. But they are attractive since they burn with less pollution. These two facts are connected. They contain oxygen--in effect they are already partly burned. Therefore they can yield less energy on complete burning. But they will complete the burning more efficiently and produce fewer pollutants from incomplete combustion.

Summary

Alternative energy source	Limitation
Lowering weights	Needs prohibitively massive weight
Electric storage batteries	Too heavy to match gasoline engine performance
Solar cells	Too little energy available
Flywheel	Needs very heavy flywheel
Other fuels that can be burned	Good, but gasoline (and related fuels like propane) has most energy of cheap, non-exotic choices

The table opposite summarizes the problems of energy sources that are alternatives to gasoline. The alternative closest to gasoline are the fuels that can be burned. All of those listed earlier can be burned in engines essentially similar to the internal combustion engine of gasoline powered automobiles.

If it is judged purely in terms of its energy content, hydrogen is far more attractive as a fuel than gasoline. But it becomes unattractive when one considers the difficulties and dangers of handling hydrogen.

Propane and other similar fuels are sold as LPG and are chemically very close to gasoline. They can be recovered from petroleum deposits and geological sites similar to them. They can and are used to propel internal combustion vehicles. The disadvantages are minor. These gases must be stored under pressure in a tank similar to those used for gas fired bar-b-q's. They are also less dense than gasoline. While a pound of LPG may hold as much energy as a pound of gasoline, that pound of LPG will occupy more space.

Alcohols are also viable as a fuel. They can only supply half as much energy as an equivalent weight of gasoline and are more expensive. But they are proving increasingly attractive for two reasons. First they burn with less pollution. Second, they *can* be made from a renewable resource, corn--but they need not be so made. For these reasons, new legislation is requiring the blending of alcohol with ordinary gasoline to produce "gasohol" in certain parts of the U.S.

Fuel cells are devices that exploit the high energy content of fuels such as hydrogen. Hydrogen's energy is usually recovered by burning the hydrogen with air and then using the resulting heat energy to power a vehicle. Fuel cells do not burn the hydrogen; they combine hydrogen with oxygen from air to produce electrical energy directly. This electrical energy can then power an electric motor that drives a vehicle. The advantage is that fuel cell powered vehicles can be three times as efficient in converting the energy stored in hydrogen to energy usable by the vehicle.

Other examples of self-moving vehicles

Airplanes

Weight of entire system is more critical than with automobiles.

Every effort to keep the machines light.
e.g. extensive use of lightweight alloys

Standard fuels

Piston engines — High octane gasoline — Allows highest efficiency engines

Aviation turbines — Kerosene — Energy content similar to gasoline, 20,000 Btu/lb

Ships

Can carry very heavy weights, so weight of fuel not so critical.

Most common fuel for large ships is petroleum derived fuel oil.

Ships and submarines can use nuclear power plants. — Too heavy for use in cars and planes!

Used in:

- Aircraft carriers
- Icebreakers
- Submarines — No need to surface for air for engines!

Similar considerations may be applied to other cases of self-moving vehicles. Weight is of critical importance in airplane design. Every extra pound must be held aloft by the engines. Therefore great effort is expended in reducing the weight of airplanes. Because of their relative low cost and very high energy content, petroleum fuels are the fuels commonly used in airplanes. Piston engine planes use a very high octane gasoline. We shall see in the chapters to come that a high octane allows the engine to recover more energy from the burning of the fuel. Aviation turbines--now used on most larger airplanes--burn kerosene, which is chemically closely related to gasoline and has a similar energy content.

Things are quite different in the case of ships. There the weight of the vehicle is far less of a concern. Nonetheless, petroleum derived fuel oil remains the the fuel most commonly used. However ships can carry power plants whose weights would preclude their use in cars and planes. That is, they may be powered by nuclear reactors. They are used in a wide array of naval vessels, including aircraft carriers, icebreakers and submarines.

In the case of submarines, they confer the added advantage that the operation of the engines does not require air to burn a fuel. Therefore a nuclear powered submarine can operate submerged without the need to risk detection by approaching the surface for air to supply the engines. In conventional submarines, the engines are marine diesel engines. They can operate underwater if the submarine remains close to the surface and draws air through a snorkel tube. Otherwise, conventional submarines use their diesel engines to generate electricity that is then stored in large batteries. The submarine can operate fully submerged by propelling itself with electric motors run by these batteries. However, as we have seen, even large batteries cannot hold much energy compared with diesel fuel. Thus a conventional submarine cannot operate very long on its battery power.

Space Shuttle

Case in which weight of fuel is most critical:

Maximum payload:
65,000 lb (=29,500 kg)

Shuttle weight at launch:
4,500,000 lb
(=2,000,000 kg)

Each pound of payload requires $\frac{4,500,000}{65,000} = 69$ lb of shuttle machinery, etc. to launch it

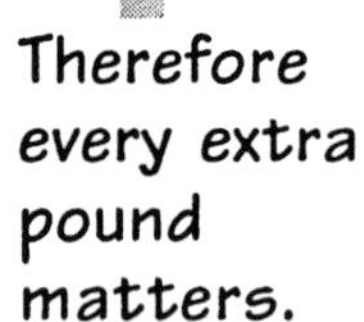

Therefore every extra pound matters.

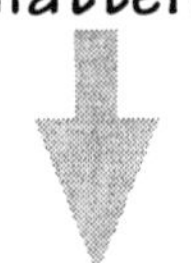

Choose fuels that have most energy for weight

Main engines

Fuel in external tank is liquid hydrogen with liquid oxygen to burn it.

Weight of the main fuel = 1,585,000 lb (=709,750 kg) is a major part of the total weight of the shuttle at launch (85%)

(Held in external tank)

- 226,000 lb hydrogen
- 1,359,000 lb oxygen

Solid booster rockets

Fuel is a solid explosive
weight = 2,230,000 lb

The case in which questions of weight become most critical is that of the space shuttle and other examples of vehicles that are to be lofted into space. At launch, the space shuttle weighs four and a half million pounds. This huge device is to used to loft a payload of 65,000 pounds into orbit. This payload is just a little more than 1% of the weight of the shuttle at launch. That is, for each pound of payload to be lofted into orbit, 4,500,000/65,000 = 69 pounds of shuttle at launch are required. Therefore the designers of the shuttle must make extraordinary efforts to eliminate any unnecessary weight. Any extra pounds that are carried with reappear as many more pounds in the weight of the shuttle at launch.

This necessity has dictated that the most energetic of fuels be used, those that would be impractically exotic for virtually any other purpose. Thus the main engines of the shuttle are fueled by hydrogen. The oxygen needed to burn this hydrogen fuel is carried with the shuttle. Both fuels are stored in liquid form at very low temperatures in the shuttle's external tank. Even these exotic fuels are not sufficient to loft the shuttle. External, solid fuel booster rockets are also used. The fuel for these external boosters is essentially a slow burning explosive. The weight of the explosive fuel comprises by far the greatest part of the weight of the whole shuttle. The shuttle, sitting on its pad awaiting launch, is essentially a huge bomb. Its launch is the miracle of the bomb exploding in a sufficiently controlled way for its payload to be lofted into the orbit desired!

Replace liquid hydrogen with gasoline as main shuttle fuel?

Expensive. Must be kept at $-253^{o}C$ to remain liquid. Forms highly explosive mixture with air in dilute or concentrated forms.

Cheap, easy to store.

Weight of hydrogen in shuttle fuel = 226,000 lb

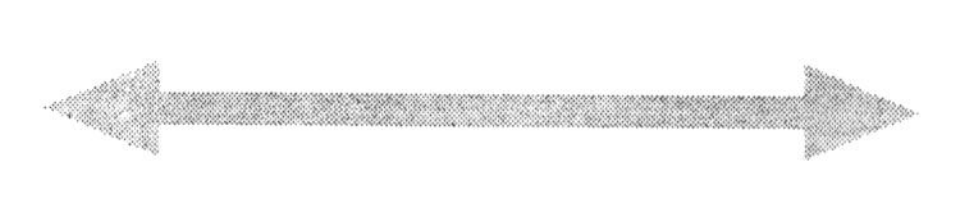

Weight of gasoline with same energy
= 226,000 x 3.05
= 689,300 lb

Energy in one pound hydrogen 61,000 Btu

Energy in one pound gasoline 20,000 Btu

3.05 times as much energy in same weight

Replace hydrogen fuel with gasoline? ← Add weight

689,300 - 226,000
= **463,300 lb**

Weight increase is 7.1 times weight of maximum payload!

Calculation assumes energy of hydrogen and gasoline can be used equally efficiently.

Is a hydrogen powered engine more efficient?

To get a sense of how critical the weight saving is when hydrogen fuel is used, it is helpful to compute the extra weight if the hydrogen fuel were replaced by gasoline. Clearly there would be much incentive to do this if it were possible. Gasoline is far cheaper and easier to handle.

To estimate how much gasoline would be needed, we calculate the weight of gasoline that has the same energy as the hydrogen fuel. Since one pound of hydrogen holds 61,000 Btu and one pound of gasoline 20,000 Btu, the same weight of hydrogen has 61,000/20,000 = 3.05 times as much energy. Therefore the 226,000 pounds of hydrogen fuel would need to be replaced with 226,000 x 3.05 = 689,300 pounds of gasoline. This would add a weight of 689,300 - 226,000 = 463,300 pounds to the shuttle.

This weight gain would be a serious problem. It is over seven times greater than the entire weight of the payload! Of course these added pounds cannot be compared directly to the weight of the payload, for the gasoline fuel is not to be lofted into space, but will be burned and discarded on the way. However it does indicate that the weight gain is significant. Also, the calculation assumes that we need merely replace the hydrogen fuel with gasoline of equal energy. That will only be so if gasoline fueled rockets can operate as efficiently as hydrogen fueled rockets. Since gasoline is less combustible, this assumption may be false.

What if the energy needed for operation does not have to be carried with vehicle?

Entirely different systems become feasible.

Suburban electric trains

Energy to run them is supplied through overhead wires. No fuel carried with train.

They use very efficient, very simple electric motors.

Two parts:

- rotor - bit that turns
- stator - bit that stays still

Compare with much less efficient much more breakdown-prone, much more complicated gasoline and diesel engines.

Energy can no longer be supplied via overhead wires.

Long distance trains

Revert to familiar petroleum fuel: diesel-electric train.

Energy source = diesel fuel burned in diesel engine. The diesel engine is used to generate electricity that runs electric motors that drive the train.

Our difficulties in storing energy have a decisive impact on our choice of technology. This is illustrated by the different types of trains. Suburban trains are released from the need to carry their energy with them as a fuel. They are supplied with electrical energy through overhead wires. The result is a machine that is technically very superior. Its motor is about as simple as one could hope for. There are two parts: a stator (the bit that stays still) and a rotor (the bit that turns and is attached to the wheels).

Compare this simplicity with gasoline or diesel engines. The latter are fearfully complicated with thousands of independently moving parts and the need for complicated maintenance. They are also ill-suited for their common applications. Consider a vehicle powered by a gasoline or diesel engine. If the vehicle comes to a halt, we must disconnect the vehicle's wheels from the engine, for otherwise the engine will stall. This is quite unlike a vehicle powered by an electric motor. When that vehicle stops so does the electric motor, which remains connected to the wheels.

Why do we continue to use gasoline and diesel engines when electric motors are so much better? We use them when we must carry our energy supply with us. Thus consider what happens with trains that travel long distances. Here it may be impractical to supply electricity through overhead wires. Then the trains must seek an alternate energy supply. The most common choice is diesel fuel. In a diesel-electric train, a diesel powered generator supplies the electricity that suburban trains get from overhead wires. Otherwise they are the same as suburban electric trains: the electric power turns electric motors which drive the train.

Answer to puzzle

The weight driven car cannot climb hills!

Energy to mount 1 meter high hill = All energy stored in raised weight

What is the other energy related reason that makes the weight drive car a complete failure? The car cannot climb hills! To see this, consider how much energy is needed for the car to climb a hill. Start with a hill one meter high. To lift the car, we must lift it together with its weight. So, in lowering the weight through its full travel of one meter, we recover almost enough energy to raise the car up the one meter hill. That might be acceptable if all we had to do was climb a one meter hill. But what happens if we have a two meter hill? The car will not have enough energy to climb it!

Assignment 5 Energy, Horses and Pyramids

1 In class we have considered many alternate forms of energy that may be used to run automobiles. We shall now consider horses as an alternate form of energy. We shall compare the energy needs of a car in city driving with those that can be supplied by a team of horses.

(a) Assume that each horse delivers one horse-power. How much energy does a single horse deliver over a period of one hour? (Answer in kJ)
(One horsepower corresponds to 0.7457 kJ of energy delivered in one second; there are 3600 seconds in an hour.)

(b) We computed that a typical car consumes 1650 kJ of energy per mile of city driving. If the average speed of this city driving is 15 miles per hour, how many kJ of energy are needed for each hour of city driving?

(c) Compare your answers to (a) and (b) to determine how many horses would be needed to pull a car around the city, if the car is to have performance matching that of an ordinary gasoline powered vehicle.

Over

2 The Great pyramid of Khufu (Cheops) in Egypt is the biggest pyramid in the world. It was built around 2500 BC and required nearly all of Egypt's male workforce to build it. We can see why such huge numbers of people were needed if we note that all the moving and lifting of stones was done by people. We can estimate how much energy was needed to raise the stones of the pyramid. We can then compare that with how much energy a single human can provide in a day. This will put a lower limit on the number of people needed to build the pyramid.

The pyramid is made from limestone and mostly filled with rubble. The total mass of the pyramid is 5.18 million metric tons and it is 147m high.

(a) Assume that the average slave working on the pyramid consumes 3,000 Calories per day (Big C Calories!) in food energy. How many kilojoules of energy is this?
(Conversion: 1 Calorie = 4.186 kJ)

(b) Virtually all of this energy is used just to keep the slave alive. Most of it is dissipated as heat. More is used as work just moving the slave and allowing the slave to find and eat food etc. Assume that just 1% of this energy is used directly in lifting stones. How much is this 1%? (Answer in kJ)

(c) Let us assume that the construction requires 30 years and that the slaves work 300 days in each year for a total of 9000 days. How much energy does each slave contribute to lifting the stones over this time period? (Answer in millions of kJ)

(d) Each metric ton of the pyramid must be raised on average one half the height of the whole pyramid. (147/2=73.5 meters). How many kJ of energy are needed to raise each ton on average? (Useful fact: 9.81 kJ will raise one metric ton by one meter.)

(e) There are 5.18 million tons of stone to be lifted to build the pyramid. How much energy overall is needed to raise these stones? (Anwer in kJ)

(f) From your answers to (c) and (e) determine how many slaves were needed to lift the stones in the pyramid.

Notice that this puts a lower limit on the number of slaves for the whole project. It assumes a healthy, well fed workforce without malingerers. We have not allowed for the energy required for many other aspects of the construction: the quarrying and carving of the stones, the moving of the stones over great distances, the energy needed to supply the workers, energy for scaffolding etc.

Chapter 6

Core Notions of the Theory of Heat Engines

What is a heat engine?	A heat engine is any device that uses heat to generate "motive power". (=work)	Does the device generate motive power/work? Test: Can it be used to lift weights?

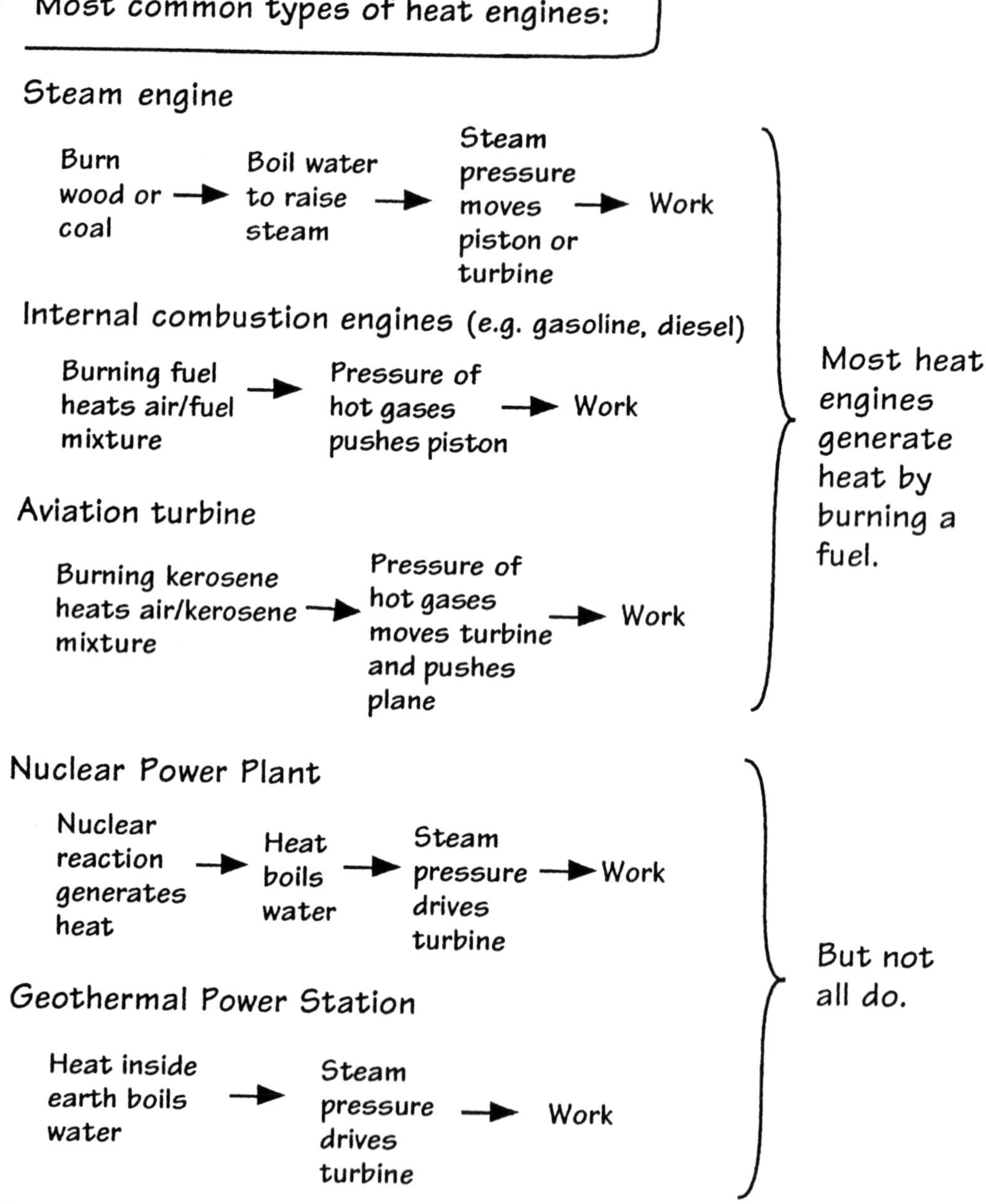

Heat engines are devices that use heat to generate motive power. From now on we shall call the motive power generated "work." Heat engines convert heat energy into work energy. It may not always be clear to you when the device uses heat to generate work. Earlier we saw a simple test: can the device be used to raise weights? If it can, then it is generating motive power or work.

Heat engines are of enormous importance to us. Transportation and the generation of electric power derive their work energy mostly from heat engines. The most familiar heat engines generate heat by burning a fuel. A steam engine burns wood or coal to boil water. The pressure of the resulting steam then pushes a piston or turns the blades of a turbine. These motions generate work: the motion of a piston may move a steam locomotive; the rotation of a turbine can run an electric generator. Internal combustion engines--such as the gasoline engines of cars or the diesel engines of trucks--are also heat engines. They burn a fuel air mixture inside a cylinder. The mixture gets very hot, expands and moves the piston which then turns the vehicle's wheels. Aviation turbines, the engines of jet planes, operate similarly. They burn kerosene in air. The mixture gets very hot, expands and pushes the plane through the air, when the hot gases are ejected in a powerful stream to the rear of the plane.

Other heat engines do not burn a fuel. In a nuclear power plant, the break-up ("fission") of Uranium 235 atoms in a reactor leads to the generation of huge amounts of heat. This heat boils water and the resulting steam drives a turbine which in turn runs an electric generator. In a geothermal power plant, steam from hot underground water turns the turbine that runs the generator.

While there are so many different types of heat engine, they are all covered by the theory of heat engines, whose central proposition is the second law of thermodynamics. In the chapters to follow, we shall see how a simple but very general analysis leads us to results of great power that cover all possible heat engines and place quite powerful restrictions on what they can do. In this chapter we will develop three core notions upon which the theory of heat engines rests.

Origins of the general theory of heat engines: Sadi Carnot, *Reflections on the motive power of fire* (1824)

First of three core notions upon which theory of heat engines is built.

Carnot's Core Notion (I)

All heat engines generate work by transferring heat from a hot place to a cold place.

The surprising part:
heat alone is not enough, you also need a cold place!

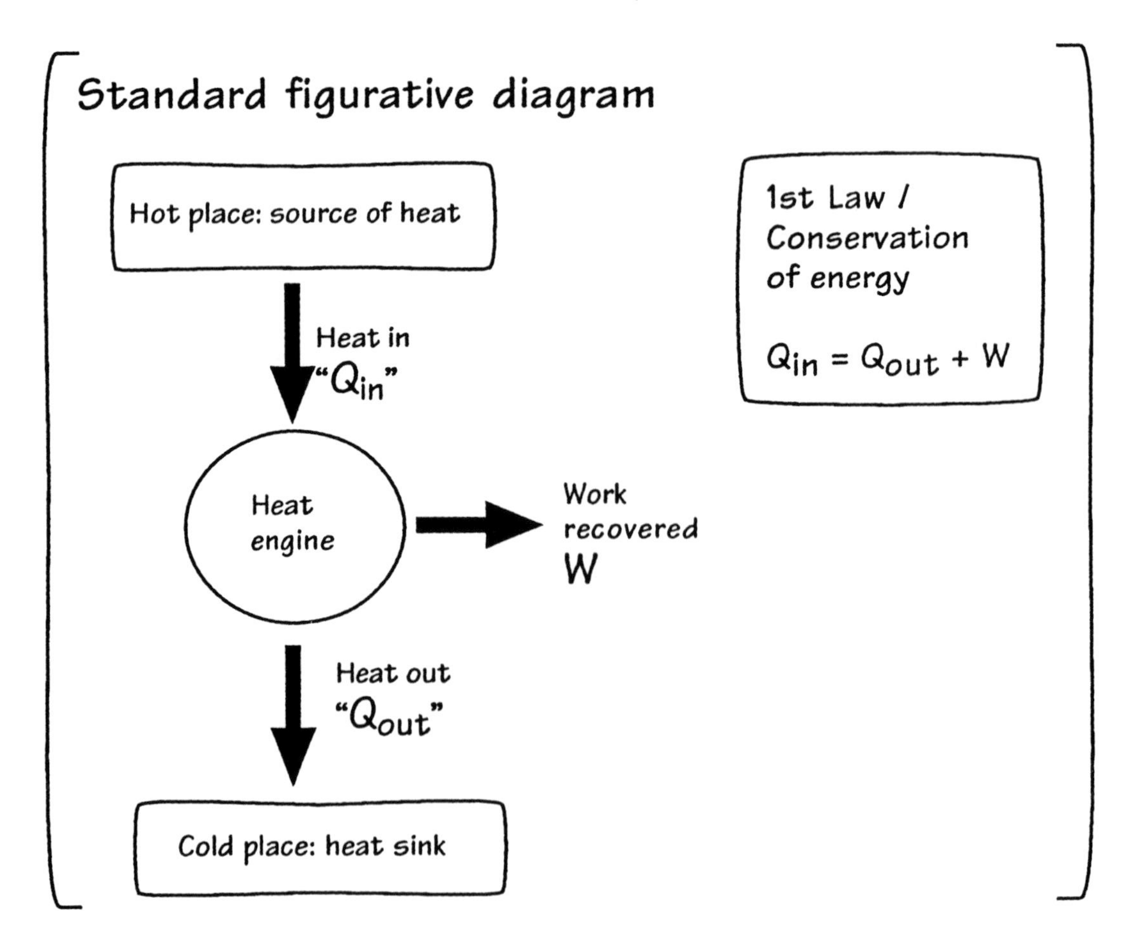

The modern theory of heat engines was first laid out in Sadi Carnot's short but brilliant work of 1824. Carnot was interested in the steam engines used so widely in the industrial revolution. He saw how engineers were working hard to devise ever more efficient engines. While the engineers sought to improve some particular steam engine, Carnot asked general questions: how much more improvement can we expect in all types of heat engines? Are some types of heat engine more efficient that others? Is it better to use heated air or steam or some other fluid? Is there some general limit beyond which no heat engine can pass, no matter how ingenious its design?

To answer these questions, Carnot developed a general theory which would apply to all heat engines, no matter what their design. The first of its three core notions is stated opposite: the operation of any heat engine necessarily involves a transfer of heat from a hot place to a cold place. This assertion is surprising. We know that a heat engine seeks to convert heat into work. Any heat discharged to a cold place represents an inefficiency; it is heat we purchased from the burning of an expensive fuel and we now fail to convert to work. This first core notion assures us that such waste must happen.

It is customary to represent the operation of any heat engine in a single diagram shown opposite. In it we see the flow of heat energy from the hot place through the engine where it divides into two parts, waste heat lost to the cold place and the useful work recovered. The amount of heat that comes in to the engine is labeled "Q_{in}". The amount of heat discharged by the engine to cold place is "Q_{out}" and the work generated is "W". This diagram gives us the theory's most general picture of a heat engine. We will use it over and over--but with particular numbers inserted in the place of Q_{in}, Q_{out} and W. For example, Q_{in} may be 100 cal, Q_{out} 60 cal and W 40 cal.

The flow of energy must be subject to the law of conservation of energy. That law tells us that the heat energy going in to the engine, Q_{in}, must equal the total energy going out. The latter is the heat discharged, Q_{out}, and work, W, and they add to Q_{out} + W. Thus the first law tells us that $Q_{in} = Q_{out} + W$. Using the numerical values above, this equation is 100 cal = 60 cal + 40 cal.

Example : Steam turbine power plant

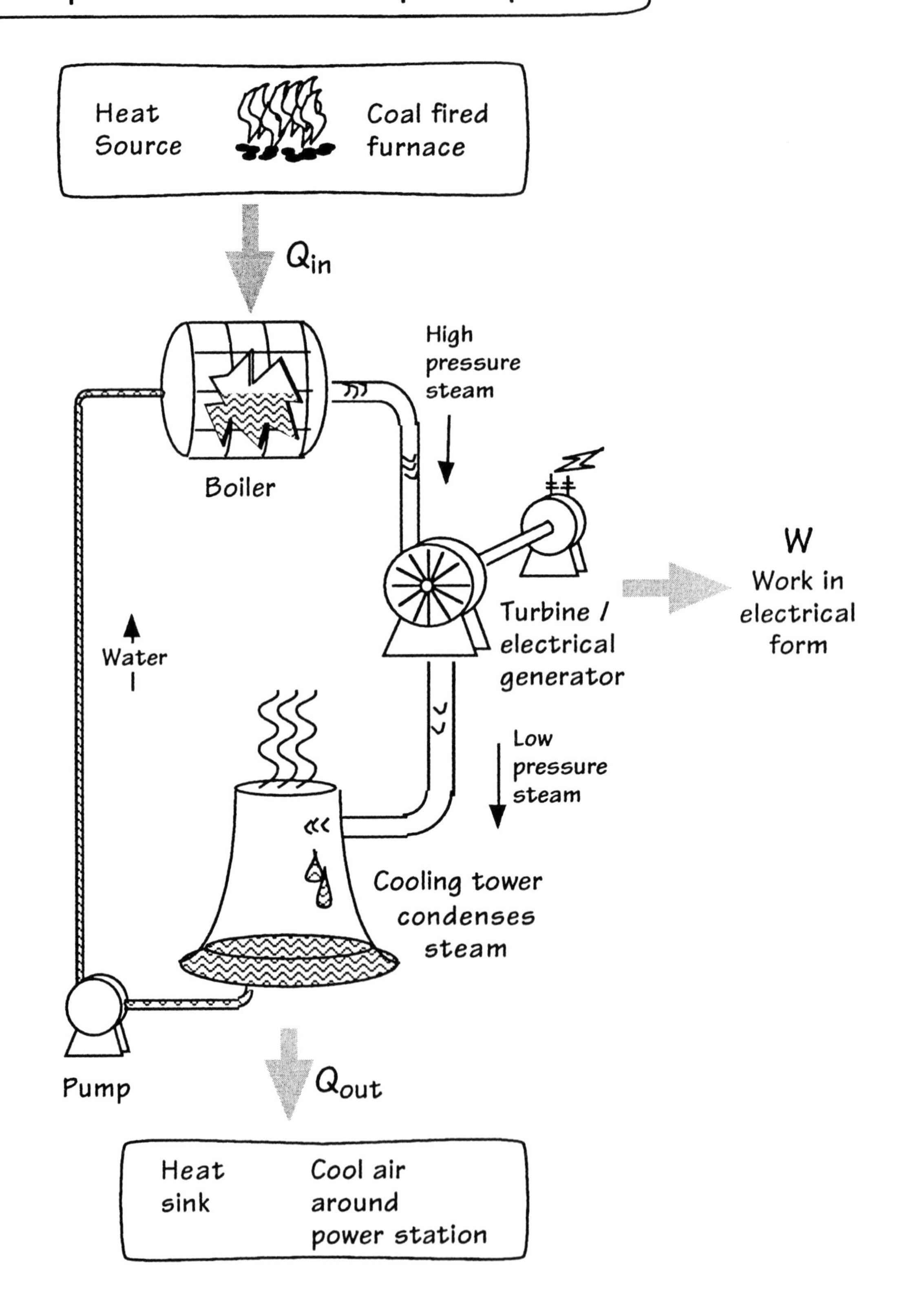

A steam turbine power plant is a heat engine that illustrates this first core notion. Power plants of this type are one of the most common ways used for generating electrical power. They operate by burning large amounts of coal to boil water. The fire in the furnace is the hot place or heat source. Q_{in} is the amount of heat it supplies to the boiler. The high pressure steam generated in the boiler is used to turn the blades of a turbine. These turning blades run a generator which produces electricity. This electrical energy is the work W. The steam loses pressure and cools in turning the turbine. The resulting low pressure steam passes to a cooling tower where it is condensed back to water. To condense the steam, a large amount of heat must be drawn out of it. The cooling tower uses cool air from the surroundings of the power station to carry off this heat. That heat is the Q_{out} discharged to the cool place or heat sink, the air around the power station. The water from the condensed steam is at a much lower pressure than the boiler. Therefore a pump must be used to force the water back into the boiler, where it can be turned into steam once again so that the cycle can continue. The pump will require some energy to return the water back to the boiler. But this energy is far less than that generated by the turbine/generator, which can easily supply it.

The first core notion is illustrated in the cooling and condensing of the steam. At this point in the operation of the heat engine, waste heat is discharged to the cool place. This discharge is essential to the operation of the heat engine. What allows it to generate work is that water expands many times in being converted to steam. This expansion is exploited to set the turbine blades in motion. If at some point the steam is not returned to its recompressed liquid state, the operation of the engine would soon cease. The condensing of the low pressure steam serves this essential function, but it can only happen through the discharge of waste heat.

Plausibility of first core notion: Cold is required to get work from heat! — Not at all obvious!

Conversion of heat energy to work energy usually requires heat expands some fluid to produce a mechanical effect. — But not always.

To continue the operation of the heat engine, we must supply more UNexpanded fluid.

Either cool hot fluid to recompress it or supply fresh, cool working fluid.

Both possibilities require a cool place.

Why Carnot found it easier to see the need for a cold place in the operation of a heat engine.

Carnot's theory was developed within the caloric theory of heat.

Heat is an indestructible fluid.

↓

It cannot be converted into work.

↓

The motion of heat produces work as a by-product of its motion.

↓

A cool place is needed if heat is to be moved.

The steam turbine power plant illustrates how the discharge of heat to a cold place *may* enable the heat engine to operate. The first core notion says more: all heat engines *must* do it. That is, any heat engine will fail to convert all the heat supplied to it to work. Some will always be lost as waste heat. We can make this idea plausible by noticing that virtually all heat engines operate by using heat to make some fluid expand: steam engines make water expand to steam; internal combustion engines heat and expand an air/fuel mixture. If a continuous cycle is to be sustained, there must also be a supply of unexpanded fluid for the heat to expand. This may be done by cooling and recycling the expanded fluid, as in the steam turbine power plant, or by supplying fresh, cool fluid--that is, fresh cool water to the steam turbine power plant. But not all heat engines operate by using heat to expand a fluid. We shall see later that ice affords an unexpected way of avoiding it. Also certain electronic devices can convert heat directly to electrical work without any intermediate steps that involve the expansion of fluids.

Therefore, suggesting that heat engines always require cold was a bold step--but one that has proved spectacularly successful. In 1824, Carnot found the suggestion far easier to make for a reason you may not expect. He was an adherent of the caloric theory of heat, a now defunct view that was then in the mainstream. According to it, heat was a fluid that could not be created or destroyed--and certainly not converted into work energy, as we now believe. Thus, in seeking the most general way to describe a heat engine, Carnot could not use the description we now give: a device that converts heat into work. That would involve the supposition that heat is destroyed. Instead he noticed that heat engines were able to produce work whenever heat moved. This became his basic characterization of a heat engine: a device that produces work as a by-product of the motion of heat. But how are we to move heat? Heat only moves from a hot place to a cold place; it leaves the fire only if the surroundings are cooler. Thus the general picture given earlier falls naturally into place. In any heat engine, heat passes from a hot place to a cold place. In the process we may generate work. But, since heat can never be destroyed, the heat flow into the engine must be balanced by a discharge of an equal flow of heat into the cold place. Because of the this emphasis on the motion of heat, Thomson talked of the dynamics of heat--thermodynamics.

The essential content of Carnot's first notion is restated as:

Second Law of Thermodynamics

"Thomson" form

No engine can produce as its sole effect the complete conversion of heat into work

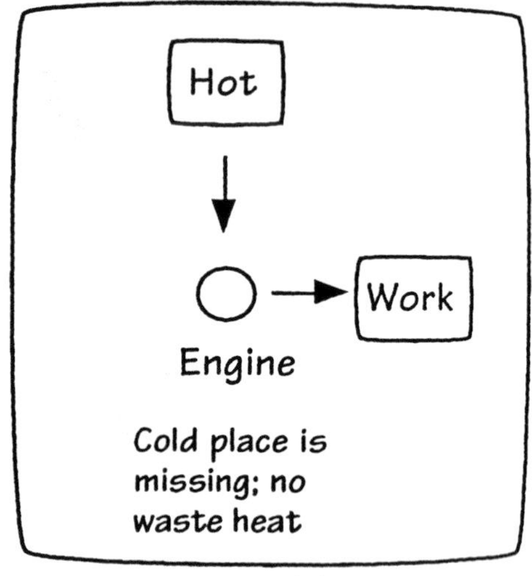

impossible

These forms are equivalent. Later we will see why.

"Clausius" form

No engine can produce as its sole effect, the transfer of heat from a cold place to a hot place

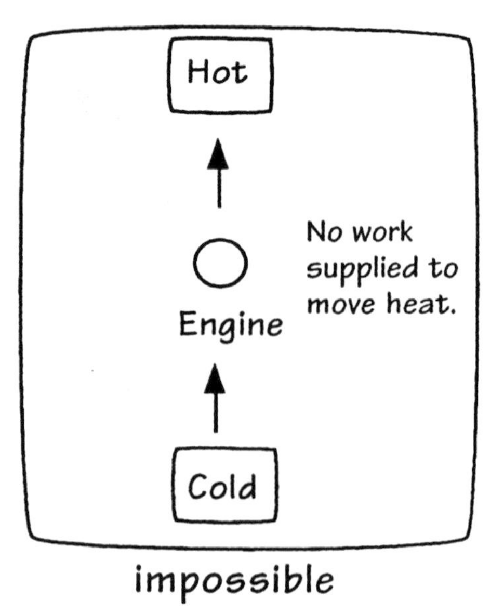

impossible

In 1824, under the influence of the caloric theory of heat, it was natural for Carnot to assume that any heat engine must discharge heat to a cold place. However, by about 1850, it became clear that heat was actually being converted into work in a heat engine. Rudolf Clausius and William Thomson began to explore how the theory of heat engines would fare under this change. They could no longer derive Carnot's notion from the properties of heat. How could they rule out the possibility of some ultra efficient heat engine converting all the heat supplied to work and discharging none as waste heat? But they found that they could not do away with Carnot's basic notion, for it was essential to developing the highly successful theory of heat engines. They soon realized that Carnot's notion actually contained a fundamental law of nature in its own right--one that was independent of the law of conservation of energy and one that was just as important.

They extracted the essential content of Carnot's first core notion and restated it in the form of a law that soon came to be known as the second law of thermodynamics. The form Thomson gave the law is closest to Carnot's original. It merely asserts that one cannot have a heat engine that exhausts no waste heat to a cold place, no matter how clever the design.

Clausius (Thomson pointed out) had given an alternate form of the same law. In Clausius' formulation, the law prohibits any machine whose sole effect is the transfer of heat from cold to hot. That is, if any machine does transfer heat from a cold to a hot place, work energy must be employed to bring about the transfer (or there must be some other equivalent compensating change elsewhere). It is not at all obvious that these two forms of the law--the Thomson and Clausius form--actually say the same thing. We shall see later that their equivalence can be proved.

The second law of thermodynamics is the basic postulate of the theory of heat engines. We shall see in the next chapter how we can deduce results of great power and generality from it concerning all heat engines.

The Thomson form of the Second Law of Thermodynamics also prohibits:

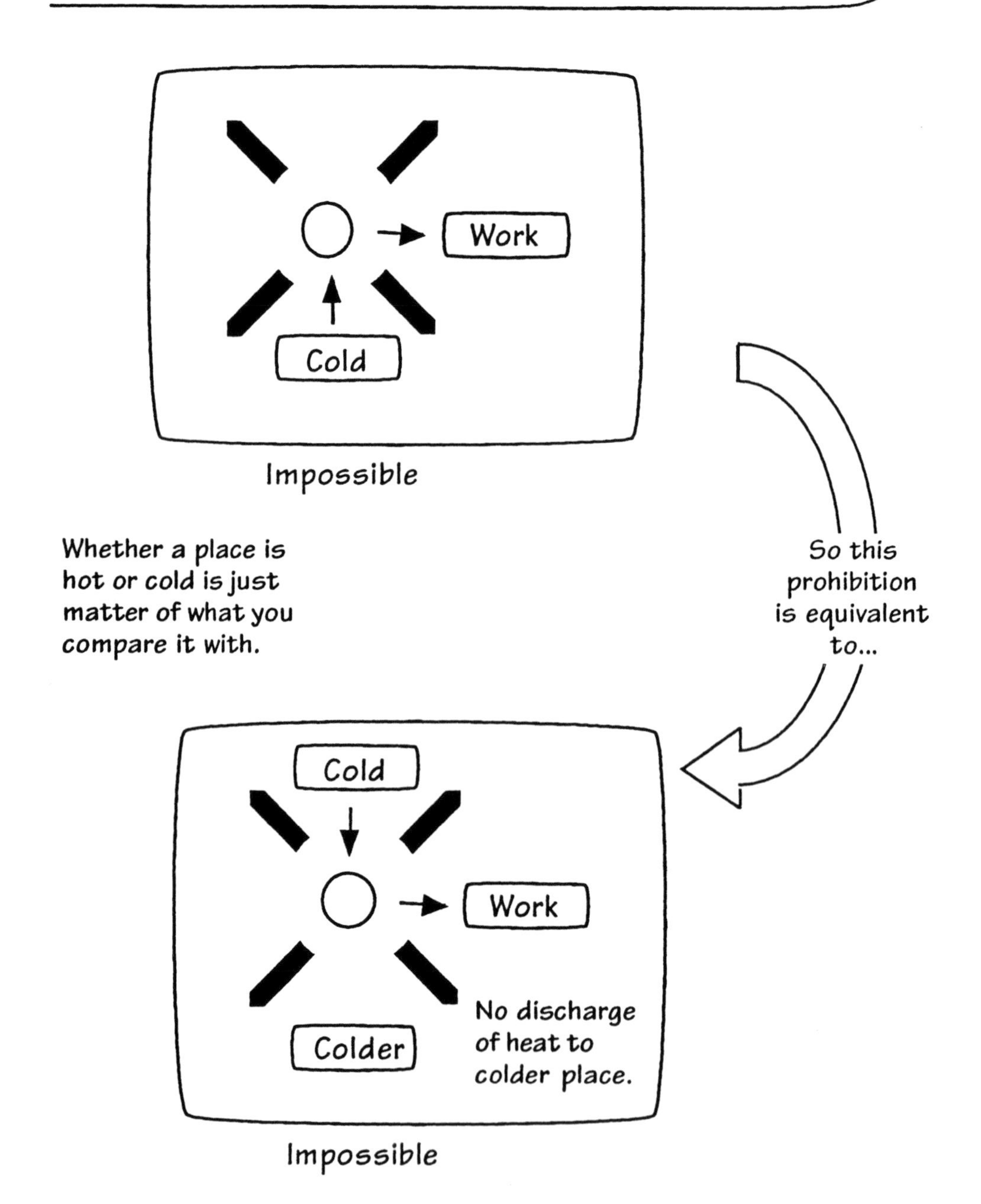

The Thomson form of the second law of thermodynamics prohibits any process whose sole effect is the complete conversion of heat into work. In the figure given earlier, this prohibition was represented as a prohibition against any process that takes heat from a hot place and converts it fully into work. That way of portraying the law has an arbitrary element. It does not matter if the place that supplies the heat is hot or cold. All that matters is that heat is drawn from it.

The reason it does not matter is that places are only hot or cold in comparison to other places. So boiling water at 100°C is hot in comparison with ice at 0°C. But it is cold in comparison with a furnace at 500°C. So the Thomson form of the law prohibits any process that extracts heat from a place and converts it fully work--whether we judge that place to be hot or cold.

Another way to see the prohibition is to imagine a prohibited process that takes heat from a cold place and converts it fully into work. That cold place might be judged as hot in comparison to another place that is even colder. So the diagram can be made to look like the earlier diagram merely by moving the cold place to the position of the hot place and inserting a yet colder place.

The prohibition shown opposite will be important to us in the next chapter when we derive consequences of the second law.

Carnot' objective:

Discover which engines get the most work from each quantity of heat.

Problem:

In some heat engines, it is hard to determine how much heat is supplied and how much is exhausted as waste heat.

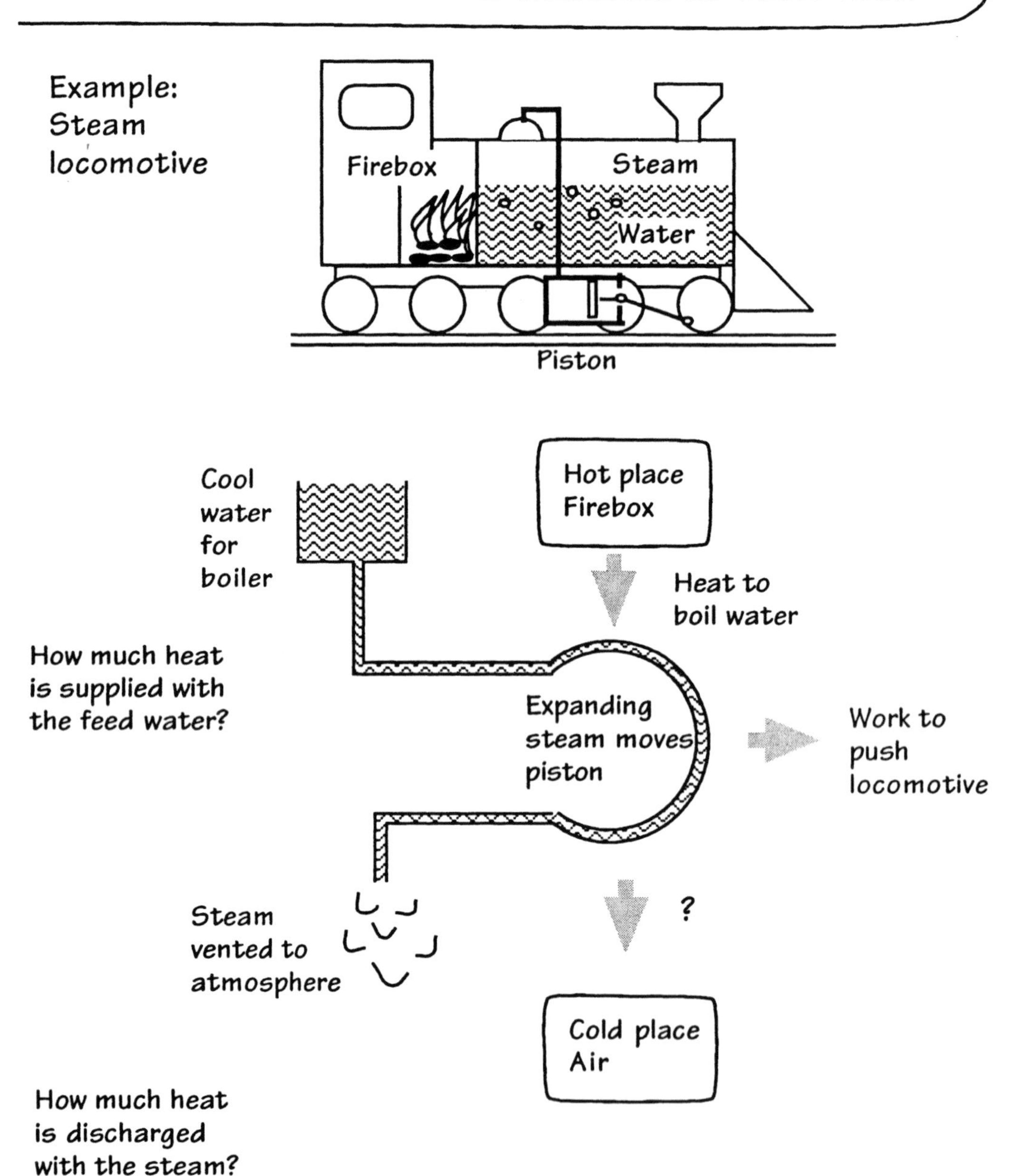

Carnot's next core notion pertains to a technical problem in the theory. One goal is to find which heat engine yields the most work for a given amount of heat. In some engine designs this proves hard to do simply because it is hard to account for exactly how much heat is supplied.

For example, consider a steam locomotive. In a steam locomotive, fire in the firebox boils water to make steam. The steam is then used to move a piston which drives the wheels of the locomotive. Unlike the steam turbine power station, once the steam has done its useful work, it is not condensed to water and reused. It is just discharged to the air.

In this design, how are we to assess how much heat is used in the engine? The problem is that the fire is not the only source of heat. There is also some heat supplied with the cool feed water. To see this, imagine that the steam locomotive refills its boiler with water at different temperatures: 20°C, 25°C, 30°C, ... The warmer the feed water, less heat need be supplied from the firebox to boil it. In effect, the warmer water contains some heat which supplements the heat supplied by the firebox and must also be considered as heat supplied to the engine in our analysis. Similarly, when the steam is exhausted to the air, this is the exhausting of waste heat to the cooler place. But exactly how much heat do we judge the steam to contain?

These are not unanswerable questions. We could decide to take 20°C as our baseline and measure the heat in the feed water and exhausted steam as the heat that would be needed to bring water at 20°C to those states. But this approach introduces a lot of complications that turn out to be quite inessential to our analysis. (Why choose 20°C? Why not choose different baseline temperatures for the feed water and exhausted steam?) It turns out that there is a better way of simplifying the analysis.

Carnot's Core Notion (II)

Analyze engines modified so that the working substance is contained in a fully closed cycle.

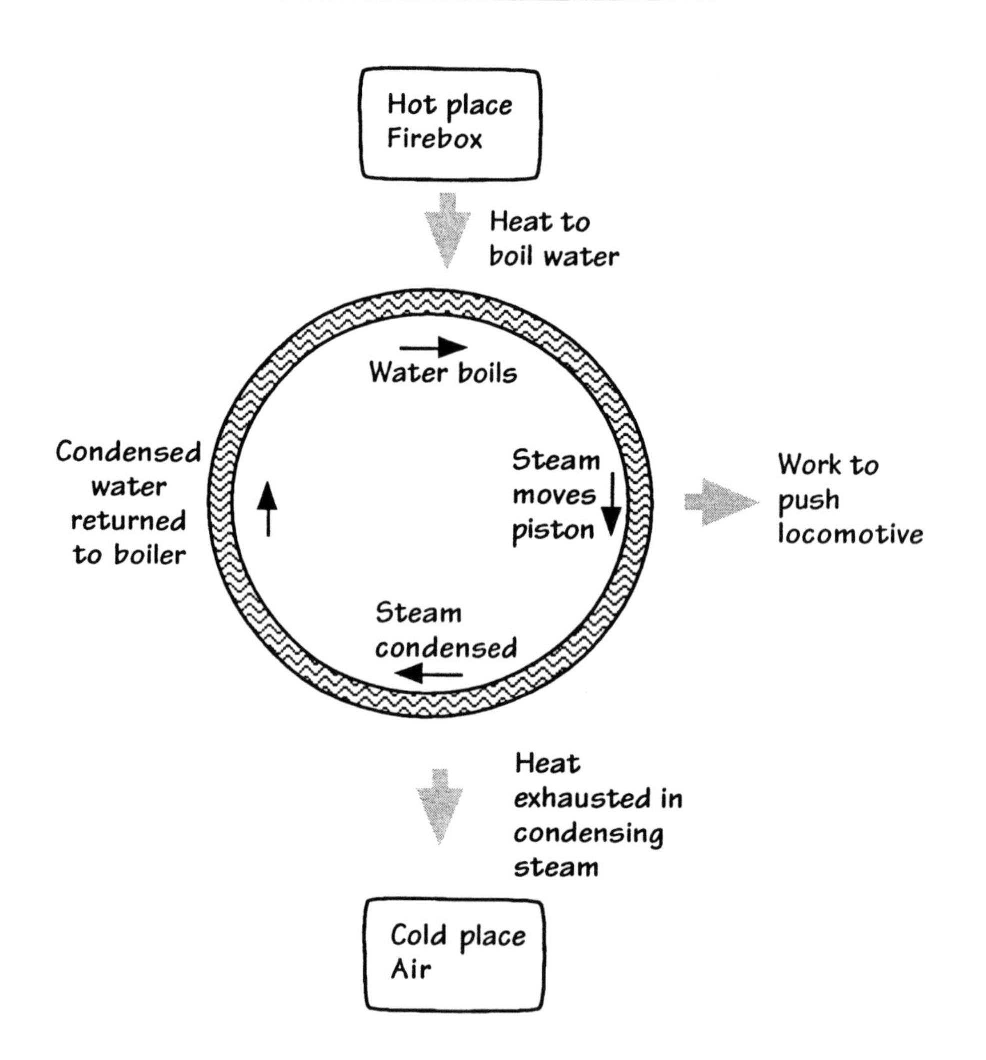

The problem described arises because our engine accepts a working substance--water--as a feed and then discharges it after it has been used. As result it is hard to see how much heat energy is supplied to the machine and how much is discharged. Carnot's simple solution is just to restrict our analysis to machines that do not accept and exhaust the working substance. That is, we will consider only heat engines that operate in a closed cycle, fully recycling all their working fluid. This is Carnot's second core notion.

At first it may seem that this notion severely restricts the scope of our analysis. Fortunately it turns out to have a far lesser effect. The design of most heat engines that accept and discharge working substance may be modified in quite small ways so that they operate with a closed cycle.

In the case of the steam engine, we would add steps in which the steam is condensed back to water and then this water would be supplied to the boiler as feedwater. In its modified form, it becomes very clear exactly how much heat is supplied to the engine and how much discharged as waste heat. The heat supplied is exactly that used to boil the water, for there is no other source of heat. The waste heat is exactly that exhausted in condensing the steam, for there is no other means to exhaust heat.

This engine that recycles its water is, of course, a different engine. It is probably not one that we would ever want to build. The condenser would have to be very large if it was to operate successfully--just as some of the largest pieces of equipment in a steam turbine power station are the cooling towers. But, thermodynamically, the ordinary steam locomotive engine and the recycling steam locomotive engine are very close. We shall see later, for example, that the recycling steam locomotive engine is one of the least efficient of heat engines because of design features common to it and the ordinary steam locomotive engine. Therefore this unhappy verdict carries over directly to ordinary steam engines.

Examples:

Geometry:	No real line is perfectly straight.	→	Perfectly straight lines
Mechanics:	All motion is impeded by friction.	→	Perfectly frictionless motion

We can build systems as close as we like to the ideal limiting case.

Ideal, limiting cases are often:

- Easy to analyze
- The case of best performance

Corresponding notion in the theory of heat engines:

Carnot's Core Notion (III)

Consider "reversible" heat engines. They can be run in both forward and reverse direction.

- They exist only as ideal, limiting case.
- We can build machines that come closer and closer in performance to a reversible engine.
- No real engine can be reversible since reversible engines operate infinitely slowly.

In most sciences it has proven very convenient to talk about things that actually cannot be. We know that no real line is perfectly straight. Every line in a house is at least slightly crooked. No matter how careful the builder, no board can be cut to a perfect straight line. In mechanics, no real motion is unimpeded by friction. No matter how smoothly the wheels of a car turn, there is always some friction in the axle that will slow it, as will air resistance. But straight lines and frictionless motions have a special place in our analyses. They are the simplest cases to deal with, so their analysis is often easy. Further, they can represent an ideal limit to performance. In plotting routes on a map of a city, the straight line route is shortest. When a car moves frictionlessly, it requires the least fuel.

There is a corresponding notion in the theory of heat engines. It is the idea of a "reversible" heat engine, one that can run in both forward and reverse direction. Carnot's third core notion asks us to consider such engines as an ideal limiting case. These engines will be of special importance to us, since they will represent the case of absolute best performance in a heat engine. They will also prove to be especially simple to analyze. But they are ideal cases. We may be able to construct sequences of engines that come ever closer to a reversible engine. But just as we can make no line perfectly straight, so we cannot make any engine that is truly reversible. One reason for this is that reversible engines actually operate infinitely slowly--so a true, reversible engine would undergo no change in any finite time!

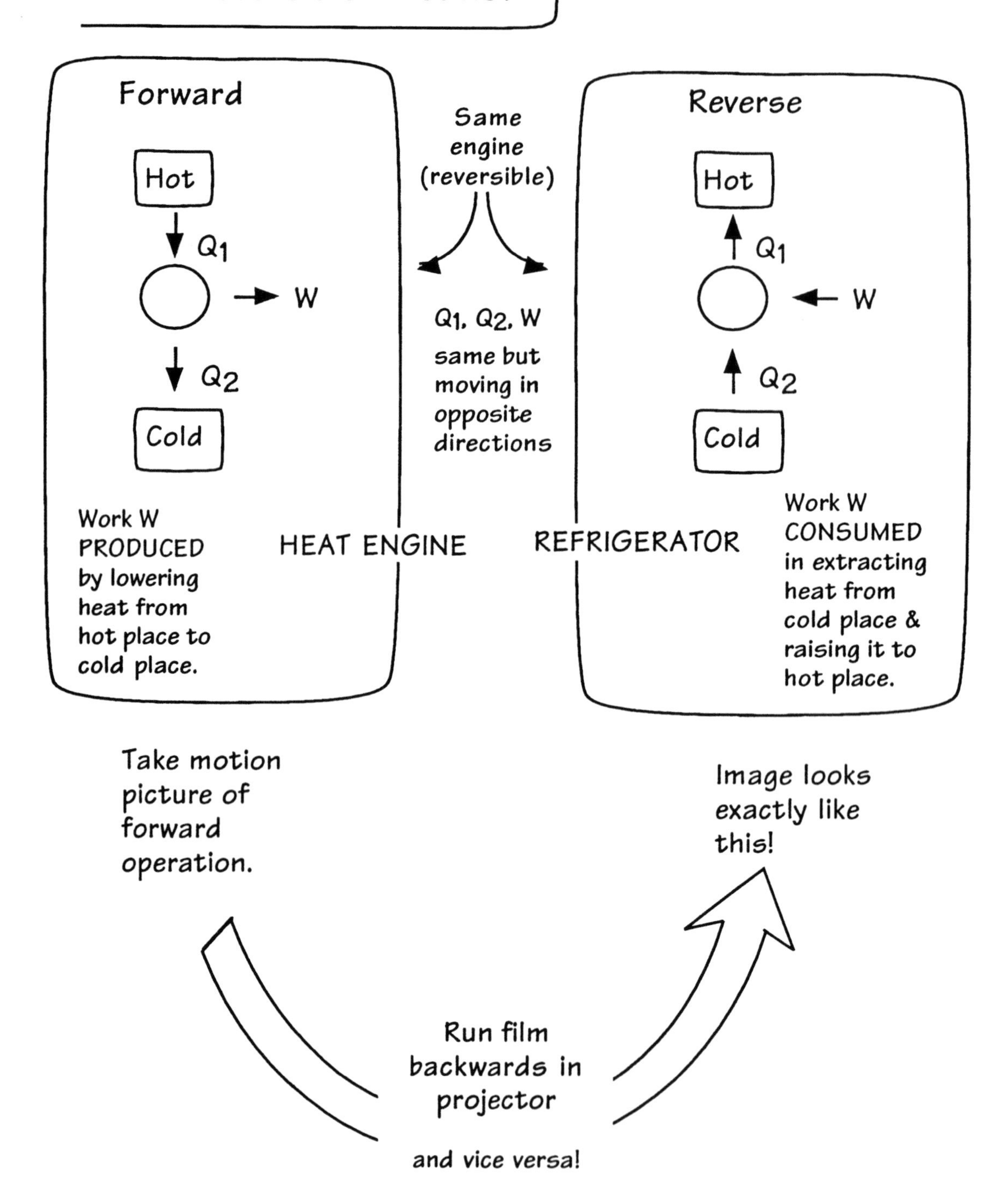
What "reversible" means:
Forward
Hot
Q_1
W
Q_2
Cold
Work W
PRODUCED
by lowering
heat from
hot place to
cold place.
HEAT ENGINE
Same
engine
(reversible)
Q_1, Q_2, W
same but
moving in
opposite
directions
Reverse
Hot
Q_1
W
Q_2
Cold
REFRIGERATOR
Work W
CONSUMED
in extracting
heat from
cold place &
raising it to
hot place.
Take motion
picture of
forward
operation.
Image looks
exactly like
this!
Run film
backwards in
projector
and vice versa!

But what does it mean for a heat engine to be reversible? It simply means that the engine can run in both forward and reversed directions. When the engine runs in the reverse direction, every change of the forward direction still takes place, but each now happens in the opposite direction.

For example, if, in forward operation, an amount of heat Q_1 is supplied by the hot place to the engine, then, in reverse operation, an equal amount of heat Q_1 will be transferred from the engine back to the hot place. Similarly, if the engine supplies work W in forward operation, it will require that the same amount of work W now be supplied to it in reverse operation.

The overall effect of running a heat engine in reverse is to undo whatever the heat engine did when operating in the forward direction. In the forward direction, the hot place loses heat energy, part of which is converted to work and part passes as waste heat to the cold place. In reverse operation, heat energy is taken from the cold place; work is supplied to the engine; and the combined energies--work and heat--are delivered to the hot place.

This last operation may seem unfamiliar. But it is actually one you know very well. That is what a refrigerator does. The cold place is the inside where the food is kept. Work is supplied to the refrigerator as electrical energy. The refrigerator extracts heat from the cold place and delivers it (with the electrical energy) as heat to the hot place. In this case, the hot place is just the room that holds the refrigerator. The heat is delivered via coils behind the refrigerator to the air. We do not normally see these coils since they face the wall behind the refrigerator. If you feel the coils, you will find they are warm as they deliver heat to the room.

The reverse operation is so complete that one can capture it vividly in the following way. Take a movie of the heat engine operating in the forward direction. It will show heat being supplied by the hot place and work generated by the engine. Run the movie in reverse. The reversed movie will show heat delivered to the hot place and the engine requiring work to be supplied to it. That reversed movie will show exactly how a reversible heat engine operates when it is run in reverse.

How can a heat engine be reversible?

It must use exclusively processes that can be reversed individually.

A process is reversible.	=	The process is always only minutely away from equilibrium.

Therefore the process proceeds infinitely slowly.

Example: Freezing/melting of ice in a pond

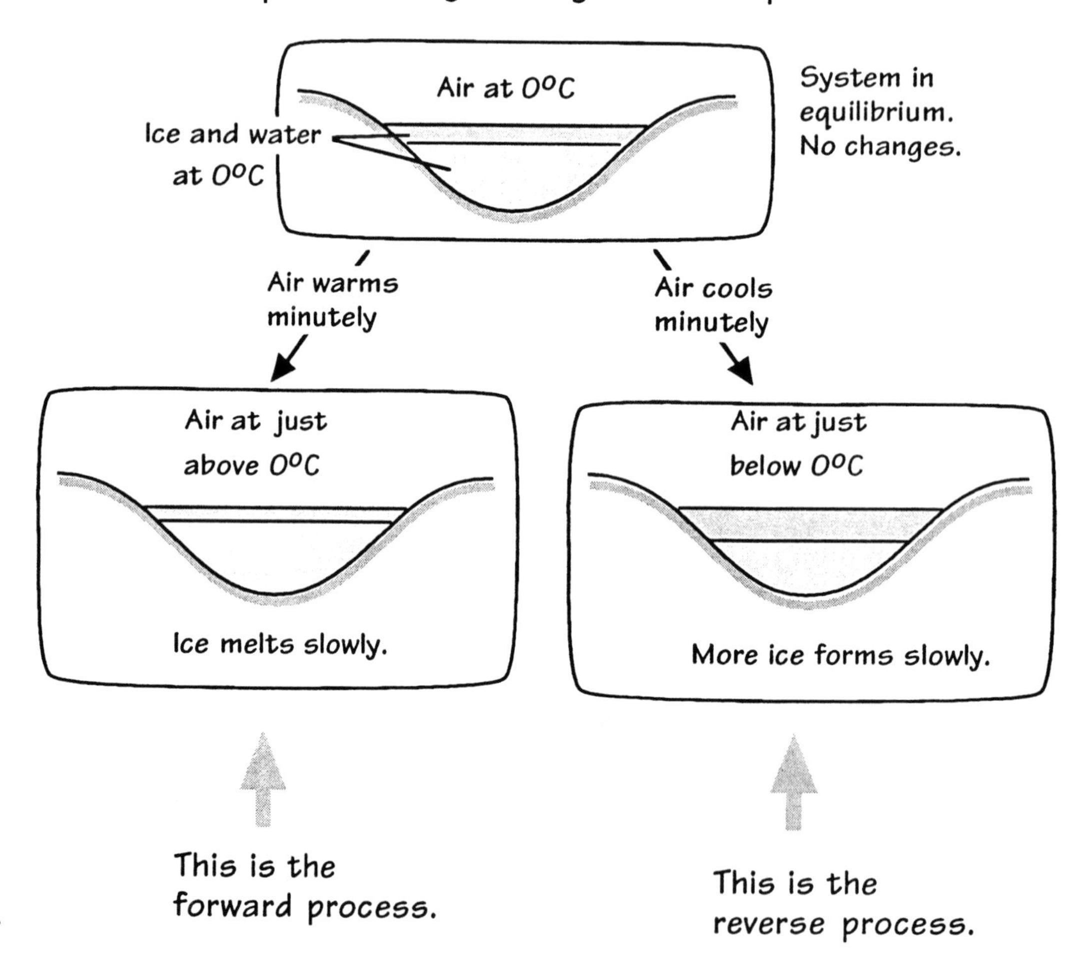

This is the forward process.

This is the reverse process.

How is it possible, even in principle, to design an engine that is reversible? Every process used in the engine must be reversible individually. If that is true, then the engine overall will be reversible.

An individual process is reversible if it is only minutely away from equilibrium. At equilibrium, the forces that drive a process in one or other direction are exactly balanced and no changes occur. To arrive at a reversible process, we allow one of these forces to grow minutely so that it starts to drive change in one direction. Because it is minutely away from equilibrium, the force can only induce minutely slow changed and the process must proceed infinitely slowly.

This is much easier to visualize in an example. Imagine a pond of water in winter. It has a layer of ice on its surface. When the air temperature is at exactly 0^oC, the freezing point of water, the air/ice/water are in perfect equilibrium. No difference of temperature seeks to move heat between the air and the pond. Nothing happens.

Now imagine that the air warms very slightly. Since the air is just warmer than the ice, heat will pass very slowly from the air to the ice/water, very slowly melting the ice. This is our forward process. If instead the air had cooled very slightly below 0^oC, then heat would pass very slowly from the ice/water and more ice would form. This is the reversed process. We can apply the movie test mentioned earlier. Imagine a movie of the ice melting run backwards. It would look just like the ice forming. It would, that is, as long as the temperature difference between the air and ice/water was minutely small so that it could be ignored in the movie.

If the temperature differences driving the processes differ greatly from the equilibrium value, the process is no longer reversible. It would be irreversible. Consider, for example, if the air warmed to 20^oC. Heat would pass rapidly from the air to the ice, so that the ice would melt rapidly. We can take a movie of this and run it backwards. It would show ice forming, but it would portray no process that could happen in nature. For it would show heat passing spontaneously from ice/water at 0^oC, a colder body, to air at 20^oC, a warmer body. That is impossible; the process is irreversible.

Irreversible and Reversible Expansion

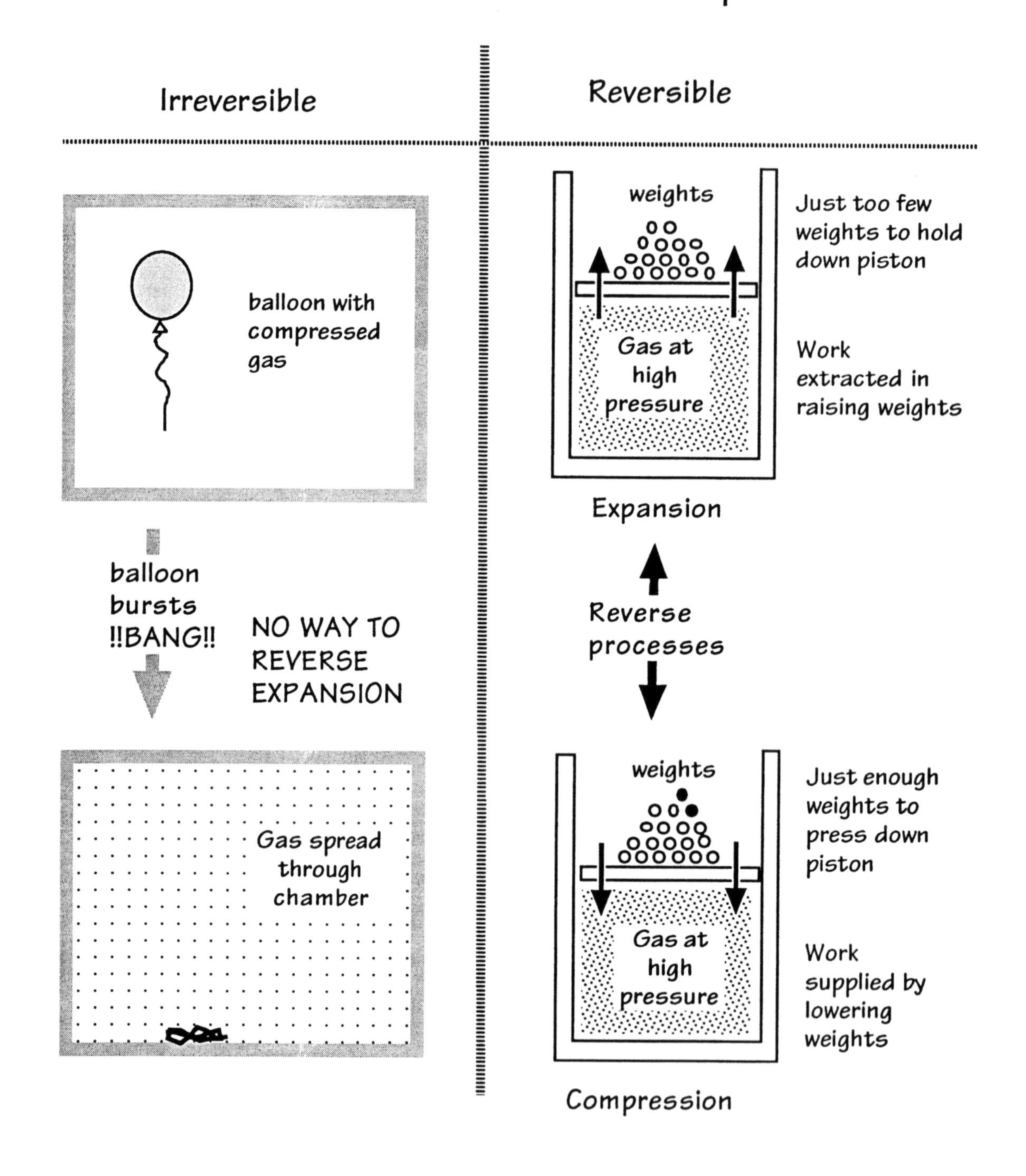

We can now apply the notion of reversibility to the two most important processes in heat engines: expansion/compression and heating/cooling.

Consider a volume of compressed gas, such as is trapped in the balloon opposite. If the balloon bursts, the gas expands to fill the chamber. The expansion is irreversible. To see this, imagine a movie taken of the bursting. If it were run in reverse, it would show the air in the chamber rushing together and recompressing spontaneously in one part of the chamber, with shreds of rubber rising up and joining together to reform the balloon. This is not a process that can happen; the bursting of the balloon is irreversible.

Consider the same volume of compressed gas held in a cylinder equipped with a piston. The gas pressure seeks to push the piston out. Imagine that the piston is kept in place by a pile of weights that exactly balance the pressure forces. The system is in equilibrium and there are no changes. Now remove a tiny portion of the weights so that it is just unable to hold down the piston. The piston will slowly begin to rise. If we then add very slightly to the weight pile, the weights will be just sufficient to overcome the pressure and the gas will be recompressed. Notice that each of these processes is the other run in reverse. If we take a movie of one and run in backwards, it will look like the other--as long as the changes in the weight pile are so tiny as to be ignorable. Therefore the piston and cylinder arrangement gives us a means for expanding the gas reversibly.

In the case of the reversible expansion, we are able to recover work: the gas pressure is used to raise weights. In the case of the irreversible expansion, no work is recovered. That is, the reversible process has more fully exploited the useful potential of the compressed gas than did the irreversible process. This hints at the result we shall see in the next chapter: reversible heat engines are those that recover the most work from heat.

Irreversible and Reversible Heating

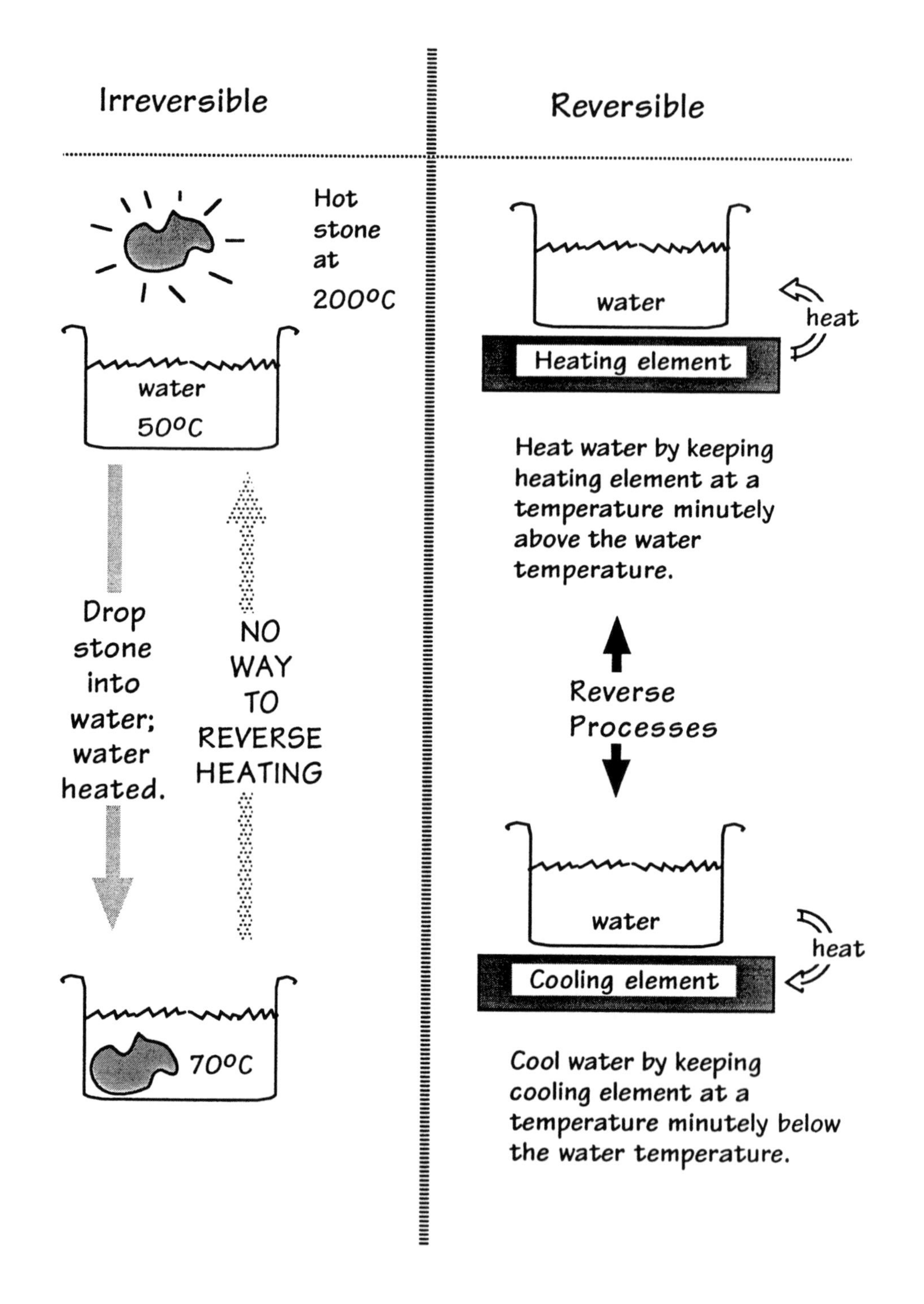

To illustrate the difference between irreversible and reversible heating, consider the heating of water in a container. If the water is just warm --say at $50^{o}C$--we can heat it by dropping in a stone that is very much hotter than it. For example, the stone may be at $200^{o}C$. Heat would pass rapidly from the stone to the water, heating the water and cooling the stone, taking both to some intermediate temperature. We can see immediately that this process is irreversible. Imagine a movie taken of it and run in reverse. It would show the water and stone starting at some intermediate temperature. Then heat would spontaneously pass from the water to the stone, cooling the stone to $50^{o}C$ and heating the stone to $200^{o}C$. This is not a process that can happen. Heat does not spontaneously pass from cold to hot. Heating water with a very hot stone is not a reversible process.

To heat the water reversibly, we must use very tiny temperature differences. Imagine that we place a heating element under the water vessel. As long as both heating element and water are at exactly $50^{o}C$, no heat will be exchanged. If we allow the heating element to warm to just above $50^{o}C$, then heat will slowly pass from the element to the water, heating the water. This heating will proceed as long as we keep the heating element just slightly warmer than the water.

This heating can be reversed. If the element were maintained at a temperature just slightly below that of the water, then heat would pass very slowly from the water to the element. The water would now cool. This cooling is the reverse of the heating, as we can see if we take a movie of the heating. That movie run backwards will look just like the cooling, as long as the temperature difference between water and element is kept minutely small and can be ignored.

Reversible heat engine:

In forward operation, heat passes from hot to cold and work is produced.

In reverse operation, heat passes from cold to hot and work is consumed.

Having trouble visualizing how an engine can operate in reverse? An example of a common device that can be run in reverse:

But it is not a *heat* engine!!

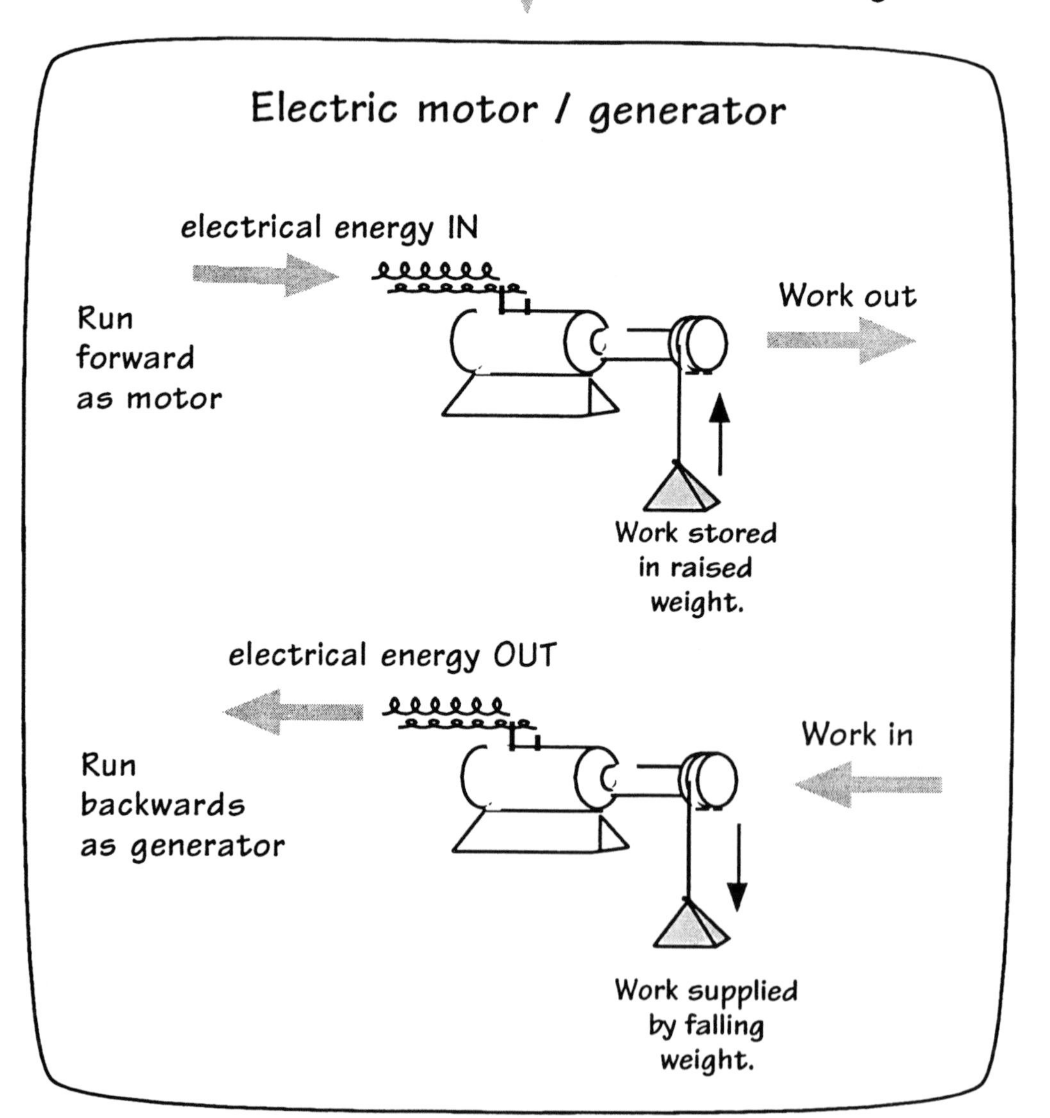

If a heat engine uses reversible processes only, then it will be reversible. This means that it can run forwards and produce work or it can run in reverse and consume work. No real heat engine can be reversible and most are quite far from reversibility. So you may find it hard to visualize a work producing engine that can be run in reverse. Here it may be helpful to recall that there are other common work producing devices that can be reversed. They are electric motors and they become electricity generators when they are run in reverse.

The forward operation is shown opposite. Electrical energy is supplied to the motor, which produces work. The work is used to raise a weight by winding a cable over a drum. The work energy is then stored in the raised weight as potential energy. In the reverse operation, the weight is lowered. The energy stored as potential energy is supplied to the motor which now functions as a generator. It converts that energy into electrical energy which is delivered through the connecting wires.

Notice this motor/generator system is NOT a heat engine. It is merely being used to illustrate the reversibility of a system.

It is a great deal more complicated to describe in detail the operation of a reversible heat engine. One example of a reversible heat engine is broken down into its individual processes and described in the appendix.

Appendix An example of a reversible heat engine

The Carnot steam cycle

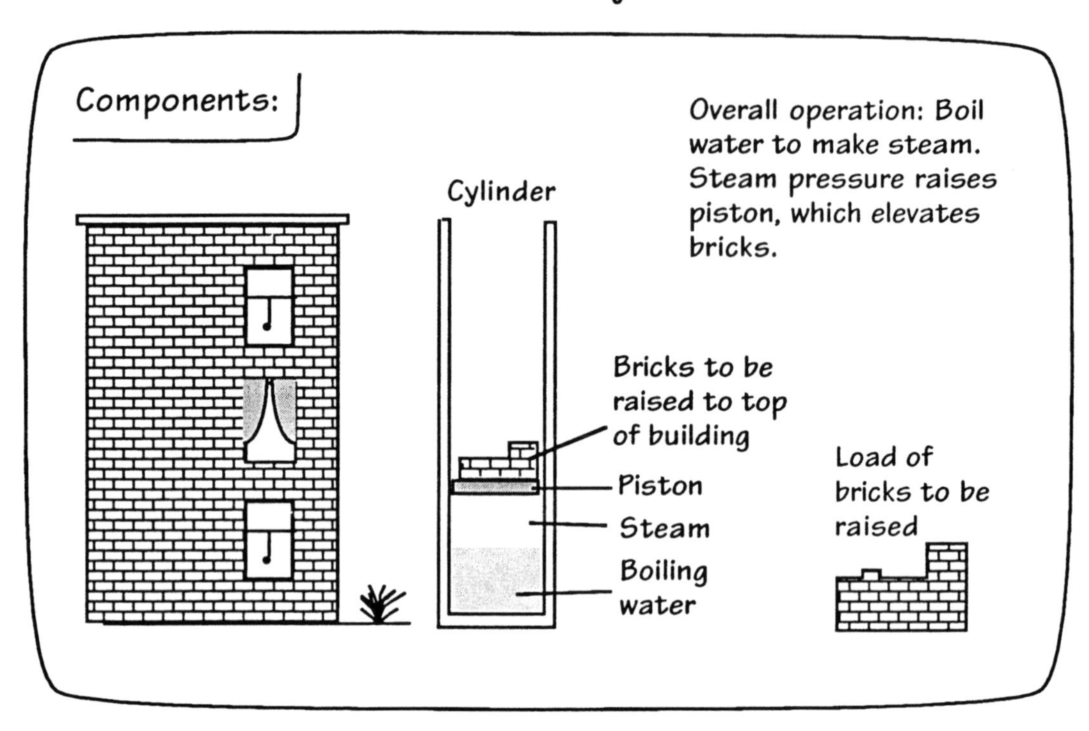

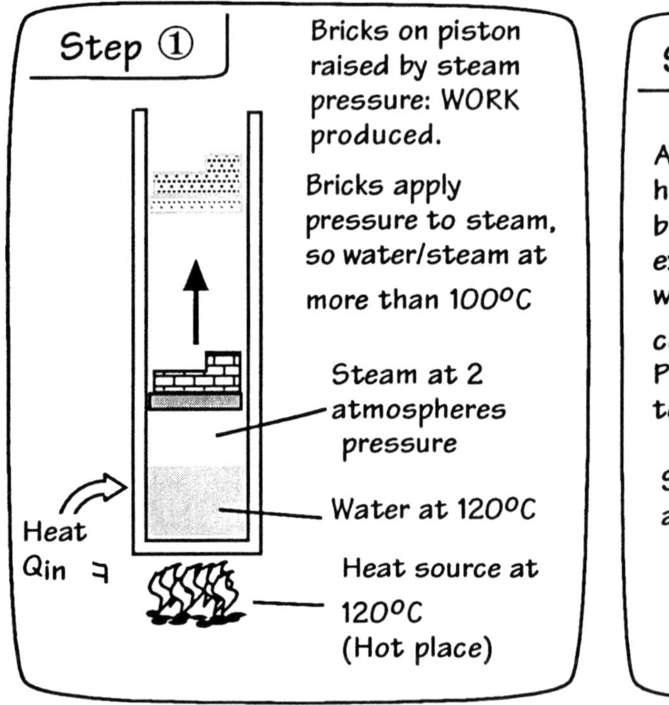

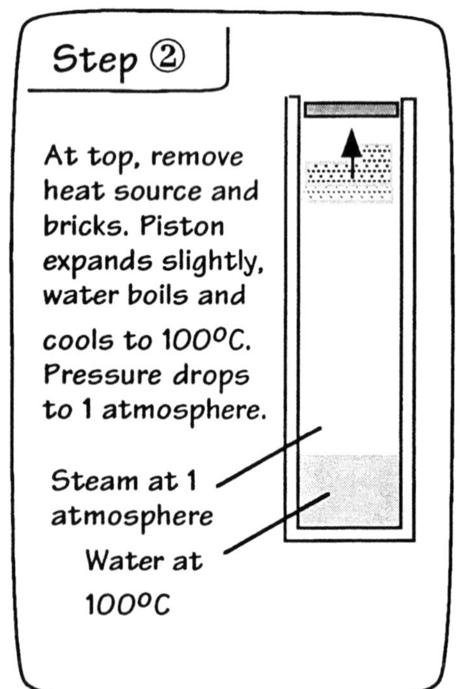

One of the simplest reversible heat engines exploits the Carnot steam cycle. Here it is used in a device for raising bricks up the side of a building. The device consists of a long cylinder with a piston. Water is boiled to make steam in the cylinder. The steam raises the piston upon which the bricks sit. This raises the bricks to the top of the building. When the bricks are removed, the piston is returned to the bottom of the cylinder and the cycle continues. The entire process can be made reversible if each individual step is reversible and we will now see how to ensure that every step is reversible.

We start with the piston at the bottom of the cylinder, loaded with bricks, and the water slowly boiling. Since the weight of the bricks presses on the steam, the pressure in the cylinder will be greater than atmospheric pressure. Let us say it is at 2 atmospheres pressure. At this higher pressure, water will boil at a higher temperature--120°C--and this will be the temperature of the water and steam. (This elevation of temperature with pressure is the secret of the fast cooking in a pressure cooker. Production of steam in the cooker increases the pressure which increases the temperature. Things cook much faster at even modestly higher temperature.)

In step one, heat Q_{in} is supplied reversibly to the water so that the steam pressure is maintained at just the level needed to slowly raise the piston with the bricks in a reversible expansion process. This is the step in the cycle in which heat is supplied and work recovered in the form of the energy needed to raise the bricks. Once the bricks are at the top of their motion, we remove the heat source.

In step two, the bricks are taken off the piston and placed on the roof of the building. As the bricks are removed from the piston, the steam pressure is no longer opposed and it can raise the piston further. The bricks are removed slowly, one at a time, so that this further expansion is reversible. As the expansion continues, further steam is needed to fill the cylinder. This steam is supplied by continued boiling of the water. But since there is no longer a heat source connected to the cylinder, this boiling will cool the water and steam and the pressure will drop. When the pressure in the cylinder drops to one atmosphere, the steam and water have cooled to 100°C.

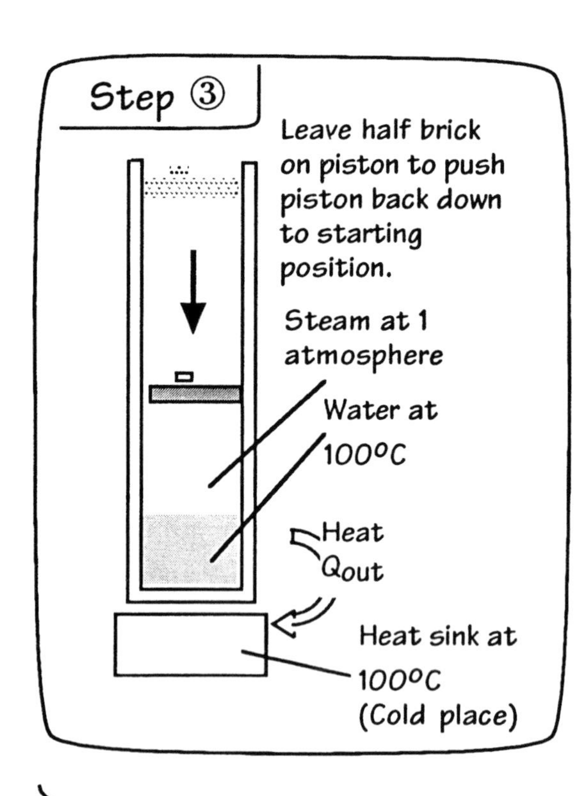
Step ③
Leave half brick on piston to push piston back down to starting position.
Steam at 1 atmosphere
Water at 100°C
Heat Q_{out}
Heat sink at 100°C (Cold place)

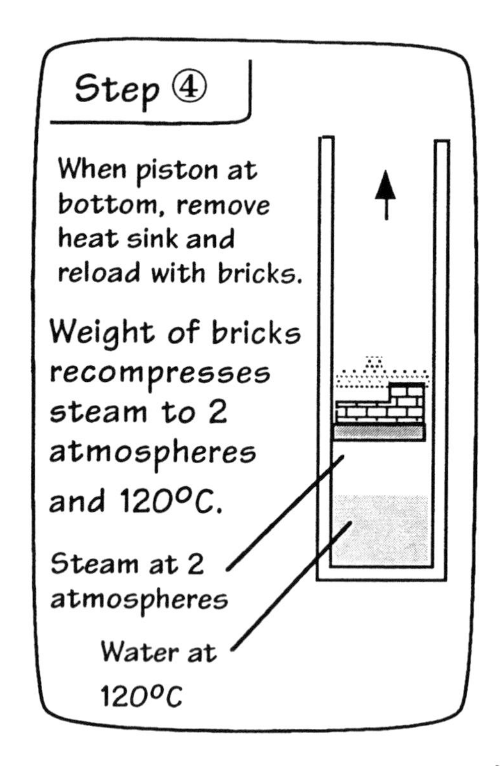
Step ④
When piston at bottom, remove heat sink and reload with bricks.
Weight of bricks recompresses steam to 2 atmospheres and 120°C.
Steam at 2 atmospheres
Water at 120°C

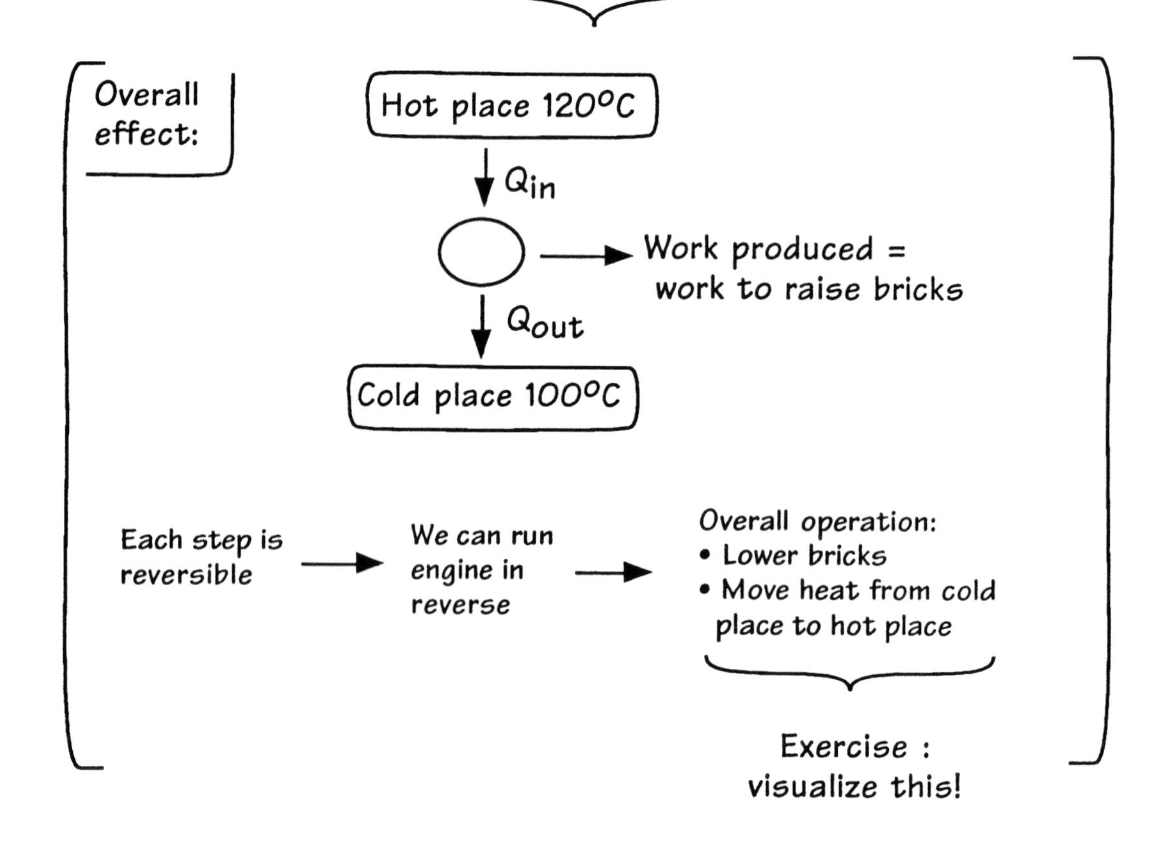
Overall effect:
Hot place 120°C
Q_{in}
Work produced = work to raise bricks
Q_{out}
Cold place 100°C
Each step is reversible
We can run engine in reverse
Overall operation:
• Lower bricks
• Move heat from cold place to hot place
Exercise : visualize this!

In step three, the piston is returned to the bottom of the cylinder. This is achieved by placing a half brick on the piston, so that the piston bears down on the cooler steam with a force slightly greater than that exerted by the steam on the piston. At the piston is slowly pressed down, the steam undergoes a reversible recompression. During this recompression, the steam filling the piston is condensed back to water and considerable heat is given off. This heat Q_{out} is released by the condensing steam at just above 100°C and is passed reversibly to a heat sink--the cold place--which is maintained at 100°C.

For step four, when the piston is nearly at the bottom, the heat sink is removed and the piston reloaded with bricks. This exerts further pressure on the steam, which is brought back to the initial pressure of two atmospheres. This recompression is done very slowly, replacing the bricks one at a time, so that the recompression is reversible. Since the heat sink has been removed, the heat given off by the condensing steam is retained and brings the steam and water back to the original temperature of 120°C. The system has now been brought back to the conditions at the start of step one. The cycle is complete.

The overall effect of this cycle is to take heat Q_{in} from the heat source at 120°C. That heat energy is split into two parts by the engine: the work energy stored in the raised bricks and the waste heat Q_{out} discharged to the heat sink at 100°C.

Since we were careful to ensure that each step was reversible, the entire operation is reversible. Running in reverse, the engine would lower, not raise, bricks. The work energy recovered in lowering the bricks would be used to take heat energy from the cold place at 100°C and deliver it along with the work energy as heat to the hot place. As an exercise, try to visualize how the reversed operation would look. Start at any step and work backwards. For example, step one, run in reverse, would involve the lowering of bricks, whose weight would compress and condense steam at 120°C leading to a discharge of heat to the hot place at 120°C.

Assignment 6: Core Notions of the Theory of Heat Engines

Anyone who has watched steam lift the lid of a pot on the stove has no doubt that heat can be used to generate work. But Carnot's first insight into heat engines seems paradoxical: heat is not enough, we also need cold; work is only recovered in heat engines as a byproduct of the passage of heat from a hot place to a cold place. To become comfortable with Carnot's idea, it helps to look at a range of examples of heat engines and to notice in every case how the presence of cold is essential to the engine's operation.

1. In each of the following examples of a heat engine, identify

--the source of heat (hot place)
--the cold place
--why the cold place is essential to the operation of the engine.

(a) Newcomen engine. In this oldest of steam engines, a fire boils water to make steam. The steam raises a piston inside a cylinder. Cold water is sprayed into the cylinder to condense the steam and form a partial vacuum, which draws the piston back. Work is recovered from the to and fro motion of the piston, which may have been used to pump water out of mines.

(b) Nuclear power plant. In a nuclear power plant, nuclear reactions make a nuclear pile very hot. Water circulated through the pile boils to make steam. The steam rotates a turbine from which work is recovered. (The work is usually used to generate electricity.) Since the steam may acquire radioactivity, it is condensed to water and returned to be boiled again.

(c) Steam locomotive: A fire in the firebox is used to boil water to make steam. The steam pushes a piston which drives the wheels of the locomotive. Rather than condense the steam, the engine will then just vent the steam to the surroundings.
(Hint: Since the steam is not condensed within the engine, it is not so obvious in this case that cold has anything to do with the operation of the engine. To see that it does, recall that a steam locomotive is supplied with water and that its operation depends on the huge increase in volume of cool water when it boils to form steam.)

(d) Gasoline car engine: A mixture of vaporized gasoline and air is burned inside a cylinder. The burning makes the burned gasoline/air mixture very hot so it expands. In expanding it pushes a piston which drives the car.
(Hint: Just as a steam locomotive is supplied with water, a car engine requires air.)

2. Heat can generate work in many ways. One would not normally think of these ways of generating work as heat engines. But since they use heat to generate work, they are heat engines and must be subject to the laws of thermodynamics just the same. In the following not so obvious examples of heat engines, identify:

--the source of heat (hot place)
--the cold place
--why the cold place is essential to the operation of the engine.

(a) A bimetal stip is made of two bonded layers of different metal. (They are commonly used in thermostats.) The two metals expand by different amounts when they are heated. As a result, a heated bimetal strip bends.

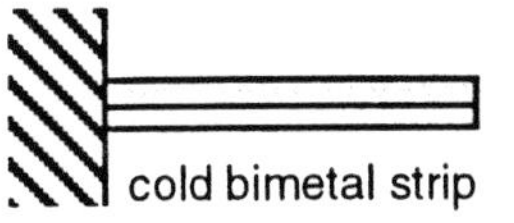

hot bimetal strip

One could use a bimetal stip to elevate weights (=perform work) by alternately heating and cooling the strip. To see how this could be done, imagine a small weight sitting on the outer edge of the bimetal strip in the figure. It would be raised when the strip is heated.

(b) A hot air ballon rises because a fire heats air trapped in the envelope. The air expands and rises, taking the balloon and payload with it. A hot air balloon making trips up and down a mountain would be a heat engine if it is used to raise weights that are dropped at the moutain top in each round trip.

(c) A waterwheel on a mountain stream is turned by the water flowing down the mountainside in a stream. It is actually part of a huge heat engine. Heat from the sun evaporates water from the oceans. The water vapor rises until it reaches the cool upper air. There it condenses into clouds and the water falls as rain into the stream. After flowing down the mountain it returns to the oceans to complete the cycle.

3. Imagine some heat engine that takes 100 cal of heat energy from a hot place, converts 40 cal into work and exhuasts 60 cal as waste heat to a cold place. The diagram for such an engine is:

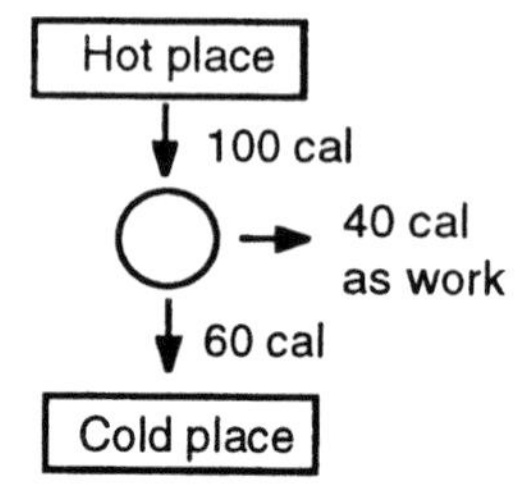

(a) Imagine also that this engine is reversible. Draw the corresponding diagram for a clone of the engine that is run in reverse.

(b) This clone engine run in reverse will consume work. The original engine run forward will supply it. What would happen if we use the work produced by the original engine to feed the clone engine? What will the total,combined effect be of the two coupled engines on:

- the total heat taken from and put into the hot place?
- the total heat taken from and put into the cold place?
- the total amount of work produced and consumed?

4. Recall that a reversible process is one that can proceed in forward or reverse directions. If the process is truly reversible, then a movie taken of it and run backwards would show a process that could happen--the reversed version of the original process. A system is in equilibrium when the forces within it are exactly balanced so that no change occurs. If we introduce a minute imbalance in these forces, then the system will undergo a very slow change. This change will be a reversible process. For example, a person eating a meal is not a reversible process. To see why, imagine a movie of someone eating, run in reverse. The process depicted is not one that is allowed in nature! The freezing/melting of ice in a pond as described earlier is a reversible process.

(a) Consider a bottle of soda. The soda contains carbon dioxide gas dissolved in it under pressure. When you open it, you hear a fizz as the pressure held is released and the carbon dioxide bubbles out of solution. Use the movie test to show that this is not a reversible process.

(b) How could one release the pressure on soda in a reversible manner? (Hint: Imagine the soda trapped in a cylinder with a piston held down by weights, such as the one discussed above in the description of irreversible and reversible expansion.)

Chapter 7

Major Results of the Theory of Heat Engines

Preview

Our suspicion from comparing reversible and irreversible heat engines:

Reversible engines are more efficient (i.e. produce more work from the same heat)

Now PROVE from second law of thermodynamics

* (Efficiency of ANY reversible heat engine) ≥ (Efficiency any other engine (i.e. irreversible) engine)

* (Efficiency of ALL reversible heat engines is the same.) Efficiency fixed solely by : Temp. of hot place & cold place

+ MORE

In the last chapter we laid out the basic framework of the theory of heat engines. We can now proceed to derive some of its basic results. We have seen that reversible heat engines provide an ideal, limiting case and this suggests that they may well perform better than other types of heat engines. By "perform better," we mean "yield more work from the same heat." We shall now see that this suspicion is well founded.

We shall be able to prove that reversible heat engines are the most efficient of all heat engines. No other heat engine can surpass them. More surprising, we will be able to prove that all heat engines have the same performance if they are working between the same hot and cold places. Notice what this means. It does not matter what design the heat engine uses, whether it operates with boiling water and steam, or with hot air, and whether the engine employs a few simple steps or very many complicated steps. If the engine is a reversible engine, then no engine is more efficient at converting heat energy into work and it will have exactly the same efficiency as any other reversible heat engine working between the same hot and cold place. A simple formula will let us compute the efficiency of these engines.

While these results all pertain to ideal engines that we cannot build, they have very real applications. First, we know that we improve the performance of our heat engines by eliminating irreversible processes, as far as we can. Second, the efficiency of a reversible engine represents an ideal beyond which no real engine can pass. It will turn out that this limit is a very restrictive one. Merely by glancing at the design of a real engine and knowing these limits we can estimate the range of its performance.

These are results of astonishing breadth and power. Moreover we shall see that we need minimal assumptions to generate them. Essentially all we need is the second law of thermodynamics, as stated in the last chapter. Once we accept it, we shall see that simple logic forces us to accept the results laid out here.

Define efficiency

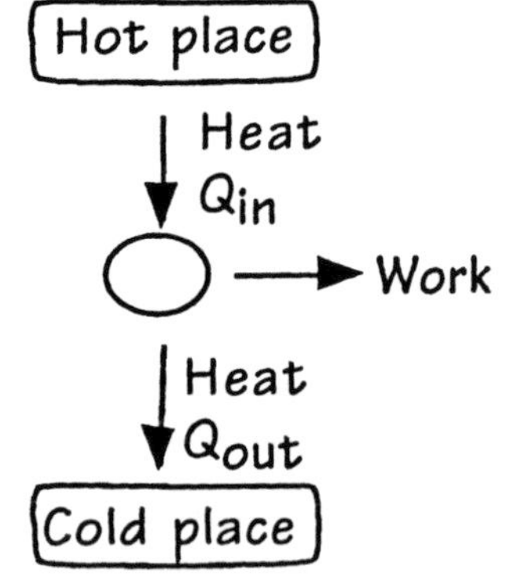

$$\text{Efficiency} = \frac{\text{Work}}{\text{Heat } Q_{in}}$$

In words: what fraction of heat supplied Q_{in} from the hot place is converted into work?

For example:

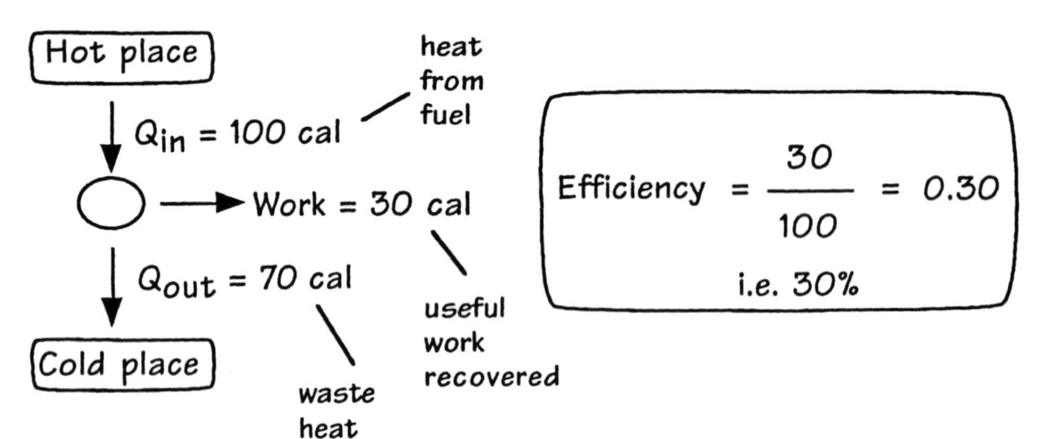

Question: Why does $\underset{Q_{in}}{100 \text{ cal}} = \underset{\text{Work}}{30 \text{ cal}} + \underset{Q_{out}}{70\text{cal}}$?

Answer: 1st law of thermodynamics. Energy is conserved

$$\text{Energy IN} = \text{Energy OUT}$$

$$Q_{in} = \text{Work} + Q_{out}$$

Before proceeding we need to settle on a definite way of measuring the performance of a heat engine. The parameter routinely used is the efficiency of the heat engine. This is just the fraction of heat supplied from the hot place that is converted into work. In terms of our standard diagram, the efficiency is just Work/Q_{in}. A numerical example is shown opposite. The engine takes 100 cal of heat energy from the hot place and converts 30 cal of it into work. 70 cal is exhausted as waste heat. Then the efficiency of the engine is 30/100 = 0.30. Expresses as a percentage, that is 30%.

Obviously the better the engine performs, the higher the efficiency; the worse it performs, the lower the efficiency.

A reminder: In the diagram, the heat supplied to the engine is 100 cal. This equals the sum of the work generated, 30 cal, and the waste heat, 70 cal. In all our diagrams this sum will hold. It just expresses the first law of thermodynamics, that energy is conserved. That is, all energy supplied to the engine must equal all energy leaving the engine (if none is retained in the engine).

How to combine heat engines

Use one heat engine to drive a reversed engine.

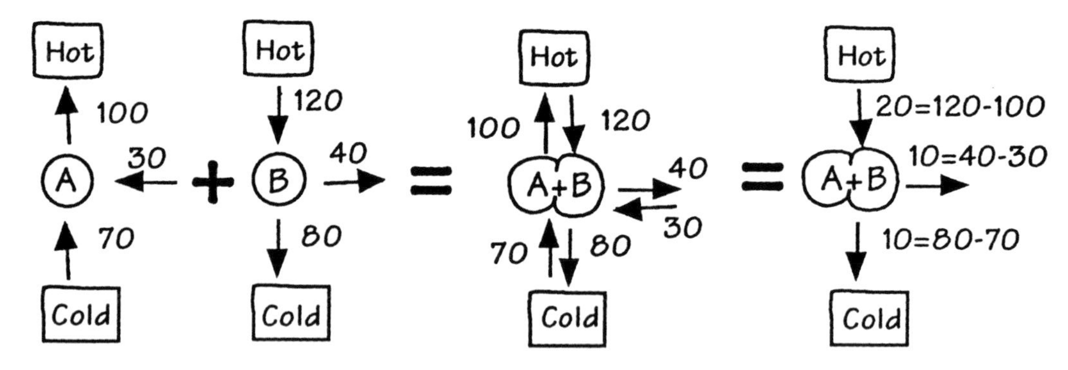

Work generated by engine B is used to operate A.

Use waste heat from one heat engine to feed another.

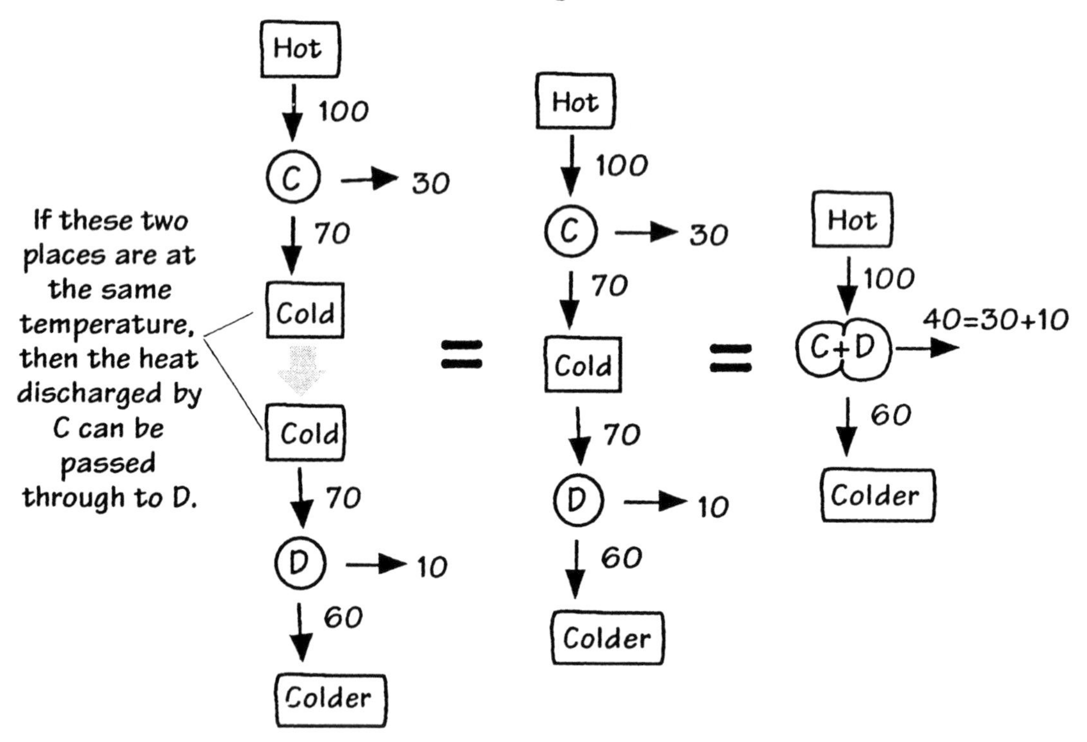

In proving our results, we will repeatedly use the following simple operation. If we have two heat engines it is often possible to combine them to build a new, bigger heat engine. The ways in which they can be combined are quite straightforward and are illustrated in the two examples shown here.

Consider engine A which is run in reverse. It consumes 30 units of work energy. The work is used to extract 70 units of heat from the cold place and to deliver 100 units of heat to the hot place. To operate, this engine needs a source of work energy. How might we secure this supply of work energy? Imagine that we had another heat engine B that operates between the same hot and cold place. As shown, this engine B extracts 120 units of heat from the hot place and recovers 40 units of work from it. We can use 30 of these 40 units of work energy to run engine A.

How might we take the work energy from engine B to engine A? If engine A were a real machine it might need the 30 units of work in the form of electrical energy, which would power some electric motor within its machinery. If B were a real engine, it might even produce this work as electrical energy. To supply work to engine A, we might only need to take the electrical power delivery cable from engine B and plug it into the electrical feed socket of engine A.

The combined effect of engine A+B is shown opposite. Engine A consumes 30 of the 40 units of work produced by engine B. That leaves a net production of work of 40-30=10 units. Similarly, Engine B extracts 120 units of heat energy from the hot place and engine A returns 100 units of heat to the hot place. So overall, 120-100=20 units of heat are extracted from the hot place. Finally, overall 80-70=10 units of energy are delivered to the cold place.

In the second example, waste heat from engine C is passed through to engine D. This can be done since we suppose that the two places marked "Cold" are at the same temperature. The net effect of the combined engine C+D is to take 100 units of heat from the place marked "Hot", deliver 60 units of heat to the place marked "Colder" and generate a total of 30+10=40 units of work. Since the 70 units of waste heat from engine C are consumed by engine D, they no longer appear in our final diagram for the combined engines C+D.

Result (I)

The efficiency of ALL reversible heat engines is the SAME when they work between the same hot and cold places.

Proof method: *"Reductio ad absurdum"*

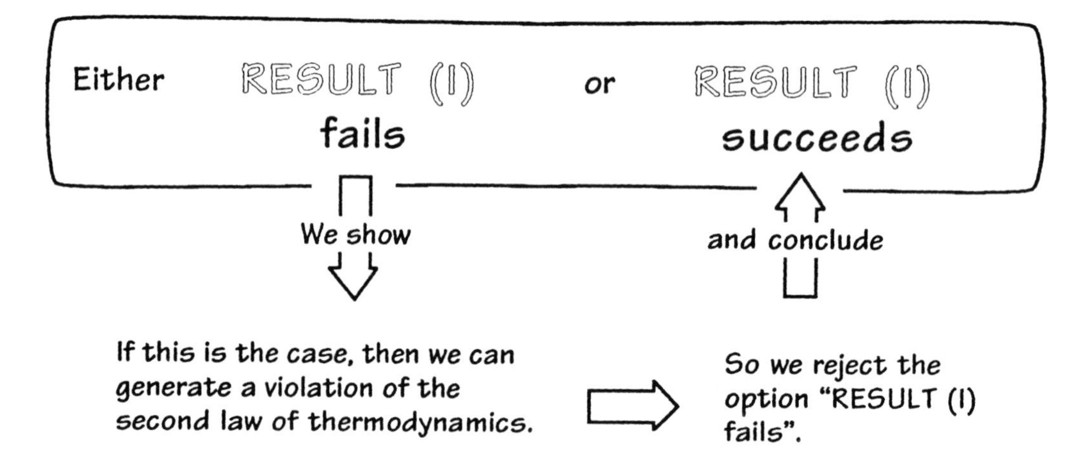

Step-by-step:

① Assume RESULT (I) fails.

② Show this contradicts the second law.

③ Accept the only other option, RESULT (I) succeeds.

Illustration in another example: is there a biggest number?

Our first major result is that all reversible heat engines, operating between the same hot and cold places, have the same efficiency.

In order to prove this, we will use a method of proof known as a "*reductio ad absurdum.*" This proof method proceeds by an indirect route. Instead of showing the result directly, we ask: "Well, if you don't believe the result, then what else might you believe?" We then show that believing anything else is impossible. You have to believe the result.

More precisely, there are only two options. Either Result (I) fails or Result (I) succeeds. We can show that the first of these options is unacceptable by assuming it as a working hypothesis and showing that we then arrive at an impossible outcome. That is, we will show that assuming the first option (Result (I) fails) allows us to build a machine that violates the second law of thermodynamics. That is an unacceptable outcome. So we must reject the option that Result (I) fails and accept the only other option: Result (I) succeeds.

The first time you see a *reductio* proof, it can be confusing. The path seems so complicated. We get the result by showing "not not the result"! But this sort of indirect reasoning can often be the easiest way to prove something and it is not so uncommon. For example, everyone knows that there is no biggest number. It's obvious. If there were a biggest number, what could it be? 1,000? That cannot be it since 1,000+1 is bigger. 1,000,000? That cannot be it since 1,000,000+1 is bigger. No number can be the biggest since that number plus one is bigger.

This argument is just a *reductio* proof of the result that there is no biggest number. For purposes of the proof, we assume the result fails; that is, we assume that there is a biggest number. We then show this conclusion allows a contradiction. That biggest number plus one is bigger than the biggest number. So we reject the option that the result fails and accept the only other choice, the result succeeds: there is no biggest number.

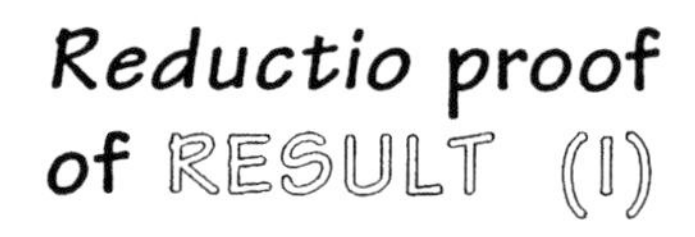

Assume Result (I) fails: There are reversible heat engines, operating between the same hot and cold places, with different efficiencies.

Consider two such engines with different efficiencies.

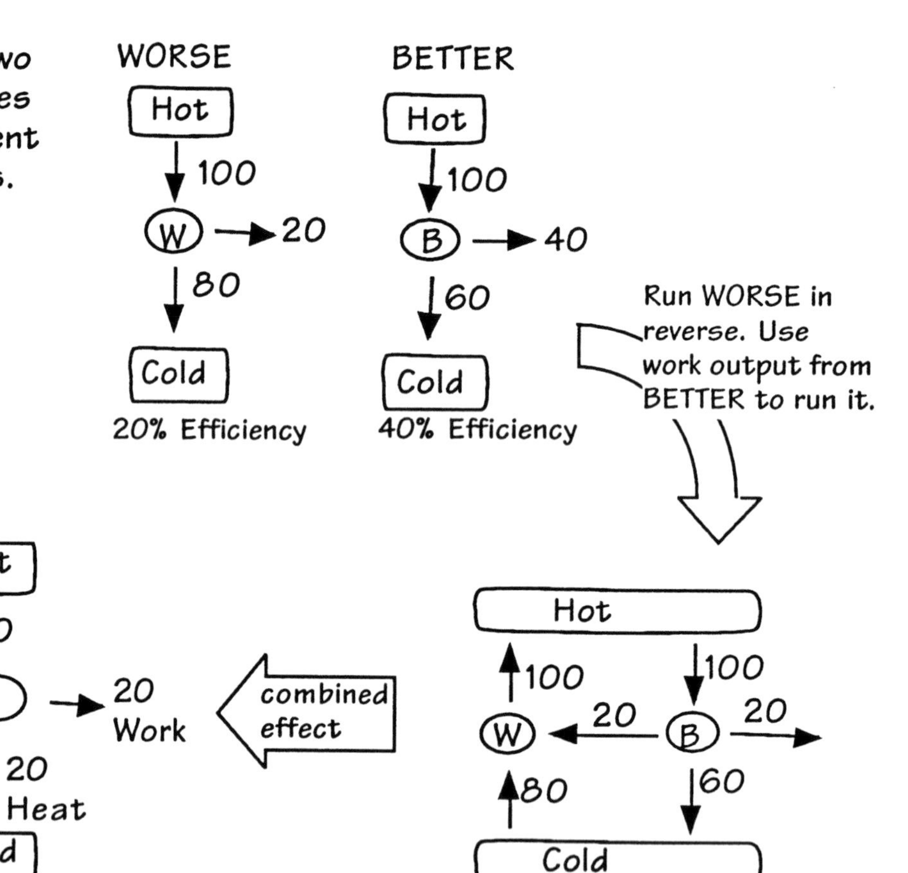

This engine extracts 20 units heat from the cold place & converts it FULLY into work !!

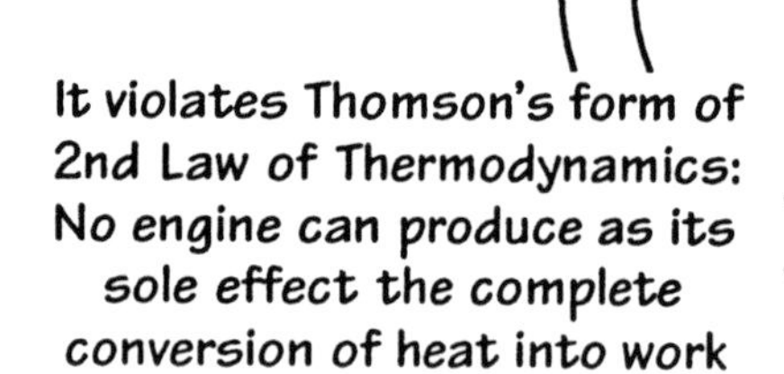

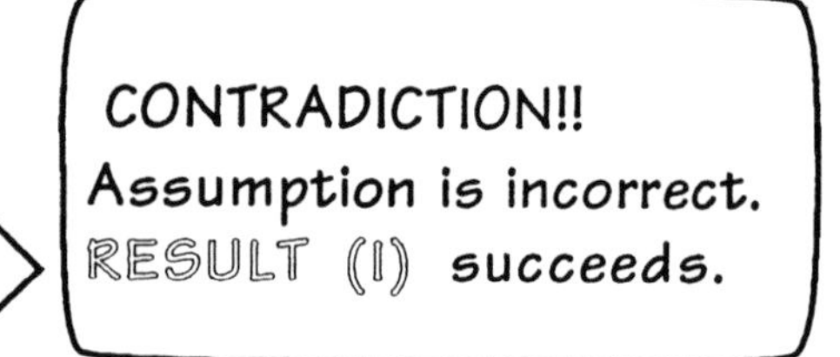

To prove Result (I) by *reductio*, we begin by assuming the opposite of what we want to prove. That is, we assume that there are reversible heat engines operating between the same hot and cold places and which have different efficiencies. We will now see that this assumption contradicts the second law of thermodynamics and, therefore, must be rejected.

To see the contradiction, consider two such heat engines with different efficiencies. For concreteness opposite, we take the efficiencies to be 20% for the engine "WORSE" and 40% for the engine "BETTER". The actual values of the efficiencies do not matter to the argument. All that matters is that they be different. Exactly the same argument could be run with any other pair of differing efficiencies.

According to our assumption, the engine WORSE is reversible. Therefore it can run in reverse. Running this way, the engine will consume 20 calories of work and use it to extract 80 calories from the cold place and deliver 100 calories of heat to the hot place. Now imagine that next to WORSE we also have the engine BETTER running in the forward direction. The engine BETTER will be able to supply the 20 calories of work WORSE needs. To supply it, the two engines will have to be coupled together. If, for example, the two engines are steam engines that deliver or consume work through rotating shafts, then the two shafts will be connected. If BETTER supplies the work needed by WORSE, there will be a work energy surplus. For every 100 calories of heat energy BETTER takes from the hot place, it will recover 40 as work energy. The figure opposite shows the net effect of coupling the two engines, 20 more than needed.

Now consider the overall transfers of heat and work for the combined engines. Every time BETTER takes 100 calories of heat from the hot place, WORSE will replace it with 100 calories, so there is no net gain or loss of heat. The combined effect of the two engines will be to extract 20 calories of heat (=80-60) from the cold place and convert it fully into 20 calories of work energy (=40-20).

This last conversion is prohibited by the second law of thermodynamics. Therefore our assumption of the failure of Result (I) has led to a contradiction. We must reject the assumption. Result (I) succeeds. Our proof is complete.

A simple variation of the proof uses Clausius' form of the Second Law of Thermodynamics.

Assume RESULT (I) fails (as before)

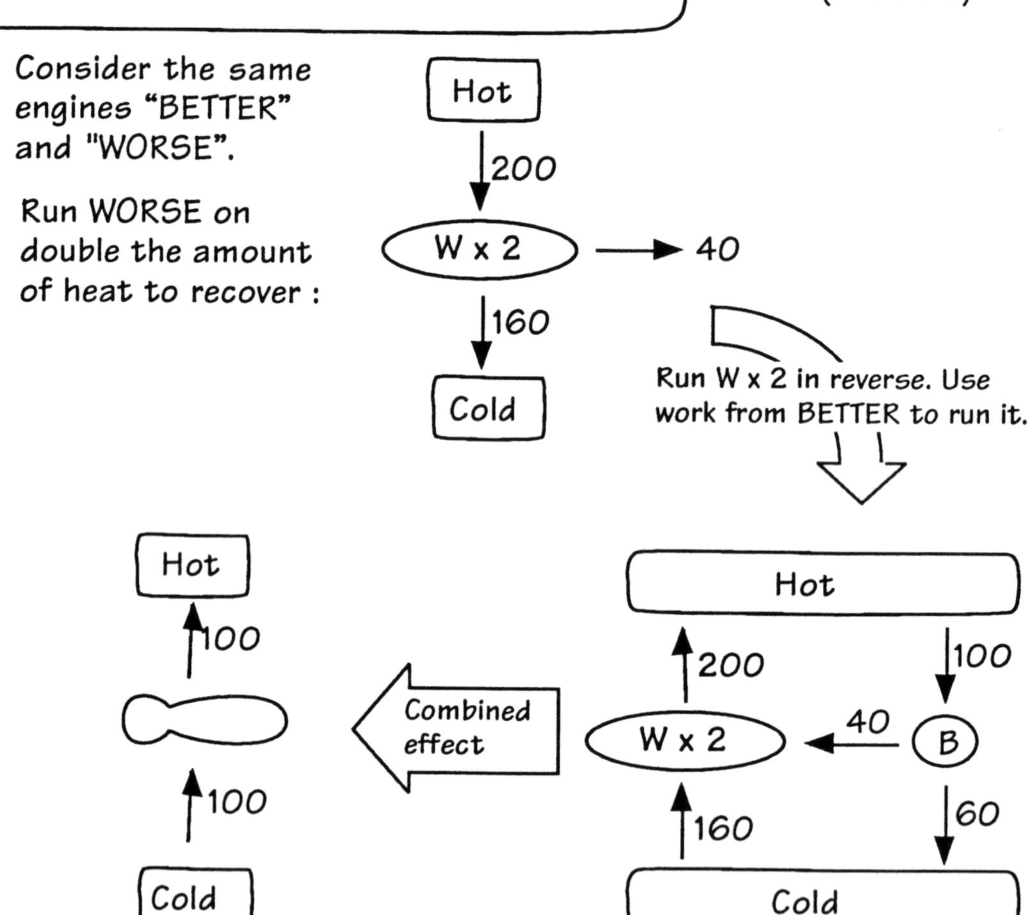

This combined engine extracts 100 calories of heat from the cold place and moves it to the hot place as it sole effect.

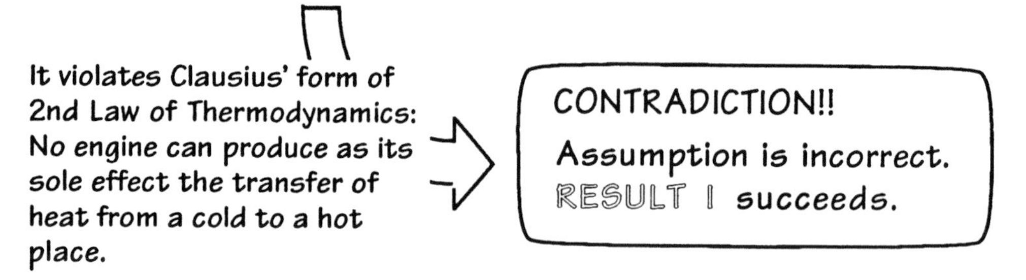

The earlier proof of Result (I) used Thomson's form of the second law of thermodynamics. With a small modification to the proof, it can also be proved using Clausius' form of the law. Of course there is no need to prove the result a second time as far as the logic is concerned. If a result is proved once, that is enough! However this second proof will give another illustration of how one proves results in the theory and it will start to give you a sense of how closely connected are the two forms of the second law of thermodynamics.

The proof begins as before by assuming that Result (I) fails. We will use the same engines, WORSE and BETTER. This time, however, we will run WORSE on double the amount of energy. That is, for every 200 calories of heat WORSE takes from the hot place, it will deliver 40 calories as work and 160 calories as waste heat. Since WORSE is a reversible engine, we can run it in reverse. The 40 calories of work that it consumes can be supplied by BETTER. When the two engines are coupled so that BETTER drives WORSE in reverse, the combined effect is as indicated opposite. There is no net production or consumption of work energy; all work energy that BETTER produces is consumed by WORSE. For every 200 calories of heat WORSE passes to the hot place, BETTER draws 100 calories out; the effect is a gain of 100 calories (=200-100) by the hot place. Similarly, the cold place loses 100 calories of heat.

Thus the overall effect of the combined engine is to extract heat from a cold place and pass it to a hot place without the need for an external supply of work to drive the combined engine. This violates Clausius' form of the second law of thermodynamics. That is, we have arrived at a contradiction. Therefore we reject the assumption and conclude that Result (I) succeeds.

Once you see how easily a result such as this is proved, it is easy to lose sight of just how profound the result is. This result is telling us that all we need to know about a reversible engine are the hot and cold places between which it operates. That then fixes its efficiency completely, no matter how the engine operates--be it with steam or hot air, freon or marshmallow. As long as the engines are reversible, they will have the same efficiency.

Result (II)

No irreversible engine is more efficient than a reversible engine, operating between the same hot and cold places.

Proof by reductio

Assume RESULT (II) fails.

There is an irreversible engine that is more efficient than a reversible engine.

Consider an irreversible engine that is more efficient than a reversible engine.

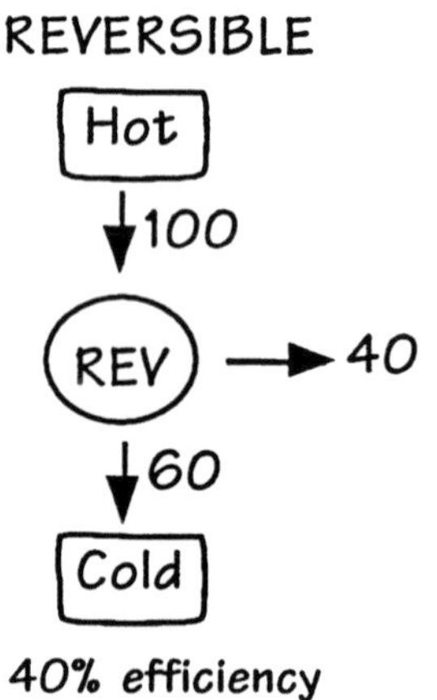

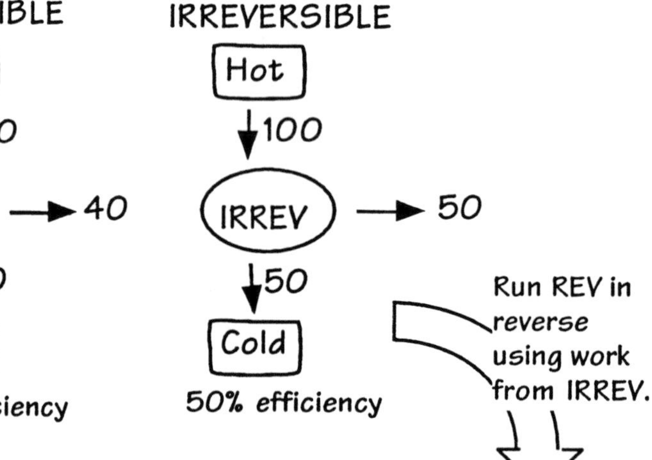

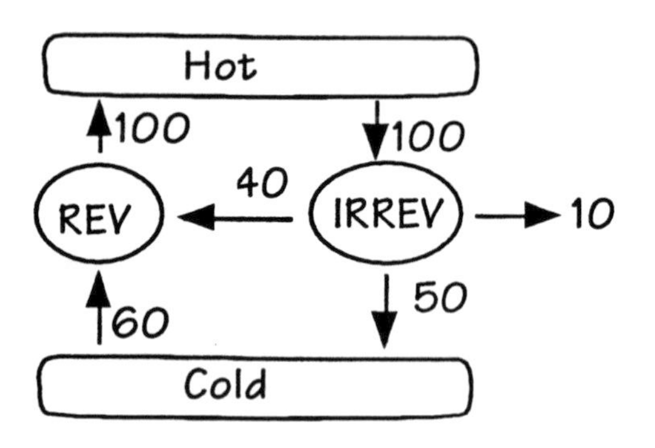

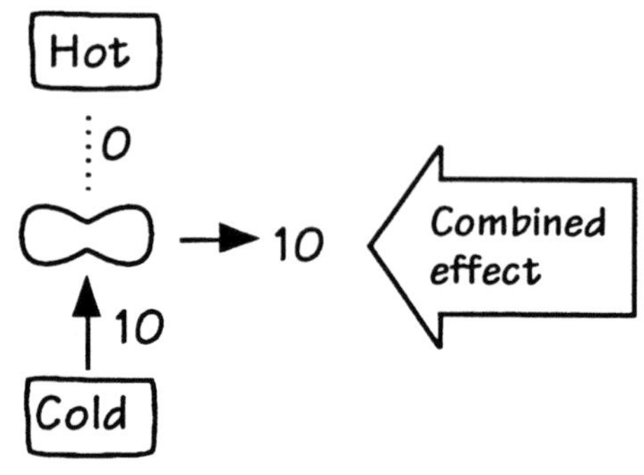

The sole effect of this combined engine is to extract 10 cal of heat from cold place & convert it FULLY to work. It violates Thomson's form of the second law of thermodynamics.

CONTRADICTION!!
Assumption is incorrect.
RESULT (II) succeeds.

Result (I) told us that all reversible engines have the same efficiency when operating between the same hot and cold places. Result (II) now assures us that this unique efficiency of a reversible engine is the best we can get, for no irreversible engine can be more efficient when working between the same hot and cold places.

The proof employs the same *reductio* technique. We assume otherwise; that is, that there is an irreversible engine that can outperform a reversible engine. We consider a pair of such engines opposite. The efficiency of IRREVERSIBLE is 50% and exceeds the efficiency of REVERSIBLE at 40%. (As before the particular numbers used are immaterial to the proof; it would proceed with any other suitable numbers.)

We now run REVERSIBLE in reverse, using the work produced by IRREVERSIBLE. The combined effect of the coupled engines is shown opposite. There is no net gain or loss of energy from the hot place. Overall, 10 calories of heat (=60-50) are taken from the cold place and converted fully into 10 calories of work energy (=50-40). This complete conversion violates the Thomson form of the second law of thermodynamics. Therefore we have a contradiction. We reject the assumption and conclude that Result (II) succeeds. The proof is complete.

How inefficient can irreversible heat engines be?

Result (II): the efficiency of irreversible engines cannot exceed that of reversible engines.

Can they be less efficient?

How much less efficient?

Proposition

The efficiency of an irreversible engine can vary anywhere from 0% up to the ideal efficiency of a reversible engine.

Proof:

We can have irreversible heat engines with 0% efficiency.

Hot
100
Cold

Irreversible heat transfer

=

Hot
100
0%
0
100
Cold

Irreversible heat engine of 0% efficiency

By combining this 0% efficiency engine with a reversible engine in any proportion, we can produce an irreversible engine with an efficiency anywhere between 0% and the efficiency of a reversible engine.

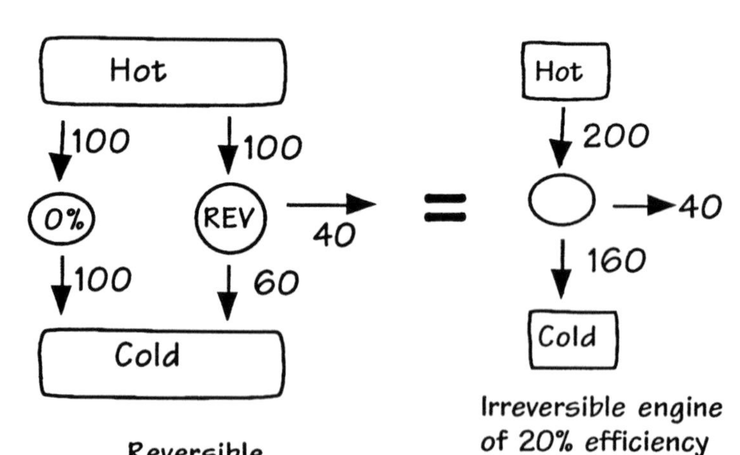

Irreversible engine of 20% efficiency

REV = Reversible engine of 40% efficiency

Result (II) assures us that no irreversible heat engine can be more efficient than a reversible heat engine. But it leaves unanswered the question of whether they can be less efficient than reversible engines. We can see quite quickly that irreversible heat engines can have any efficiency less than that of the reversible engine--this is stated as a proposition opposite.

The proof is quite simple. It is trivial to build an irreversible heat engine that has 0% efficiency. The simplest such device is merely a heat conducting rod that connects the hot and cold places. By its construction 100% of the heat supplied by the hot place is discharged as waste heat to the cold place. No work is recovered. (If you feel that this does not count as a heat engine since it has no moving parts, you may consider more complicated cases. For example, imagine any honest-to-goodness heat engine. Use all the work it produces to rub a disk against a brake pad, so that all the work the engine produces is converted to heat; then conduct this heat to the cold place. This engine-brake system combined will have 0% efficiency, since it produces no work.)

Once we have an engine with 0% efficiency, we can combine it with a reversible engine to produce a combined engine with an efficiency anywhere between 0% and that of the reversible engine. Imagine, for example, that we have a reversible engine "REV" whose efficiency is 40%. We take the heat drawn from the hot place and divide it as we please between REV and a 0% efficiency engine. Opposite, we show the case of 200 calories divided into equal parts of 100 calories and the part distributed to each machine. The combined machine has an overall efficiency of 20%. (It delivers 40 calories of work from 200 calories of heat; 40/200 = 0.20, which corresponds to 20%.) By dividing the heat drawn from the hot place in different proportions between the REV and the 0% efficiency engine we can make engines whose overall efficiency are anywhere between 0% and that of REV.

Calculating the efficiency of a reversible heat engine.

A reversible heat engine runs between 300°C and 200°C.

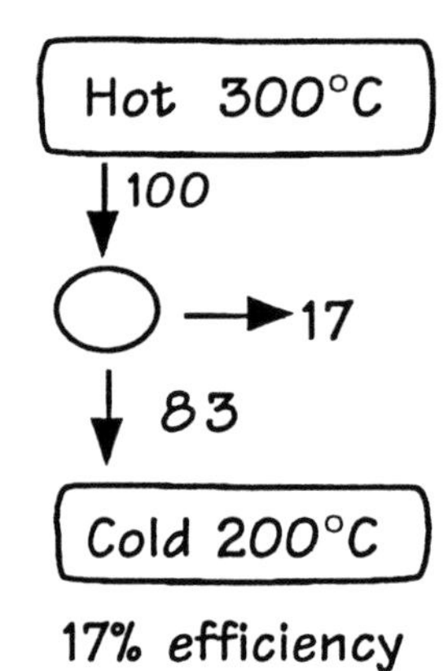

Result (I): All reversible heat engines operating between the same hot and cold places have same the same efficiency.

Therefore the 17% efficiency of this heat engine is fixed by the temperatures of the hot and cold places.

Result (III)

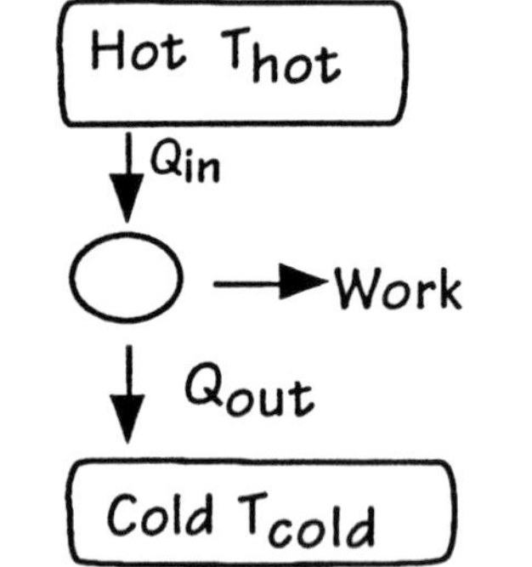

For all reversible heat engines operating between a hot place at temperature T_{hot} and a cold place at temperature T_{cold}.

$$\text{Efficiency } \text{“}\eta\text{”} = \frac{\text{Work}}{Q_{in}} = 1 - \frac{T_{cold}}{T_{hot}}$$

The Greek letter "eta" is the standard symbol used for efficiency.

BEWARE !
Temperatures for Result (III) are ALWAYS measured on the absolute scale, e.g. degrees Kelvin, written as "K"

For example:

T_{hot} = 300°C = 573 K

T_{cold} = 200°C = 473 K

$$\text{Efficiency } \eta = 1 - \frac{473}{573} = 1 - 0.83 = 0.17$$

i.e. 17%

Result (I) tells us that the efficiency of a reversible engine is fixed once the hot place and cold place are fixed. In effect, this tells us that once the *temperatures* of the hot and cold places are fixed, then the efficiency of the reversible engine is also fixed. For the temperatures of these places are what matters for the operation of a reversible heat engine. Thus, if the hot and cold places are at 300°C and 200°C, then the efficiency of the reversible heat engine turns out to be 17%. The result that lets us calculate this percentage is Result (III), which is stated opposite.

There are two things to bear in mind when using the formula of Result (III). First, it applies only to *reversible* heat engines. Second, the temperatures must be measured on the absolute scale in degrees Kelvin. The example shows how the formula is applied to the case of the hot and cold places at 300°C and 200°C. When these temperatures are converted to degrees Kelvin they become 573 K and 473 K. (We will see shortly how to convert to degrees Kelvin.) Inserting these values into the formula of Result (III) gives an efficiency of 0.17, that is, 17%.

Result (III) is presented here without proof, since the proof is a little more involved than the proofs of Results (I) and (II). The proof does use essentially the same methods as used to prove Results (I) and (II).

The formula of Result (III) is the most useful formulae in the theory of heat engines. We will use it a lot. It is worth committing it to memory-- and not hard to do since the formula is so simple.

Why T_{hot} and T_{cold} in Result (III) must be measured in an absolute temperature scale (degrees Kelvin).

The problem with using an ordinary scale:

$$\eta = 1 - \frac{T_{cold}}{T_{hot}} = 1 - \frac{0}{300} = 1$$

(T_{cold} = 0°C, T_{hot} = 300°C)

100% efficiency!! (violates Thomson's form of second law)

$$\eta = 1 - \frac{T_{cold}}{T_{hot}} = 1 - \frac{-100}{0} = ?$$

(T_{cold} = -100°C, T_{hot} = 0°C)

Undefined. ("You cannot divide by zero.")

Solution: Use a temperature scale in which no real system can have a temperature of zero degrees.

Zero degrees on absolute temperature scale (0 degrees Kelvin) = Unattainably coldest temperature

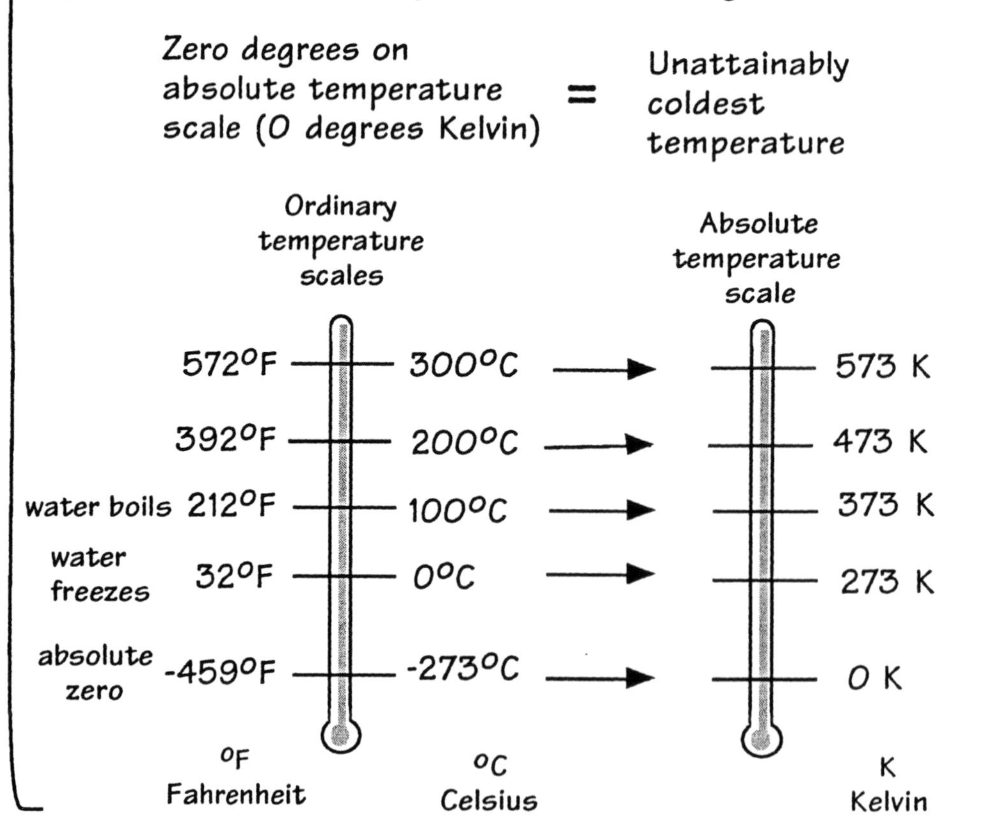

In using Result (III), it is essential that the temperatures be measured in an absolute temperature scale. The Kelvin scale is the absolute scale most commonly used.

To see why we need to use an absolute scale, we need only start to use Result (III) with ordinary temperatures. Whenever we try to use a temperature of 0°C, we have problems. For example, if the cold place happens to be at 0°C, the freezing point of water, an application of Result (III) leads to an efficiency of 1, that is, 100%. This violates Thomson's form of the second law of thermodynamics. Worse, if we can devise a heat engine that derives heat from a source at 0°C, then, when we try to use Result (III), we end up trying to divide by zero. That leads to no result at all, since division by zero is undefined.

To escape these problems we use an absolute temperature scale. The zero temperature of this scale is so cold that no real body can be cooled to it. (This unattainability of absolute zero is actually the third law of thermodynamics--but that is another story!) Therefore, if we use an absolute scale, we will never run into a problem from using a zero in Result (III).

Some typical values on the Kelvin scale are shown opposite. For example, absolute zero (0 K) corresponds to -273°C or -459°F. The freezing point of water is 273 K, which corresponds to 0°C or 32°F.

The easiest conversion to calculate is between degrees Kelvin and degrees Celsius. To convert from degrees Celsius to degrees Kelvin, you simply add 273; to convert from degrees Kelvin to degrees Celsius, you subtract 273.

Rules for the conversion of temperatures between different scales.

To convert °C to K	Add 273	0°C = 273 K 100°C = 373 K
To convert K to °C	Subtract 273	
To convert °C to °F	Multiply by 1.8 and add 32	(°F) = (°C) x 1.8 + 32 212 = 100 x 1.8 + 32 212°F = 100°C
To convert °F to °C	Subtract 32 and divide by 1.8	(°C) = [(°F) - 32] / 1.8 100 = [212 -32] / 1.8 100°C = 212°F

Because of the different temperature scales that are used, from time to time you will need to convert temperatures from one scale to another. The simplest conversion is the one you will probably do most often and it has already been described: between degrees Celsius and degrees Kelvin. This conversion is effected simply by adding or subtracting 273 as indicated opposite.

The conversion between degrees Celsius and degrees Fahrenheit is a little more complicated and involves two operations. Review the samples opposite to be sure you understand how to carry out the conversion.

Note that these conversions may be chained to produce conversions not on the table. For example, to convert 1000°F to degrees Kelvin, you would proceed in two steps. First you would convert 1000°F to degrees Celsius. It is (1000 - 32)/1.8 = 968/1.8 = 538°C. You would then convert this 538°C into degrees Kelvin. It is 538 + 273 =811 K.

Some typical values for efficiency η of a reversible heat engine.

T_{hot} \ T_{cold}	2000°C	1000°C	500°C	300°C	100°C	0°C
2000°C 2273 K 3632 °F	0%	44%	66%	75%	84%	88%
1000°C 1273 K 1832 °F		0%	39%	55%	71%	79%
500°C 773 K 932 °F			0%	26%	52%	65%
300°C 573 K 572°F				0%	35%	52%
100°C 373 K 212°F					0%	27%
0°C 273 K 32°F						0%

Why are these squares blank?

The table shows the result of applying Result (III) to many pairs of temperatures T_{hot} and T_{cold}. The rows are labeled with the values of T_{hot} and the columns with the values of T_{cold}. To find the efficiency of an engine operating between T_{hot} and T_{cold}, just look at the box in the table where the relevant row and column intersect.

For example, if T_{hot} is 2000°C and T_{cold} is 100°C, then we look at the row labeled 2000°C and the column labeled 100°C. They intersect on a box with an efficiency of 84%. For review, let us see how this value was calculated. The two temperatures, 2000°C and 100°C, are converted to degrees Kelvin: 2273 K and 373 K. They are then inserted into Result (III) to give us efficiency $\eta = 1 - 373/2273 = 1 - 0.16 = 0.84$, that is 84%.

We can read off some very important properties of Result (III) from the table. The efficiency of a reversible heat engine increases as the temperature T_{cold} of the cold place drops. To see this, look at the first row of the table for which T_{hot} is 2000°C. As T_{cold} drops, the efficiency increases from 0% through 66% to 88%. Similarly, the efficiency of a heat engine increases as we increase T_{hot}, the temperature of the hot place. To see this, look at the last column for which T_{cold} is fixed at 0°C. Reading from the bottom, as the temperature T_{hot} increases, the efficiency increases from 0% through 52% to 88%.

The moral: the efficiency of reversible heat engine is maximized by maximizing the temperature difference between T_{hot} and T_{cold}.

A sample problem

A reversible heat engine operates between the temperatures of 1000°C and 300°C. What is its efficiency? For each 100 calories of heat supplied by the hot place, how much work is recovered? How much heat is exhausted as waste heat?

① Represent the problem in a diagram. Convert temperatures to degrees Kelvin.

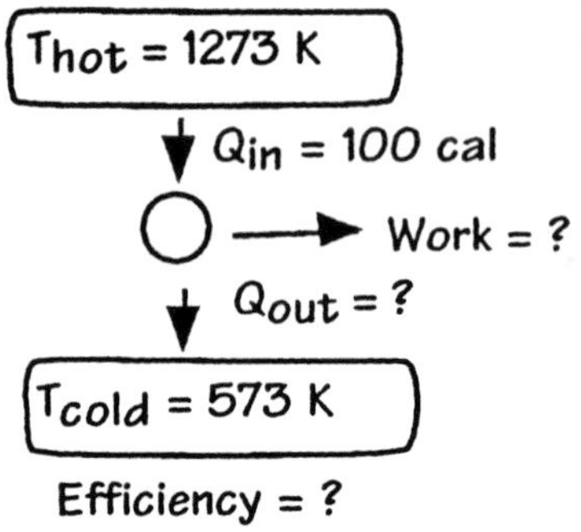

$T_{hot} = 1273$ K

$Q_{in} = 100$ cal

Work = ?

Q_{out} = ?

$T_{cold} = 573$ K

Efficiency = ?

1000°C -> 1273 K

300°C -> 573 K

② Compute the efficiency using Result (III).

$$\text{Efficiency} = 1 - \frac{T_{cold}}{T_{hot}} = 1 - \frac{573}{1273}$$

$$= 1 - 0.45 = 0.55 \qquad 55\%$$

③ Compute Work from Q_{in} and the efficiency.

$$\text{Efficiency} = \frac{\text{Work}}{Q_{in}} = \frac{\text{Work}}{100} = 0.55$$

$$\text{So, Work} = 0.55 \times 100 = 55 \text{ cal}$$

④ Compute the waste heat Q_{out} from Q_{in} and Work using the first law of thermodynamics.

$$Q_{in} = \text{Work} + Q_{out}$$

$$\text{So, } 100 = 55 + Q_{out}$$

$$\text{So, } Q_{out} = 100 - 55 = 45 \text{ cal}$$

This sample problem illustrates how we can use Result (III) to compute the various quantities associated with the operation of a reversible heat engine.

The first step is to represent all the information we have in the usual diagram. This gives us a chance to see the problem at a glance and to assess if we have enough information to solve it. Notice that the temperatures are converted to degrees Kelvin by adding 273. For example, 1000^oC becomes 1000 + 273 = 1273 K. In the second step, the temperatures are substituted into Result (III) and the efficiency of the engine computed.

In the third step, we compute the amount of work produced by using the definition of efficiency. Recall that efficiency was defined as Work/Q_{in}. This gives us Work/100 = 0.55. What value can Work have if it is to satisfy this condition? That is, what divided by 100 gives 0.55? The answer is 55.

In the fourth step we compute the waste heat from the first law of thermodynamics. We can "talk out" the calculation in this way. The engine is supplied with 100 cal of heat (Q_{in}). We know that 55 cal are converted into Work. The first law tells us that energy cannot be created or destroyed. So the remaining 100 - 55 = 45 cal must go somewhere. That somewhere is the waste heat, Q_{out}.

This calculation applies to a reversible engine--one that can never exist, since reversible engines are idealizations. But it also tells us something about real engines. For the performance of a reversible engine is the best possible. Therefore any real engine will not perform as well. To see what this means, imagine that we have a fancy nuclear power station whose reactor supplies heat at the very high temperature of 1000^oC and discharges waste heat at 300^oC. Our calculation tells us that the best efficiency such a power station can possibly achieve is 55%. In all likelihood it will not come close to that. Thus on the basis of this short calculation, we know that such a power plant will exhaust at least half of its energy as waste heat--even though we know virtually nothing about the details of its design! I think it amazing that a tiny calculation can tell us so much!

Why do reversible heat engines perform so well?

Each irreversible process = The loss of an opportunity to recover work

Heating

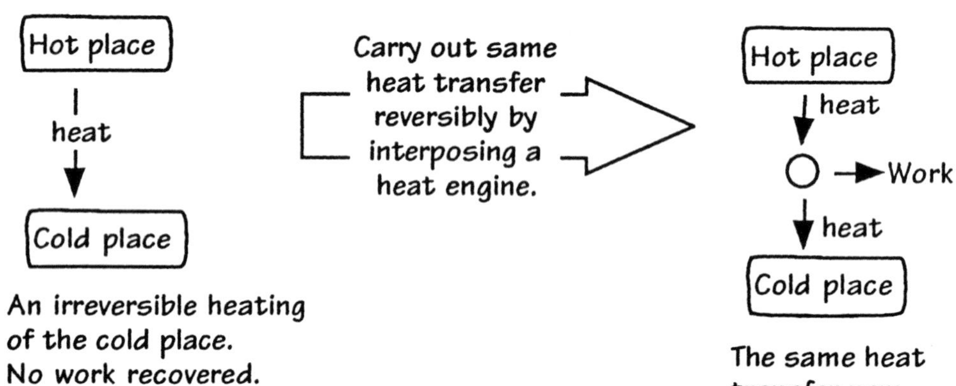

An irreversible heating of the cold place.
No work recovered.

The same heat transfer now generates work.

Expansion

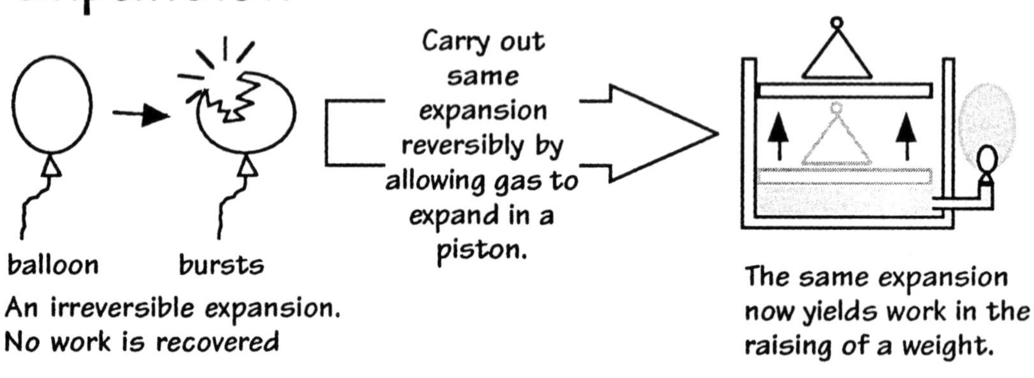

An irreversible expansion.
No work is recovered

The same expansion now yields work in the raising of a weight.

Moral for heat engine design.

To maximize the efficiency of the engine,

seek to eliminate every irreversible process or replace it by a reversible process.

Reversible heat engines are the most efficient of all heat engines. There is a physical reasons for this special performance of reversible engines. Every time there is an irreversible process, we lose a chance to convert a form of energy into work energy. So a reversible engine avoids all these lost opportunities by employing no irreversible processes and can give us the maximum work.

To see how irreversible processes are a lost opportunity to recover work, consider the two processes used most in heat engines: heating and expansion.

In an irreversible heating, we simply allow heat to pass directly from the hot place to a cold place. For example, we may allow a large coal fire to heat the air. We can see the loss of opportunity to recover work if we imagine this same heat transfer mediated reversibly by a heat engine. Then, some of the heat energy to be transferred to the cold place will be redirected into work energy. For example, if we take that same coal fire and place it in a power station, we will still heat the air around the fire through the waste heat from the power station; but we will also recover useful work through the intermediate of the electricity generated.

Similarly, consider the irreversible expansion of a gas under pressure. For example, consider the bursting of a balloon. No work is recovered. If we carry out the same expansion reversibly, then we can recover work. For example, we may connect the balloon to a cylinder fitted with a piston. As the gas expands from the balloon into the cylinder, it will raise the weighted piston and produce work.

We can combine these considerations into an important piece of advice for the design of a heat engine. If you wish to maximize the efficiency of the engine, you should seek to make it as close to a reversible engine as possible. That is, you should seek to eliminate every irreversible process or replace it with a reversible process.

The early history of steam engines

Improvements in performance by eliminating or replacing irreversible processes.

Newcomen engine. Earliest steam engine (early 1700s).

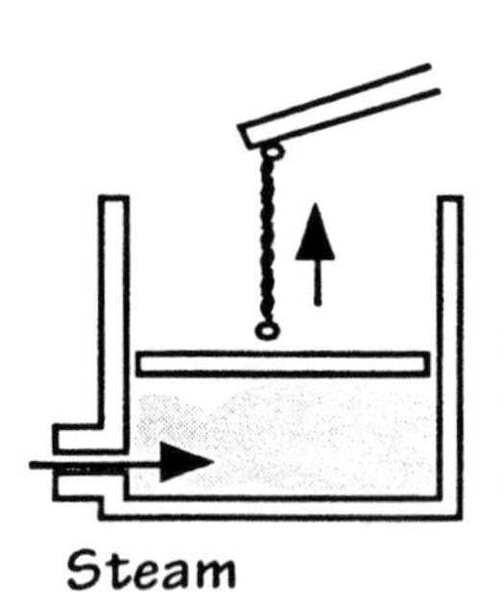

Piston raised by counter-weights and steam.

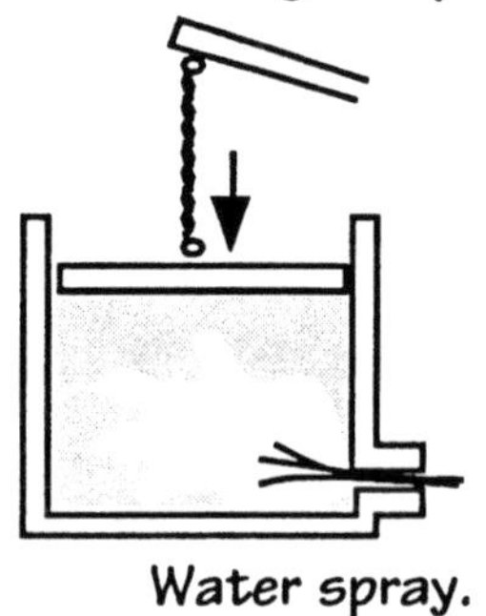

Work producing stroke: Steam condenses and forms partial vacuum which draws piston down.

James Watt (1765): In each cycle, the steam must reheat the cooled walls of the cylinder. This requires 8 times as much steam as is needed to fill the cylinder!

Watt's improved steam engine.

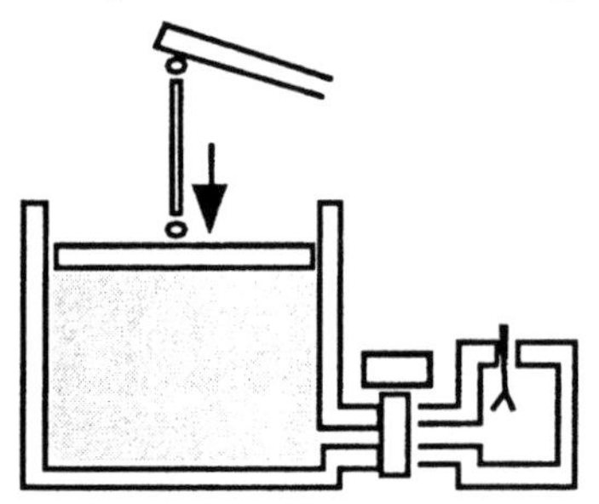

Condense steam in a separate chamber. Main cylinder remains hot.

Watt's improvement:	Reduce steam consumption by eliminate the irreversible heating and cooling of main cylinder.

The early history of the steam engine provides a nice illustration of how one can improve the performance of a heat engine by eliminating or replacing irreversible processes. The earliest widely used steam engine was the Newcomen engine. In its operating cycle, the piston was raised in the cylinder by counterweights and entering steam. Water was then sprayed into the steam filled cylinder. That would condense the steam, forming a partial vacuum, which would then draw the piston down. This downward stroke was the principal source of work in the engine. It did not rely on the pressure of steam to push; it depended on the formation of a partial vacuum to pull. For this reason, the Newcomen engine is sometimes called a vacuum engine.

The Newcomen engine was notoriously inefficient; it required huge amounts of steam to operate. James Watt undertook the task of systematically improving the efficiency of the steam engine. He identified the cooling of the walls of the main cylinder with the water spray as a major cause of inefficiency. He determined that eight times as much steam was needed to reheat the cylinder in each cycle as was needed merely to fill it with steam. In Watt's improved design, the main cylinder was kept warm at all times. Steam was condensed in a separate condenser. At the appropriate time in the operating cycle, a valve to the condenser was opened to allow the steam to escape and condense, thereby forming the partial vacuum that would pull the piston down.

In our terms, what Watt had identified were two irreversible processes: an irreversible heating of the cool walls of the cylinder when it was filled with steam and an irreversible cooling of the hot walls of the cylinder when the water spray was admitted. Each represented a transfer of heat from hot to cold without any recovery of work. Watt improved the efficiency of the engine by simply eliminating these irreversible processes.

Of course this early period came before Carnot's theory, so that people like Watt could not explicitly employ notions like reversibility. Rather his improvements were made using engineering intuition and what was then known about heat and work. Here Watt was well informed. He had been an assistant to Joseph Black, who was one of the leading theorists of the 18th century in the theory of heat and instrumental in the development of the caloric or material theory of heat.

Original operation of Watt's engine

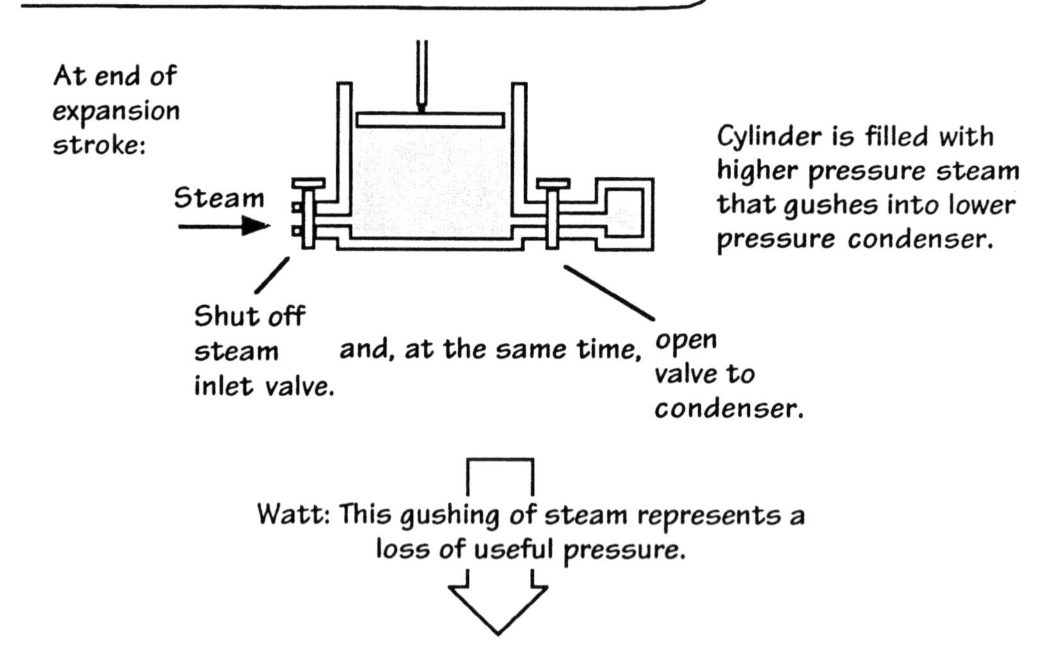

Watt: This gushing of steam represents a loss of useful pressure.

Improved operation of Watt's engine

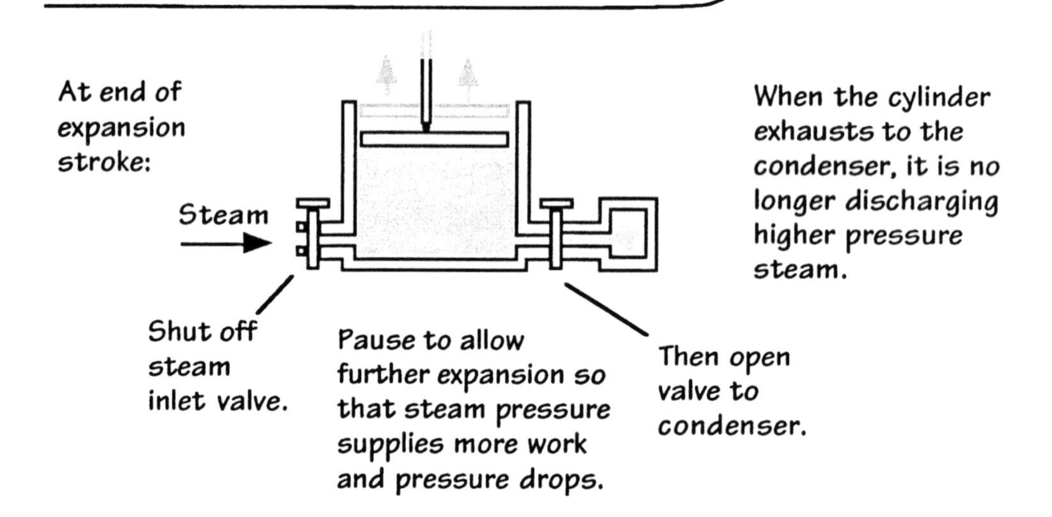

Watt's improvement:	Replace	irreversible expansion of higher pressure steam into condenser	with	reversible expansion of steam in cylinder.

The further development of Watt's engine gives another illustration of this type, but now concerned with an irreversible expansion. The obvious operation of Watt's engine was as follows: Fill the cylinder with steam. When it is full and the piston at the furthest point of its travel, the steam supply is shut off and, at the same time, the valve to the condenser opened. Now, at this point in the cycle, the piston is completely filled with steam at the highest pressure available. The cool condenser is at a much lower pressure. So when the valve to the condenser is opened, this higher pressure steam gushes into the condenser.

Watt noticed that this was an inefficiency in the operation. For the pressure released in this gushing could, in principle, be employed to do useful work. As it turned out, it was easy to arrange for it to do useful work. All that Watt had to do was to time the cycle slightly differently. That is, he would arrange to have the steam supply shut off before the piston had reached the end of its travel and before the valve to the condenser was opened. Thus, when the steam supply was shut off, the cylinder would be filled with steam at the higher pressure. But this higher pressure would not be wasted. It would be used to move the piston to the end of its travel providing useful work. During this last phase of expansion, no further steam is supplied to the cylinder, so the steam pressure would drop. When the valve to the condenser is opened, steam at much lower pressure passes to the condenser.

Watt's improvement amounts to the replacing of an irreversible process by a reversible process--or, more precisely, one that is closer to being reversible. The gushing of the steam in the original operation is an irreversible expansion, akin to the bursting of a balloon. When that same pressure is used to move the piston, the process has moved closer to an ideal reversible expansion in which the pressure of the expanding gas is almost exactly balanced by the restraining force of the piston. The result, again, is that more work is recovered from the same amount of steam. Since the amount of steam generated is directly proportional to the heat supplied from the fire, this means that Watt's engines are more efficient. Since the amount of heat supplied to the engine is proportional to the fuel burned and the fuel burned proportional to the fuel budget, Watt's customers were happy too!

Assignment 7: The Efficiency of Reversible Heat Engines

1 (a) Reversible heat engine A works between temperatures T_{Hot}=900°C=1173K and T_{Cold}=500°C=773K. What is its efficiency? If Q_{IN} is 100 cal, what is the Work generated and Q_{OUT} (measured in cal.)?

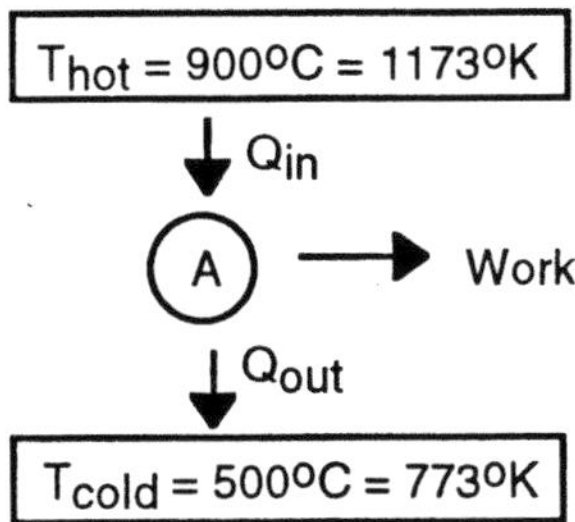

(b) What engine is the most efficient of all heat engines that operate between 900°C and 500°C? What is the range of efficiencies of irreversible engines that operate between 900°C and 500°C?

(c) Reversible heat engine B works between temperatures T_{hot}=500°C=773K and T_{Cold}=200°C=473K. What is its efficiency? If Q_{IN} is 100 cal, what is the Work generated and Q_{OUT} (measured in cal.)?

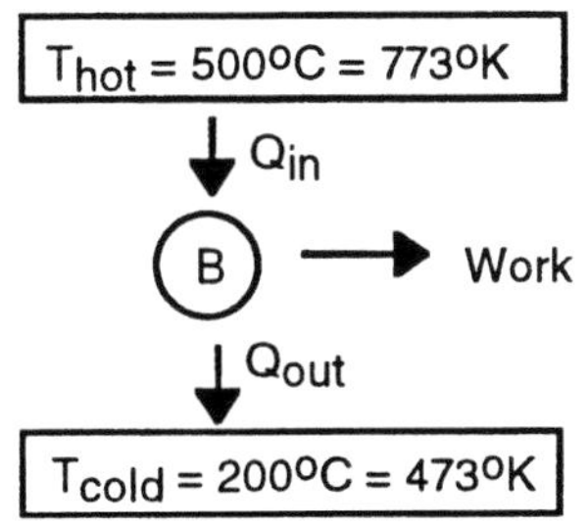

2 Imagine that 100 cal of heat are fed as Q_{IN} into engine A of question 1 and that A's waste heat is fed into engine B as B's feed.

(a) How much work is produced by engine A? How much work is produced by engine B? Sum these two to compute the total work produced by the combined engine.

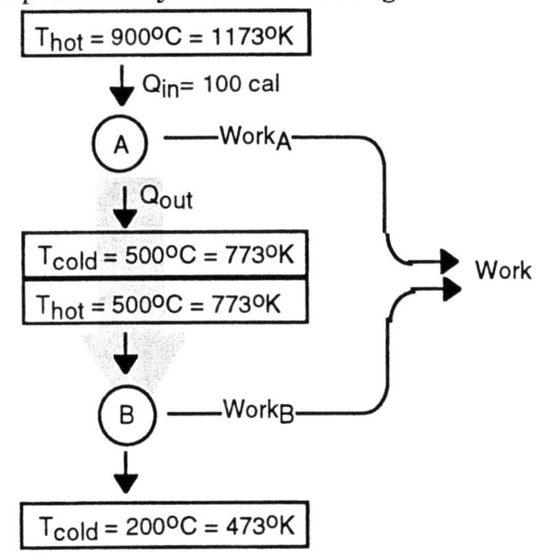

(b) Now imagine that the two joined engines A and B form a single big engine working between the temperatures 900°C and 200°C. From your answer to (a), compute the efficiency of the combined engine. (i.e. compute the fraction of heat entering from the 900°C hot place that is eventually converted into work.)

(c) The combined heat engine of 2(b) is just a reversible engine operating between the temperatures of 900°C and 200°C. Therefore Result (III) can be used to compute its efficiency directly. Using Result (III), what is the efficiency of the combined engine? How does this result compare with the efficiency computed in 2(b)?

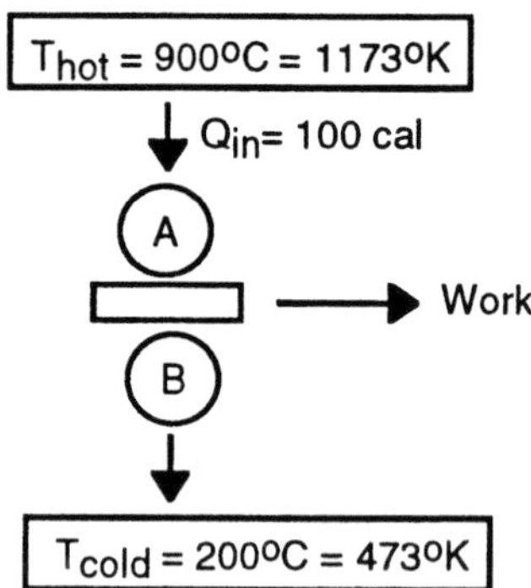

3 There are two versions of the second law of thermodynamics named after the scientists who gave the formulations:

"Clausius" No engine can produce as its sole effect the transfer of heat from a cold place to a hot place.

"Thomson" No engine can produce as its sole effect the complete conversion of heat into work.

In this question you will show that the two are equivalent by showing that whenever "Thomson" is violated, so is "Clausius" and whenever "Clausius" is violated, so is "Thomson".

(a) Assume that you have an engine "UnClausius" whose sole effect is to transfer heat from a cold place to a hot place. Show that this engine can be coupled with an ordinary reversible engine in a way that violates the Thomson form of the law.

(b)Assume you have an engine "UnThomson" whose sole effect is to convert heat into work. Show that this can be coupled with an ordinary reversible engine in a way that violates the Clausius form of the law.

Hint: It is easier to assume concrete forms for the engines such as

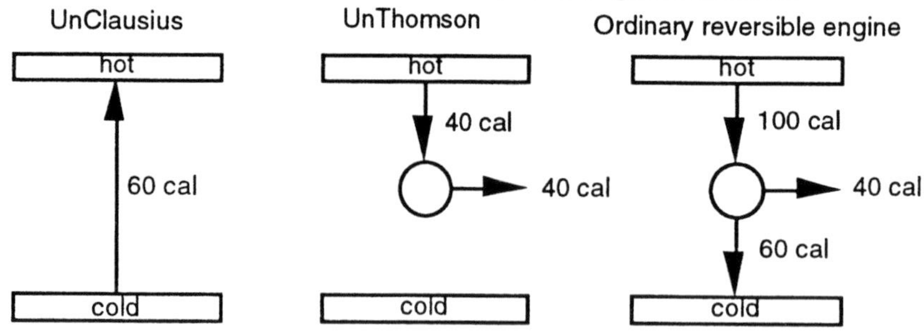

Chapter 8

Applications of the Theory of Heat Engines

Major result
of the last chapter

Efficiency of reversible engine operating between T_{hot} and T_{cold} $= \dfrac{\text{Work}}{Q_{in}} = 1 - \dfrac{T_{cold}}{T_{hot}}$

But reversible engines are theoretical toys. e.g. they operate infinitely slowly. What is the practical significance of this result?

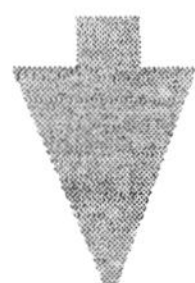

- Reversible engines are the most efficient.

  We can improve the efficiency of real engines by eliminating irreversible processes e.g. irreversible heating, expansion.

- The formula gives the MAXIMUM efficiency of any engine.

  This limit often proves much lower than expected.

- The theory applies to all means that we have of building heat engines.

  The theory turns out to tell us a great deal more about our physical world than we expect.

The theory of heat engines lets us calculate the efficiency of a reversible heat engine, once we know the temperatures between which it operates. The formula, repeated opposite, is very simple, but it will prove to be one of the most useful results of the theory, as far as our investigations of real heat engines is concerned. At first blush, this seems impossible. Reversible heat engines are devices that can never exist in the real world. They are the inventions of pure theory. If we could build one, it would operate infinitely slowly. That is, it would never actually do anything.

The importance of this result for real engines is that it represents an upper limit for their performance. We can improve the performance of a real heat engine by making it more like a reversible engine, that is, by eliminating any irreversible processes it might contain. We have already seen how such modifications produced great improvements in efficiency in the earliest steam engines. Further, imagine we have a real heat engine and know the temperatures between which it operates. Then we can use the formula to compute the efficiency of a reversible engine operating between the same temperatures. This efficiency is the upper limit for the efficiency of any heat engine operating between those temperatures, no matter how fancy its design. We shall see in this chapter that this upper limit can be far lower than we would expect. This theoretic limit proves to be one of the major boundaries against which heat engine designers must push.

Finally, the theory of heat engines turns out to apply to far more than just heat engines. We shall see that, with a slight modification, it turns into a theory of air-conditioners and refrigerators. Far less obvious is that it places surprising restrictions on the properties of matter. We can design a heat engine with any working substance we please. But the resulting engine must obey the laws of thermodynamics. We shall see that this can only happen if the behavior of matter under heating and cooling is constrained in suitable ways. This will allow us to give thermodynamic reasons for why it *has* to be possible to skate on ice.

Old fashioned steam engine
(e.g. steam locomotive)

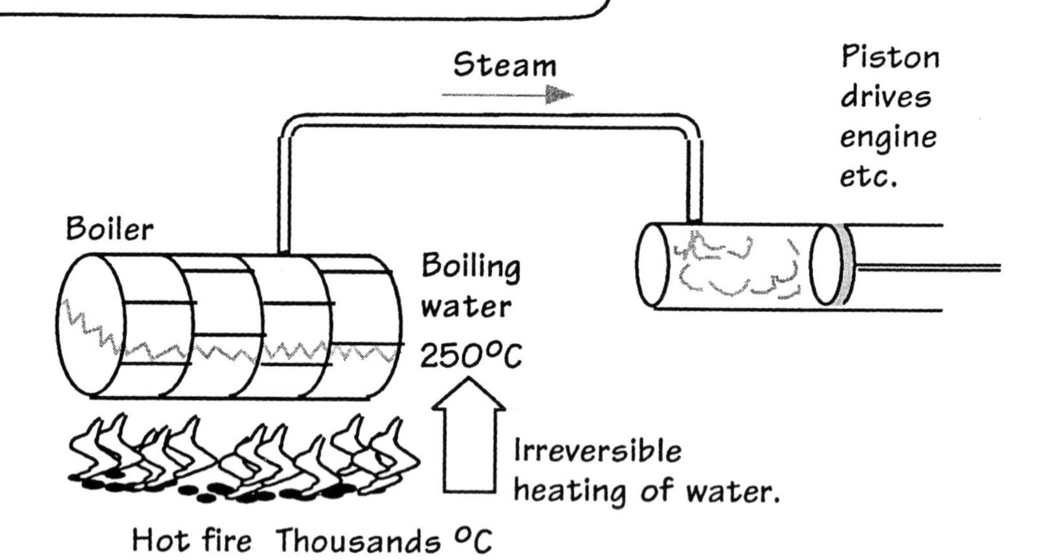

Major source of irreversibility and inefficiency in engine:

Hot fire at thousands °C irreversibly heats boiling water at hundreds °C

Maximum temperature of water limited by the ability of the boiler to contain the steam pressure.

Maximum efficiency of engine after this irreversible heating:

Ideal efficiency of reversible engine working between

T_{hot} = 250°C = 523 K (steam temperature.)

T_{cold} = 50°C = 323 K

$$\text{Efficiency} = 1 - \frac{323}{523} = 0.38 \qquad 38\%$$

This is the best possible efficiency.
In real engines, the efficiency will be lower.

In its early history, great improvements were made in the efficiency of steam engines. The efficiency of the earliest Newcomen engines was less than 1/2%. Watt improved this to at least 2 to 3%. That enabled a four to six fold reduction in fuel needs--impressive by any standard! But there were serious limits to how much improvement could be achieved. Inherent in the design of the old fashioned steam engine is a major flaw. They contain a very significant irreversible heating.

The source of heat energy is a fire that burns at thousands of degrees Celsius. That energy is transferred irreversibly to water boiling at several hundred degrees only in the boiler. In this one transfer, the opportunity to convert the bulk of the heat into work is lost. The obvious solution would be to increase the temperature of the water as close as possible to that of the fire. What prevents this is the pressure of the steam generated by the boiling water. As the temperature of the water is raised, that pressure increases rapidly and this pressure must be safely contained by the walls of the boiler. A very high pressure for a boiler would be 200 atmospheres, that is, 3,000 pounds per square inch. At this pressure, the water temperature is still only at 370°C--far short of the fire temperature. Watt himself ran his engines at very low pressures, distrusting the technology of his time to build a system that could contain higher pressures.

We can use our formula for the efficiency of a reversible heat engine to estimate the maximum efficiency of a steam engine even allowing for the loss due to the irreversible heating. We can ask this question: once the heat is at the lower temperature of the water in the boiler, what is the best efficiency that we can expect? Let us take a more modest 250°C as the boiler temperature. (It still requires a boiler pressure of about 40 atmospheres or 600 pounds per square inch.) We assume that we exhaust our heat at 50°C. The calculation opposite shows that the maximum efficiency of our engine is 38%. Clearly the real efficiency will be much less. We have not allowed for many other inefficiencies: the heat from the fire that escapes in the chimney, other irreversible processes, mechanical friction in the piston and connecting gears, etc. In practice, traditional steam engines could achieve an efficiency of about 10%. Modern steam turbines in power plants, whose steam temperatures can go over 500°C, achieve efficiencies of 50%--but only through extreme measures.

Internal Combustion Engines
(e.g. gasoline auto engines, diesel truck engines)

Irreversible heating of working substance by fuel eliminated. Far more efficient.

Cylinder filled with highly flammable fuel-air mixture.

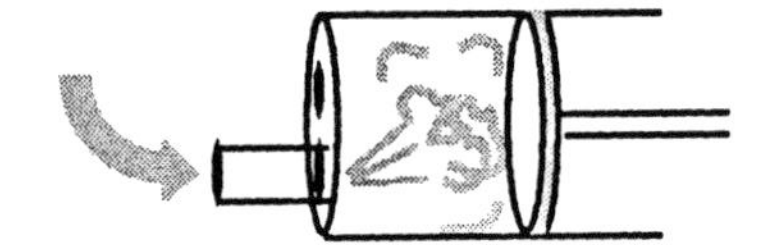

Mixture burns, gets very hot, expands and presses on piston.

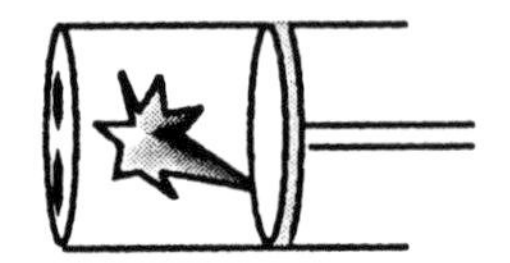

Flame *internal* to cylinder

When the fuel burns, it heats directly what is in the cylinder.

Typical temperature during combustion: 5000°F ≈ 2760°C
(gasoline engine)

Far higher than attainable in simple steam engine!

Ideal efficiency
of reversible engine running between:

T_{hot} = 2760°C = 3033 K
T_{cold} = 50°C = 323 K

$$\text{Efficiency} = 1 - \frac{323}{3033} = 1 - 0.11 = 0.89 \quad 89\%$$

But actual efficiency is lower!

With the steam engine, one struggles to find a way to transfer the heat from the fire into the water and steam without a huge drop in temperature. This problem represents the major obstacle to high efficiency. It can solved by a simple change in engine design. Instead of having the fire external to the stuff that pushes the piston (the steam), it is possible to design a fuel burning heat engine with the fire right inside the stuff that pushes the piston. These engines are the "internal combustion engines." They include the familiar automobile gasoline engine and truck diesel engines.

In a steam engine, the piston in the cylinder is moved by steam that has been heated by a fire elsewhere. In an internal combustion engine, the fire is in the cylinder itself. A mixture of fuel and air is drawn into the cylinder. It is ignited and burns explosively. The high temperature of the flame is communicated directly to the burnt fuel-air mixture, which proceeds to expand and press on the piston. Since the stuff that pushes the piston is, in effect, the flame itself, there is no irreversible loss in heating it! The effect is a dramatic improvement in efficiency.

A century ago, steam was king. It drove locomotives across the country and powered the great ships. Now one scarcely finds a steam engine in these uses. The demise of steam is due to the far higher efficiency achieved by the internal combustion engine. Steam engines tend now only to be used when one wishes to burn a fuel that cannot be burned easily in an internal combustion engine. Many electric power stations burn coal as a fuel. Coal cannot easily be burned internally. Therefore such power stations burn it in external combustion engines, where it is used to raise steam which turn turbines that drive electric generators.

The ideal efficiency of an internal combustion engine is given by our formula as in the calculation opposite. Because there is no wasteful, irreversible heating, an internal combustion engine can approach this ideal efficiency far more closely than could a steam engine. Nevertheless, many other factors reduce the performance of an internal combustion engine below this theoretical ideal. We can understand some of these factors if we look at the operation of one of the most popular internal combustion engines.

Four stroke, spark ignition, gasoline engine.

Most commonly used in automobiles.

"Otto cycle"

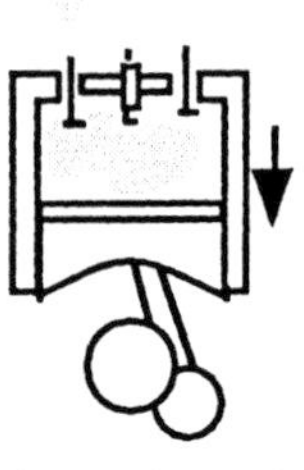

1. Induction
Gasoline-air mixture drawn into cylinder

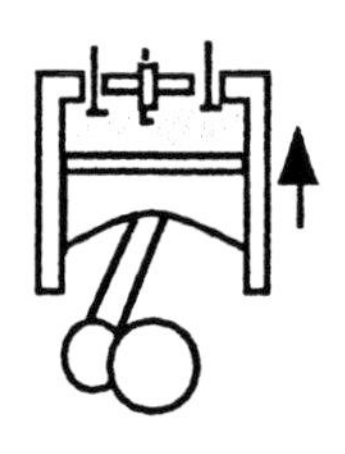

2. Compression
Mixture compressed

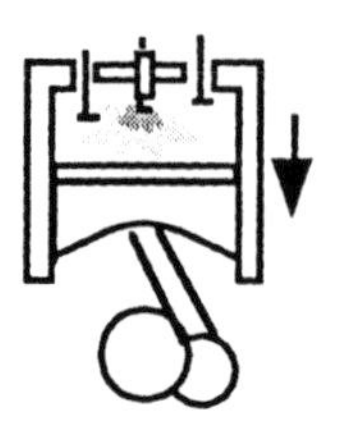

3. Power
Spark ignites mixture which burns and presses on piston

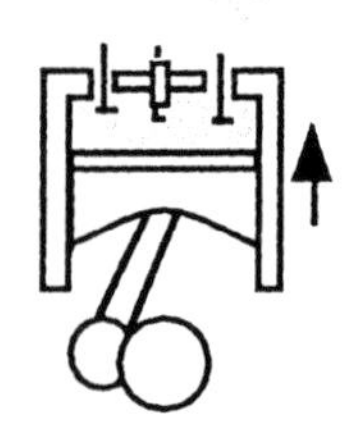

4. Exhaust
Burnt fuel driven out of cylinder

Efficiency of (near) reversibly operated Otto cycle:

Efficiency fixed by the compression ratio.

Compression ratio	Efficiency
7	49%
9	53%
11	57%

In this example, the compression ration is 10.

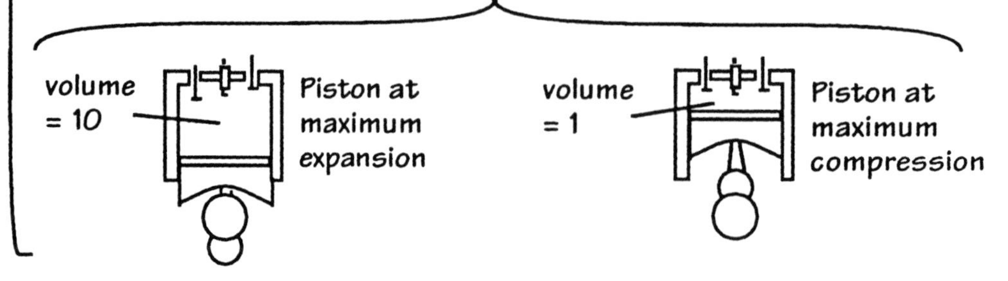

The gasoline engine most commonly used in automobiles is a four stroke, spark ignition engine. The pistons in the cylinders repeatedly go through a cycle of 4 strokes that comprise the "Otto cycle." In the induction stroke, a mixture of vaporized gasoline in air is drawn into the cylinder. In the compression stroke, this mixture is compressed. A spark ignites the mixture which expands explosively in the power stroke. Finally the burnt fuel is exhausted on the fourth stroke.

A typical car engine may have four, six or eight of these cylinders. Under normal operating conditions, each cylinder will execute a cycle of four strokes as often as 50 times a second. The operation of a diesel engine is similar. However a diesel engine does not employ a spark to ignite the fuel. Heat generated during compression of the mixture is sufficient to ignite it.

This cycle cannot operate at the ideal efficiencies computed earlier. For example, in the power stroke, a huge expansion would be needed to allow the hot gases to expand reversibly so that they cool from the thousands of degrees temperature of the flame to the cool ambient temperatures. Instead, the hot gases are vented irreversibly after there has been much less expansion. If we analyze the operation of the Otto cycle assuming reversible processes everywhere we can, we find that the efficiency of the cycle is fixed by the compression ratio. The higher the compression ratio, the greater the efficiency. Some sample values are given in the table opposite for compression ratios commonly found in automobile engines.

The compression ratio is simply the ratio of the volumes of the cylinder when the piston is at maximum expansion and contraction. It measures how much the fuel mixture is compressed in the compression stroke. The more the fuel mixture is compressed, the more its heat will be concentrated into a smaller space on burning and the higher the maximum temperature. We have already seen that the higher the temperature of the heat source, the more efficient the engine. So we should not be surprised to find the efficiency of the Otto cycle increasing with compression ratio.

A real gasoline engine does not achieve these efficiencies. Its efficiency is reduced further by mechanical losses, such as friction and in the energy needed to pump its lubricating oil and coolant. Actual efficiencies of gasoline engines are closer to 25% to 30%.

What limits increases in compression ratio?

Engine designers stop at modest compression ratios of 10-12 in ordinary car engines.

Lower compression ratios

Higher compression ratios

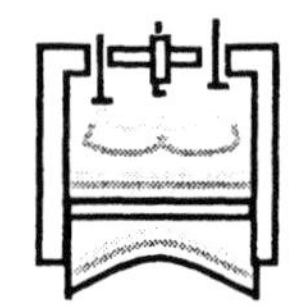

Smooth combustion

Burning gases exert smooth pressure on piston. Closer to reversible expansion.

Detonation "knocking", "pinging"

Violent detonation, shock waves in gases. Piston not driven smoothly. Closer to irreversible expansion. Felt as loss of power from engine.

Onset of detonation determined by quality of fuel.

Worst anti-knock characteristics

Best anti-knock characteristics

Pure n-heptane

89% iso-octane 11% n-heptane

Pure iso-octane

0% iso-octane

89% iso-octane

100% iso-octane

"Octane number" of a gasoline

A gasoline has octane number 89 if it has the same antiknock characteristics as a mixture of 89% iso-octane in n-heptane.

The gasoline need not have any iso-octane in it at all to merit an octane number of 89!

We can improve the efficiency of a gasoline engine by increasing its compression ratio. So why do car engine designers stop at a modest compression ratio of 10 to 12 for ordinary car engines? The reason is a breakdown in smooth running of the engine at higher compression ratios. At lower compression ratios, the fuel-air mixture burns smoothly, exerting a smooth pressure and transferring much of its energy to the piston. This process is closer to a reversible expansion. At higher compression ratios, detonation sets in. The fuel-air mixture burns with explosive shock waves that do not exert a smooth pressure on the piston. The result is an inefficient transfer of energy to the piston and a process closer to an irreversible expansion.

Detonation is sometimes called "knocking" or "pinging." Most of us have heard it, although we may not have known what it was. It sounds like someone shaking a bag of coins and the driver will feel a marked loss of power from the engine. It is most likely to happen with the car engine under heavy load but running at low speed, such as when trying to accelerate up a hill in too high a gear.

The compression ratio at which detonation sets in is determined by the quality of the fuel burned. Two chemicals are used as an industry standard to measure anti-knock qualities in a gasoline. Iso-octane has very good anti-knock qualities and is assigned a 100 "octane number." n-heptane has very poor anti-knock qualities and is assigned a 0 octane number. Mixtures have intermediate qualities. Thus an 89% mixture of iso-octane in n-heptane will have anti-knock characteristics approaching that of pure iso-octane. It has an octane number of 89. More generally, any gasoline at all that has the same anti-knock qualities as this mixture will have an octane number of 89 assigned to it. To get this number, it need not have any iso-octane in it all; it just has to behave the same as the 89% mixture.

Since a high octane number is good, many people think that a higher octane number must be better, so they buy the highest octane number gasoline they can find for their cars. This can be a waste of money. Once the fuel quality is sufficient to prevent detonation, further improvement in octane number confers no further advantage. If your car runs without knocking on an 89 octane number fuel, then you gain nothing as far as detonation is concerned by paying more for a fuel with an octane rating of 94.

Refrigerators, air-conditioners and heat pumps

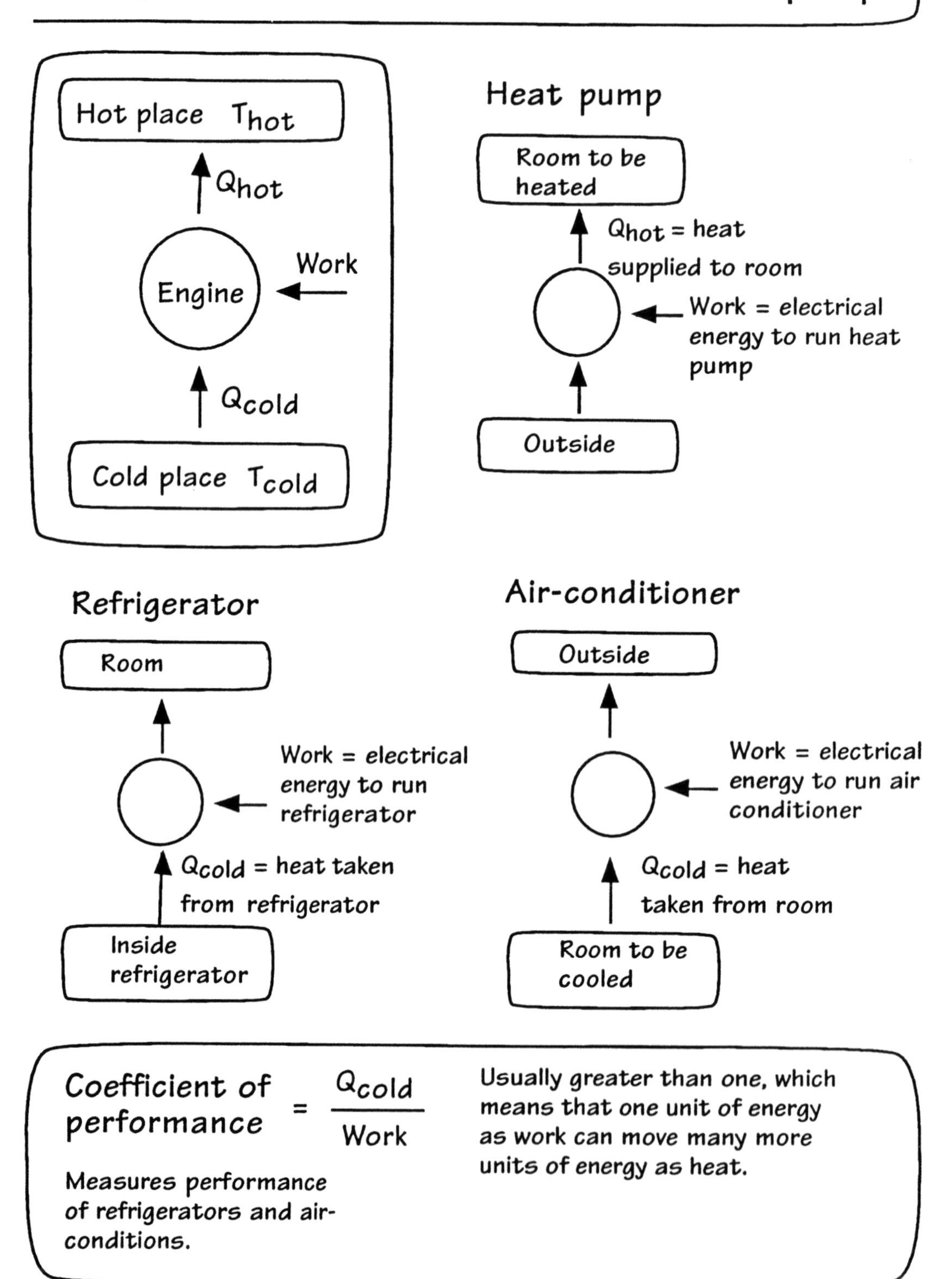

A reversible heat engine run in reverse uses work to move heat from a cold place to a hot place. That, of course, is just what a refrigerator does. It uses work, supplied as electrical energy, to move heat out of a cool compartment into the room that holds the refrigerator. Of course our real refrigerators are not the the same as reversible heat engines, for the latter are ideal devices that operate infinitely slowly. But, as the connection to reversible heat engines suggests, the theory of refrigerators is essentially just that of heat engines with the directions of heat and energy flow reversed.

The basic diagram used in the theory of refrigerators is shown opposite. It differs from that of heat engines only in the direction of the arrows and some slight relabeling. As in the case of heat engines, we will have the law of conservation of energy giving us Q_{hot} = Work + Q_{cold}. This means that Q_{hot} will always be greater than Q_{cold}.

The same theory applies to air-conditioners, too, for what they do is move heat from a room to be cooled to the outside. It also applies to heat pumps which are used for room heating in cool weather. They use the same type of machinery as air-conditioners, but they move the heat from the outside into the room.

In the case of the heat pump, the two quantities that we are most concerned with are Q_{hot} and Work, for they represent the heat delivered to the room and the electrical cost of moving it. In the case of refrigerators and air-conditioners, we are most concerned with Q_{cold} and Work, for they represent the heat extracted from the place we wish to cool and the electrical cost of doing it. In this last case in which we are concerned with Q_{cold} rather than Q_{hot}, there is a standard measure used to gage how well the machine is operating. The coefficient of performance tells us how much heat the machine extracts for each unit of work. A coefficient of performance of 5 tells us that the machine can extract 5 units of energy for every unit of work energy supplied to power the refrigerator or air-conditioner. In general, in all these devices, one unit of work energy can move several units of heat energy. This leverage makes heat pumps very attractive as devices for room heating. One unit of electrical energy fed into an electric heater supplies just one unit of heat. But the same unit of energy fed into a heat pump can supply many units of heat energy.

Standard Results From the Theory of Heat Engines Apply in Corresponding Form in the Theory of Refrigerators.

Real refrigerators cannot be run in reverse. [ideal limit] ⇒ Reversible refrigerator = heat engine when run in reverse

Result (I) The performance of all reversible refrigerators is the same when they work between the same hot and cold temperatures.

Result (II) No irreversible refrigerator performs better than a reversible refrigerator (working between the same hot and cold temperatures).

Result (III) For a reversible refrigerator

$$\frac{\text{Work}}{Q_{hot}} = 1 - \frac{T_{cold}}{T_{hot}}$$

equivalent to

$$\text{Coefficient of performance} = \frac{Q_{cold}}{\text{Work}} = \frac{1}{\frac{T_{hot}}{T_{cold}} - 1}$$

The theory of refrigerators now develops in almost exactly the same way as the theory of heat engines. We introduce the notion of the reversible refrigerator. This is a refrigerator that employs reversible processes only, so that it can exist only as an idealization. It can run in reverse and when run in reverse becomes a heat engine. It provides the case of the best performing refrigerator, just as the reversible heat engine supplied the case of the most efficient heat engine.

Using the same methods as before, we arrive at the three results opposite. For example, Result (I) is proved by a *reductio* argument. We assume that reversible refrigerators can perform differently. We show this assumptions allows us to construct a device from coupled refrigerators that violates the second law of thermodynamics.

These three results correspond exactly to the three result listed earlier in the theory of heat engines. In the place of the notion of efficiency, we have the notion of performance. We assess the performance of a refrigerator by seeing how much heat it moves for each unit of work supplied. So "better performance" in the context of Result (II) refers to moving more heat for the same amount of work supplied.

The formula in Result (III) is exactly the same as that in the theory of heat engines (after we replace Q_{in} by its analog Q_{hot}). The formula is a little less informative in the theory of refrigerators since Work/Q_{hot} has no simple interpretation. (In the theory of heat engines, the analogous quantity Work/Q_{in} is the efficiency of conversion of the heat Q_{in} to Work. In a refrigerator, we no longer seek this conversion, so the quantity Work/Q_{hot} measures nothing of direct interest to us.) With a little algebra, the formula in Result (III) can be converted into a formula for the coefficient of performance, as shown opposite. Both formulae have the same physical content, but the formula for the coefficient of performance supplies us direct access to the most useful measure of performance of a refrigerator. By substituting values for T_{hot} and T_{cold} into the expression, we find that the performance of a reversible refrigerator improves as the two temperatures come closer together. This improvement in performance is what we would expect, since, as the two temperatures approach, the refrigerator must not work as hard to move heat energy between them.

How does a refrigerator work?

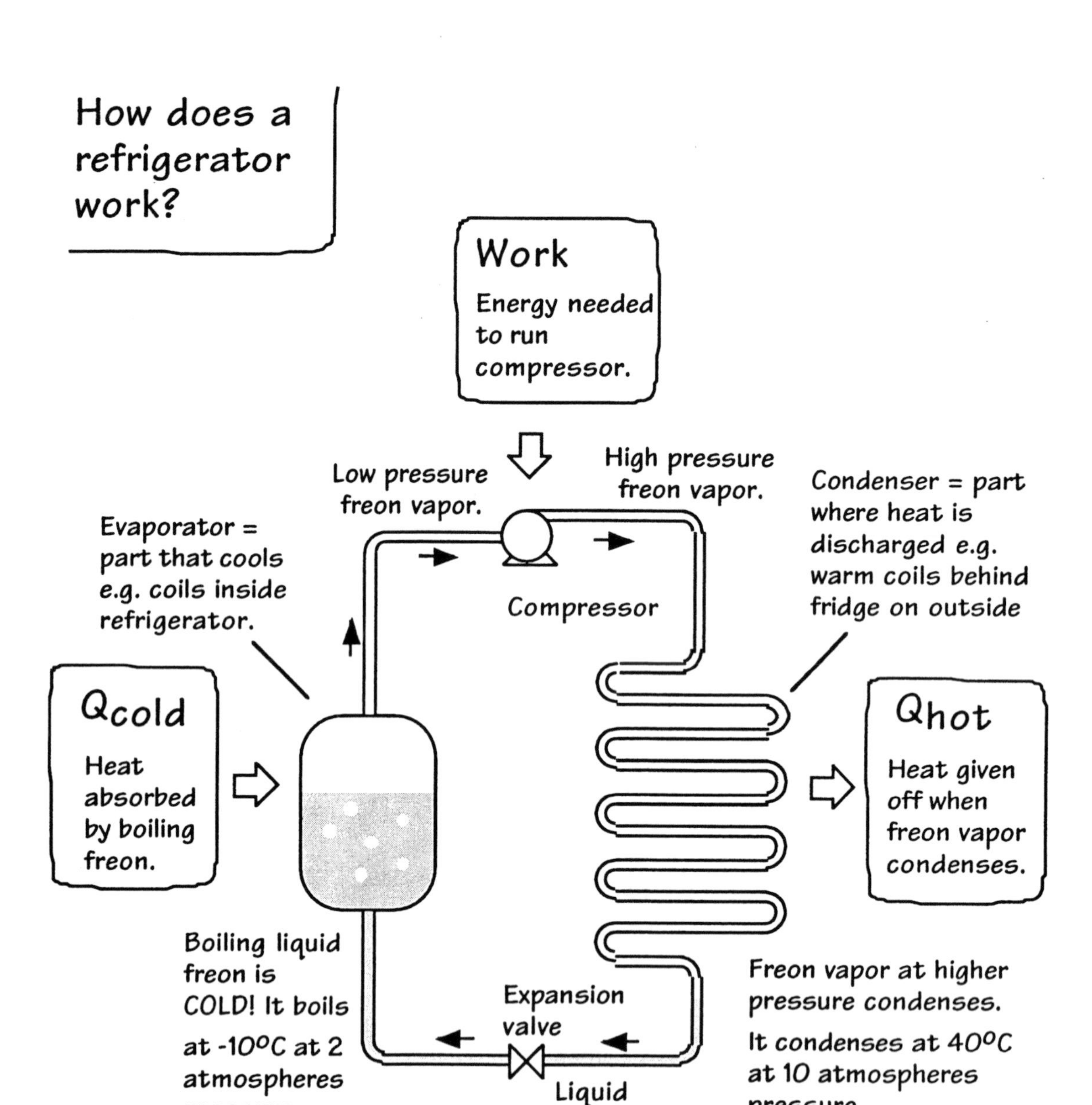

While we are all familiar with refrigerators as machines that can make things cold, it may be helpful to describe precisely how a refrigerator is able to cool things. The mechanism described here is the same used in air-conditioners and heat pumps.

A refrigerator cools using the same physical effect that makes you cold when you step out of the shower. Your skin is covered with a film of water. The water evaporates and, as it evaporates, absorbs a large amount of heat, so your skin cools. A refrigerator uses liquid freon instead of water, since liquid freon boils at a very low temperature. The evaporator of a refrigerator contains boiling liquid freon which absorbs large amounts of heat at very low temperatures as it boils. If the freon is under 2 atmospheres of pressure, it will boil at -10°C, so a freon based refrigerator can easily maintain temperatures below freezing. In a domestic refrigerator, the evaporator is often visible as aluminum coils that encase the freezer. They are typically coated with ice and a moistened finger will freeze to them. The heat absorbed by the boiling freon is Q_{cold}.

To keep operating, a refrigerator requires a supply of liquid freon. The remainder of the machinery in the refrigerator ensures that supply is available. Once the liquid freon boils, it is converted into a vapor. The machinery converts that vapor back into a liquid to resupply the evaporator. A compressor compresses the vapor to a higher pressure. The energy needed to operate the compressor is the Work required by the refrigerator. In most domestic refrigerators, this work is supplied as electrical energy. At higher pressures, freon vapor will condense. At 10 atmospheres pressure, for example, it will condense at 40°C, which is above ordinary room temperature. So the compressed vapor is passed through long coils that are surrounded by the air of the room. Since the room will be cooler than the 40°C needed to condense the vapor, the freon condenses and gives off a lot of heat to the room air. This heat is Q_{hot}. The condensing coils are usually found on the rear of the refrigerator hidden from sight. They will feel warm to the touch as they discharge their heat. Room air must be able to circulate over these coils in order to carry off the heat needed to condense the freon. Otherwise the condenser will overheat and the freon will no longer condense. The condensed liquid freon passes through an expansion valve to the lower pressure evaporator and the cycle is complete.

An Application: The performance of an air-conditioner

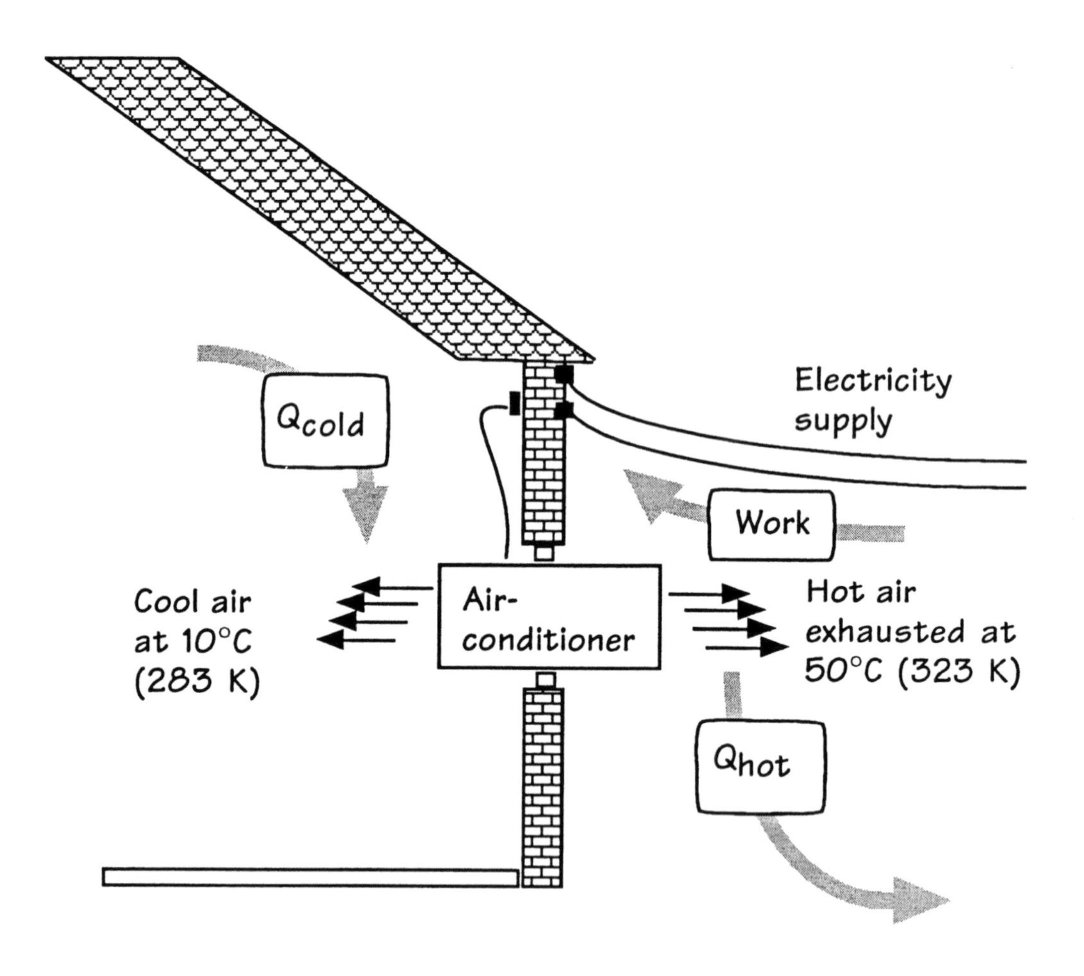

Coefficient of performance of a *reversible* air conditioner $= \dfrac{Q_{cold}}{\text{Work}} = \dfrac{1}{\dfrac{T_{hot}}{T_{cold}} - 1} = \dfrac{1}{\dfrac{323}{283} - 1} = 7$

For real air-conditioners expect less performance, e.g. 4.

The reversible refrigerators of our general theory are idealizations that cannot exist. However, our knowledge of them provides important information about real refrigerators, because reversible refrigerators provide an upper limit on the performance of any real refrigerator--and this upper limit proves to be a very restrictive one. To illustrate how our knowledge of these idealizations of theory can provide useful, practical results, we will look briefly at the economics of a form of refrigeration: air-conditioning. We will discover that air-conditioning by traditionally designed equipment must be expensive.

A home air-conditioner is shown schematically opposite. We will assume that the air-conditioner supplies cool air to the house at 10^oC (just modestly cool) and that it exhausts its heat in a warm air current at 50^oC (just warm enough to allow the machine to function). The heat absorbed by the air-conditioner through the cool air is Q_{cold}. The heat exhausted with the warm air is Q_{hot}. The Work needed to move the heat from inside the house to the outside is supplied as electrical energy.

Let us compute the best possible performance for an air-conditioner operating under such circumstances. This best performance would be provided by a reversible machine. Opposite we calculate the coefficient of performance of such a reversible air-conditioner using the formula of Result (III). (Notice that temperatures are converted to absolute degrees Kelvin!) The result is a coefficient of performance of 7. That means that every Btu of electrical energy supplied would be able to move 7 Btu of heat from the house. Of course the leverage is nice--one moves seven. But one can already start to suspect that this will be a costly operation even with an ideal machine. Air-conditioners must move a lot of heat energy to cool a house. Even a seventh of this amount might be large especially if that amount will appear on our electricity bill.

The coefficient of performance of 7 is for the ideal, reversible air-conditioner. A real air-conditioner cannot perform at this ideal level. A good air-conditioner, operating under these same conditions, may well achieve half this performance. For the calculation to follow, let us assume that the real air-conditioner has a coefficient of performance of four.

Estimate daily operating cost of an air-conditioner on a summer's day

Pittsburgh latitude: Solar energy falling on one square foot on an average sunny day in June	= 2000 Btu
Solar energy in one day falling on a modest sized house of 1600 square feet roof area	= 2000 x 1600 = 3,200,000 Btu
Some of this heat will be reflected, some re-radiated. Therefore assume that the air-conditioner must move only 1/3rd of this heat load.	Q_{cold} = 3,200,000/3 = 1,066, 000 Btu (Heat to be removed from house.)
Conversion factor: 3413 Btu = 1 kWh	Q_{cold} = 1,066,000/3413 = 312 kWh
Coefficient of performance: $\frac{Q_{cold}}{Work} = 4$	$Work = \frac{312}{4} = 78$ kWh (Electrical energy needed each day to run air-conditioner)
Electricity costs approximately $0.06 per kWh	Daily electricity cost = 78 x 0.06 = $4.68

Let us now convert the performance of this air-conditioner into a daily operating cost for air-conditioning a house in summer. To do this, we need to estimate how much heat the air-conditioner needs to move. This figure will vary very much from house to house, so we can only hope for a very rough estimate.

Sunlight is the major source of the heat taken up by a house in summer. There are other sources, such as heat entering from warm surrounding air by conduction through glass or with air leaks. In this first, rough calculation, let us consider just the burden due to solar energy. About 2,000 Btu of solar energy fall on each square foot on an average sunny day in June in the Pittsburgh latitudes. A modest sized house might expose an area of 40'x40' = 1600 square feet to the sun, so that the total amount of solar energy falling on it is 2,000 x 1,600 = 3,200,000 Btu each day. Not all of this heat enters the house. Much of it will be reflected, more of it will be blocked by roof insulation and more reradiated at night. Let us assume that only a third enters the house and must be moved by the air-conditioner. This third amounts to 1,066,000 Btu, which is equivalent to 312 kWh of energy. This figure is the Q_{cold} for the air-conditioner.

We have already estimated that the coefficient of performance of the air-conditioner is 4. Thus 1 kWh of electrical energy is needed to move 4 kWh in Q_{cold}. Therefore the air-conditioning of the house will require 312/4 = 78 kWh each day. A typical cost of electricity is 6 cents per kWh--you can read this directly off your electricity bill! Thus the daily operating cost of this air-conditioner would be 0.06x78=$4.68. This daily cost rapidly grows to be a significant expense. It corresponds to a monthly cost of $140, for example.

This calculation shows us that air-conditioning will be expensive. Clearly the figures in the above calculation are very rough. Houses with better or worse insulation, more or less shade, better or worse performing air-conditioners will generate different results. But this calculation shows us the types of cost that we should expect.

Coefficient of **P**erformance

$$= \frac{Q_{cold} \text{ (measured in Btu)}}{\text{Work (measured in Btu)}}$$

The measure most commonly used in theoretical calculations.

and

They both measure exactly the same quantity. They differ only in using different unit.

Energy **E**fficiency **R**atio

$$= \frac{Q_{cold} \text{ (measured in Btu)}}{\text{Work (measured in Wh)}}$$

The measure most commonly used in the air-conditioning industry.

Conversion:
1 Wh = 3.413 Btu

$$\text{EER} = 3.413 \times \text{CoP}$$

In theoretical analyses of refrigerators and air-conditioners, the coefficient of performance is most commonly used as an index of effectiveness. It is simply a ratio of two energies: Q_{cold}, the heat energy extracted from the cold place and Work, the energy used in extracting it. Since it is a simple ratio, any unit of energy may be used: Btu, calories, Joules--whatever you like. The main thing is that the *same* unit be used for both quantities. Then the coefficient of performance will always come out correctly. We use this freedom in choosing units when we compare the heights of people as a ratio. If Jane at 6 feet is twice the height of Joe at 3 feet, then their height ratio is two. We get the same ratio if we measure their heights in different units--as long as we use the same unit for both. Jane at 2 yards is still twice the height of Joe at one yard.

The practice in the air-conditioning industry, unfortunately, is not to use the same units for Q_{cold} and Work. Since Q_{cold} is a quantity of heat energy, it is measured in Btu, a unit commonly used for measuring heat. Since Work is usually supplied as electrical energy, it is measured as Wh, that is, Watt-hours, a unit commonly used for electrical energy. If the same ratio of Q_{cold}/Work is computed using these two different units, the result is an index known as the Energy Efficiency Ratio, or just EER. The EER is the same quantity as the coefficient of performance; it is just measured in funny units. Since 1Wh is 3.413 Btu, we can easily convert between the two. Multiply the coefficient of performance by 3.413 to get the EER; divide the EER by 3.413 to get the coefficient of performance.

The EERs for air-conditioners have been increasing in recent years. That means they are becoming more efficient. Older models might have EERs as low as 6. The newer models have EERs in the range of 10 to 14. This range corresponds to coefficients of performance in the range 10/3.413 to 14/3.413, that is 2.9 to 4.1. Recall that we estimated the coefficient of performance of a reversible air-conditioner as 7. This corresponds to an EER of 3.413x7=23.9. So, as we should expect, real air-conditioners fall short of the ideal performance of a reversible device. An EER of 24 provides an upper limit that we would not expect real devices to surpass.

Surprising power of thermodynamic reasoning

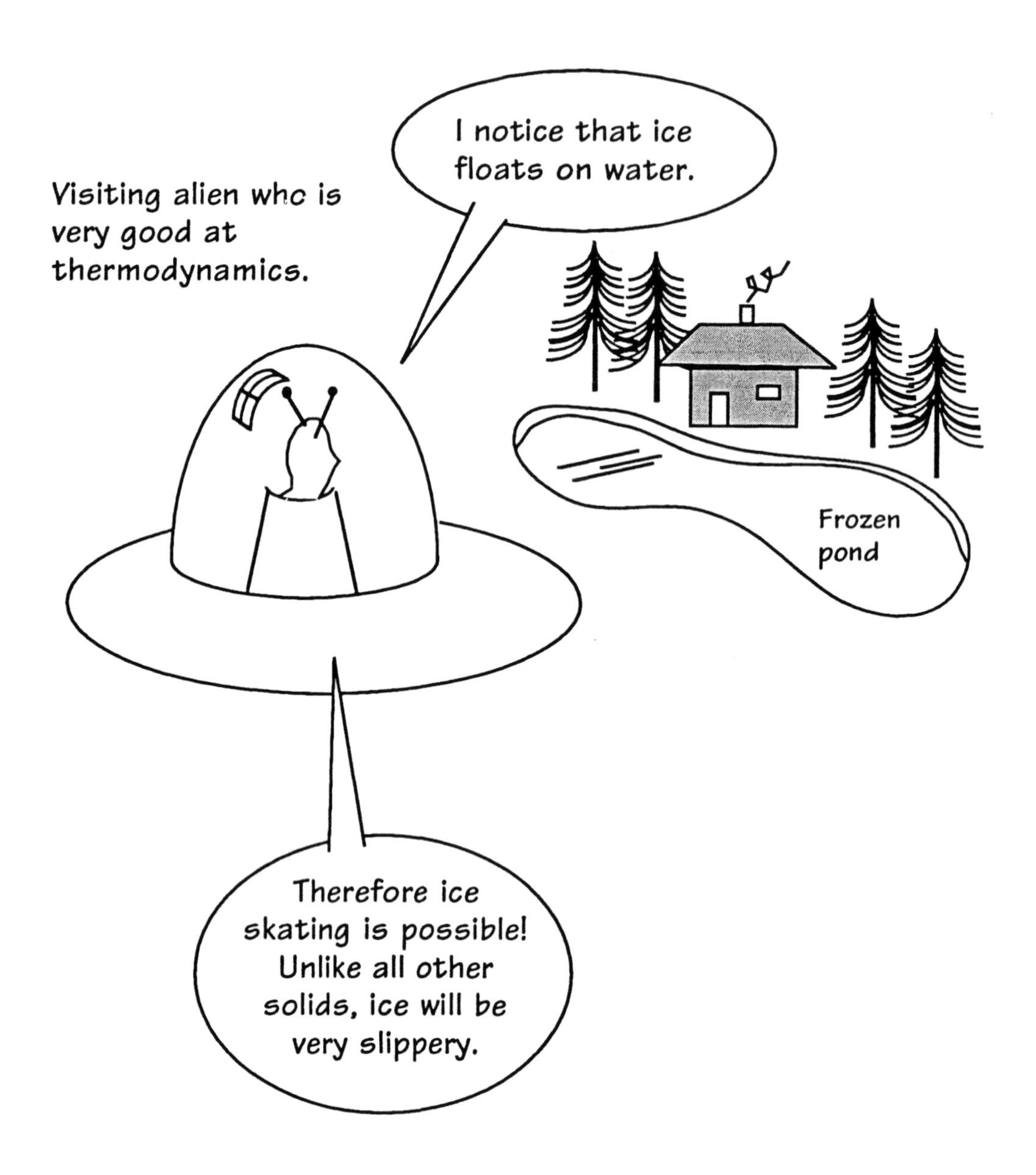

It is a sparkling midwinter's day in the mountains. A small cabin nestles on the shores of a pond. A weary alien has traveled greater distances than we can count to make first contact with planet earth. As the spaceship swoops low over the mountain, the alien sees water for the first time. The alien notices that ice--frozen water--is floating on the surface of the pond. The alien, who is very good at thermodynamics, is able to conclude from this fact that ice skating is possible and that ice is very slippery.

This is the hidden power of thermodynamics. So far, we have treated the theory as if it were a general theory of heat engines and refrigerators. But it is much more. Thermodynamics can tell us about many hidden relationships in the physical world in domains that seem quite remote from heat engines. In particular, it call tell us of many interrelationships in the properties of different forms of matter. If one is an adept in thermodynamics, one can see a new significance in the fact that ice floats on water. It entails that ice must be very slippery, so that ice skating is possible.

I hope that the possibility of such an inference seems quite mysterious to you, for then you have sensed the power of thermodynamics. We shall see now, however, that there is no magic in the inference. Once we start to tease it out, the inference will turn into a fairly ordinary piece of thermodynamics. And so it is with conjuring, as well. The conjurer makes an elephant appear on stage. It seems like a miracle until one discovers the clever combination of ordinary trickery used to carry out the effect.

What the alien has noticed

The theory of heat engines teaches us about a lot more than heat engines! It prescribes the behavior of any substance that can be used in a heat engine.

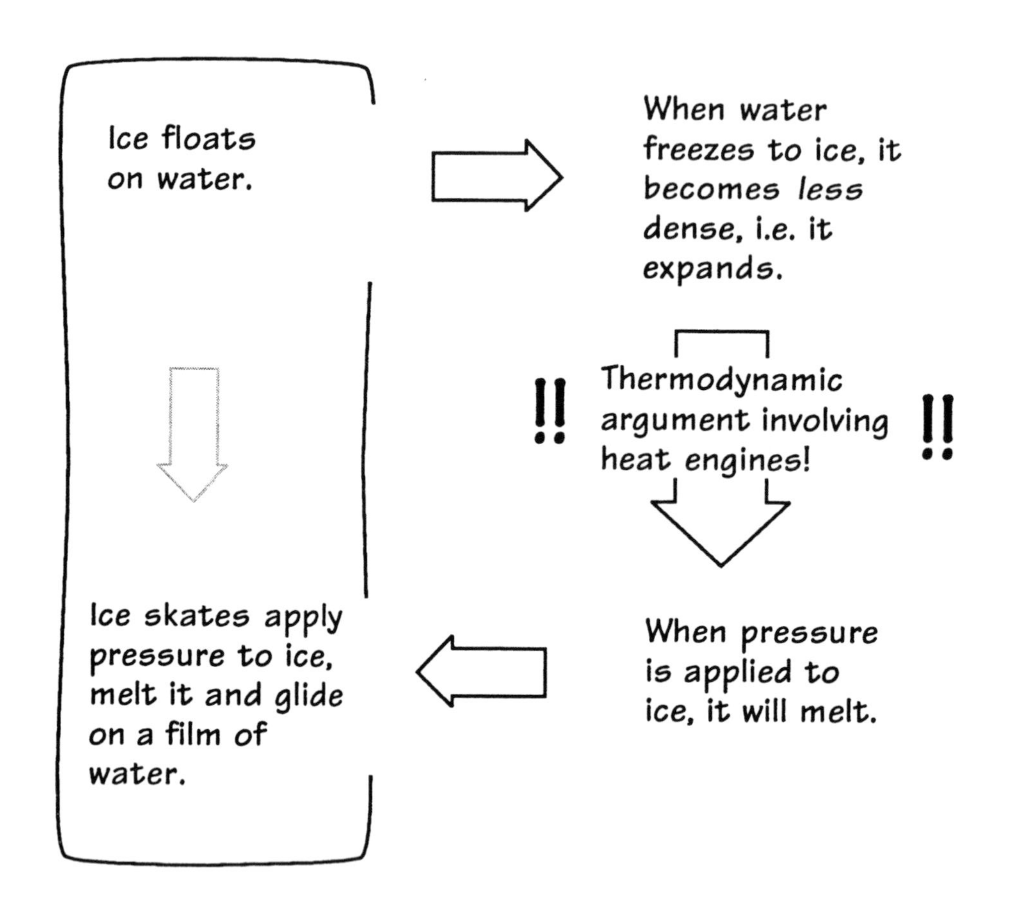

Thin film of water under blade lets skate move smoothly.

Let us now reveal the intermediate steps in the alien's inference. We start with the observation that ice floats on water. It follows that ice is less dense than water. (This is the reason cork floats on water and iron sinks: the cork is less dense than water and the iron more dense.) If ice is less dense than water, then water must expand on freezing.

Now comes the thermodynamic part: we can use the expansion of freezing water to build a heat engine. We will see below that this heat engine will allow us to violate the laws of thermodynamics--unless we assume that ice melts when we apply pressure to it. Since we do wish to retain the laws of thermodynamics, we conclude that ice does melt under pressure. We will see how this inference proceeds in the pages to follow.

The real work is now behind us. It is easy to see how this melting of ice under pressure enables ice skating. Imagine the blade of an ice skater. With the full weight of the skater bearing down on the blade, the pressure immediately under the blade becomes quite high. This high pressure melts the ice immediately under the blade, so that the blade is separated from the ice by a film of water. This film of water lubricates the blade which can then slide freely over the ice--ice skating is possible. This same effect makes skiing possible. It even makes ice cubes hard to pick up. The effect of grabbing an ice cube is to form a layer of water where you press on it, allowing the ice cube to slip easily out of your grasp.

The melting of ice under pressure explains why we can only skate on ice. Ice is unique in melting under pressure. Most other substances tend to solidify if one increases the pressure. Thus trying to skate on cork is a waste of time. Cork does not melt under pressure--even though it does float on water.

Water has two properties that make is unique: it expands on freezing, where other substances contract; it melts under pressure, where others solidify. Thermodynamics reveals the connection between the two properties. If the laws of thermodynamics are to be retained, then the first property forces the second to hold.

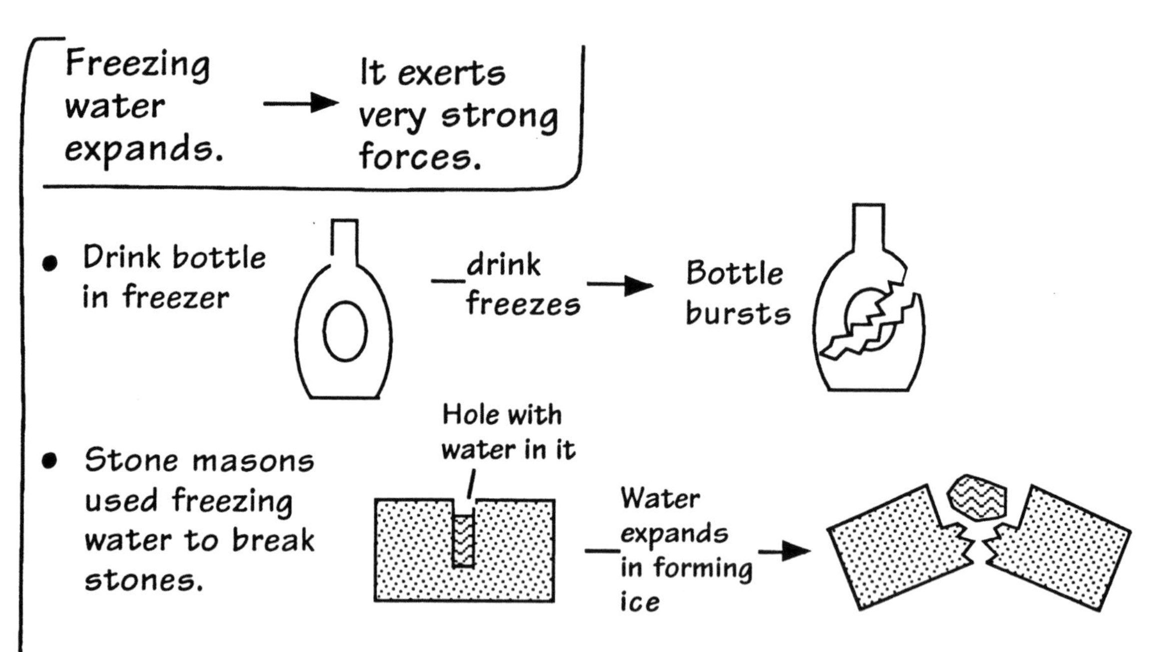
Freezing water expands.
It exerts very strong forces.
Drink bottle in freezer
drink freezes
Bottle bursts
Hole with water in it
Stone masons used freezing water to break stones.
Water expands in forming ice
Water pipes freeze and burst in winter.
Freezing water expands and opens cracks in road.

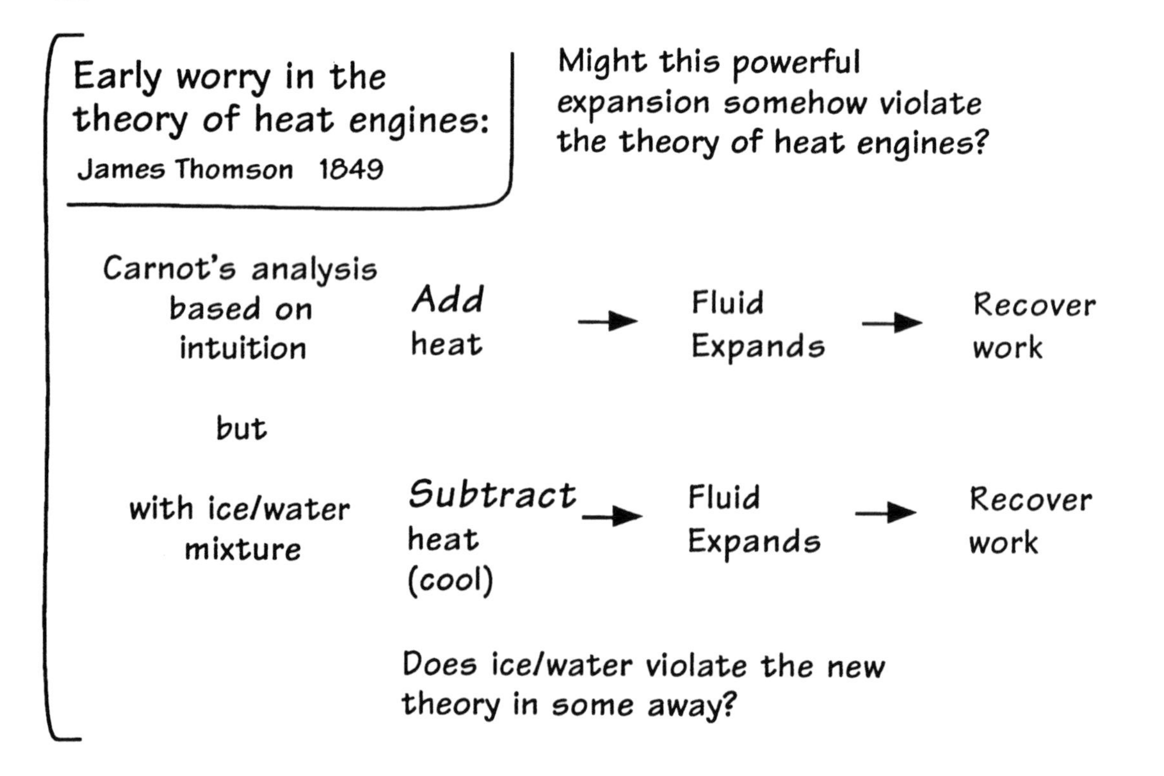
Early worry in the theory of heat engines:
James Thomson 1849
Might this powerful expansion somehow violate the theory of heat engines?
Carnot's analysis based on intuition
Add heat
Fluid Expands
Recover work
but
with ice/water mixture
Subtract heat (cool)
Fluid Expands
Recover work
Does ice/water violate the new theory in some away?

The story of the alien is a fable, of course. But something not so different actually did happen at the time of the discovery of thermodynamics around 1850. It had long been known that freezing water expands and exerts powerful forces in expanding. This is something we are familiar with today. A drink freezing in a glass bottle will burst even a very strong bottle. Exposed water pipes in winter can freeze and burst. Every winter our roads are broken up by the expansion of freezing water that has seeped into cracks. Stone masons even used to use this effect to break stone blocks.

The early researchers in thermodynamics wondered if these forces might somehow be used in a heat engine. However they were concerned about the way that this effect might bear on the theory. The theory was built on simple intuitions about how heat is used in generating work. We expect that heat added to some fluid will cause it to expand. In expanding it will supply work. We must then cool the fluid, discharging the heat as waste, in order to return it to its initial state so that the engine cycle can continue.

An ice/water mixture is incompatible with these intuitions. It does not expand if we heat it; it expands if we *cool* it. This is just the reverse of what we expected. Might there not be some way of using this odd effect in a heat engine so that the heat engine operates in a way that violates the laws of thermodynamics?

William Thomson's elder brother James Thomson addressed this worry in 1849. He showed how the expansion of freezing water could be incorporated into a heat engine and then showed that the engine was no threat to thermodynamics--as long as we assume that ice melts when placed under pressure. James Thomson's analysis was actually carried out within the older caloric theory of heat. But the essential points of his analysis carry over directly into the modern theory and they are reproduced below.

A simple ice/water heat engine that violates both laws of thermodynamics?

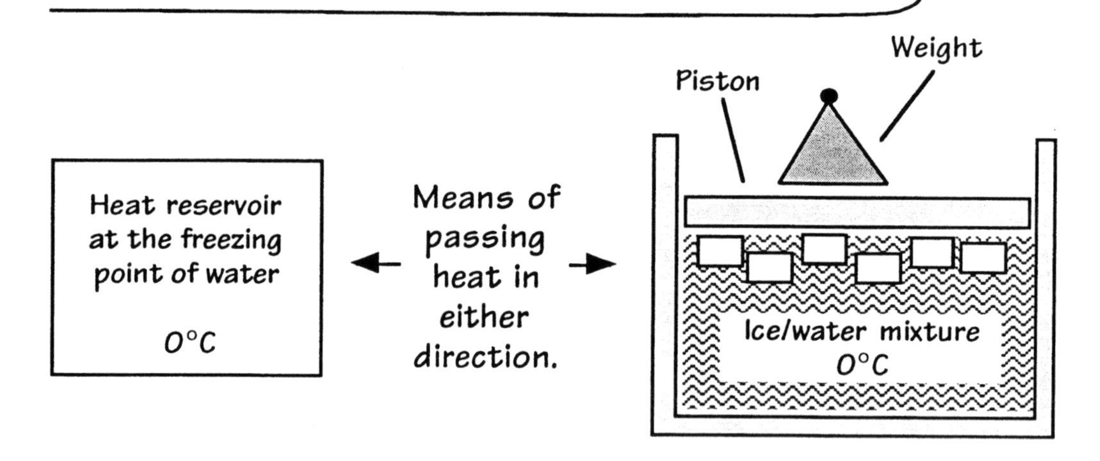

Work generating step in cycle:

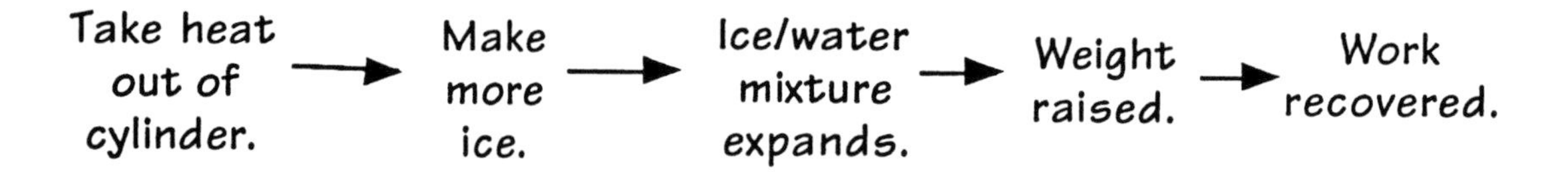

Let us now proceed with the design of an ice/water heat engine that will appear to violate the laws of thermodynamics. It will use a cylinder filled with an ice/water mixture. A piston will fit snugly over the mixture and carry a weight. When heat is extracted from the ice water mixture, some of the water will freeze and expand. We will be careful to keep the cylinder filled with an ice/water mixture and never solid ice. The result is that the formation of new ice merely makes the existing pieces of ice grow a little in size. The mixture will remain a mobile slurry able to conform to the shape of the cylinder as the mixture expands and the piston in raised.

As the ice/water mixture expands and raises the piston, work is supplied through the raising of the weight sitting on the piston.

To complete the engine, we will need a place to discard the heat extracted and to supply heat if needed. For simplicity, let us make the engine a reversible engine. Thus the heat extracted will pass reversibly to a heat reservoir at the same temperature. We will need some means to move the heat to and from the reservoir. A heat conducting metal rod would suffice.

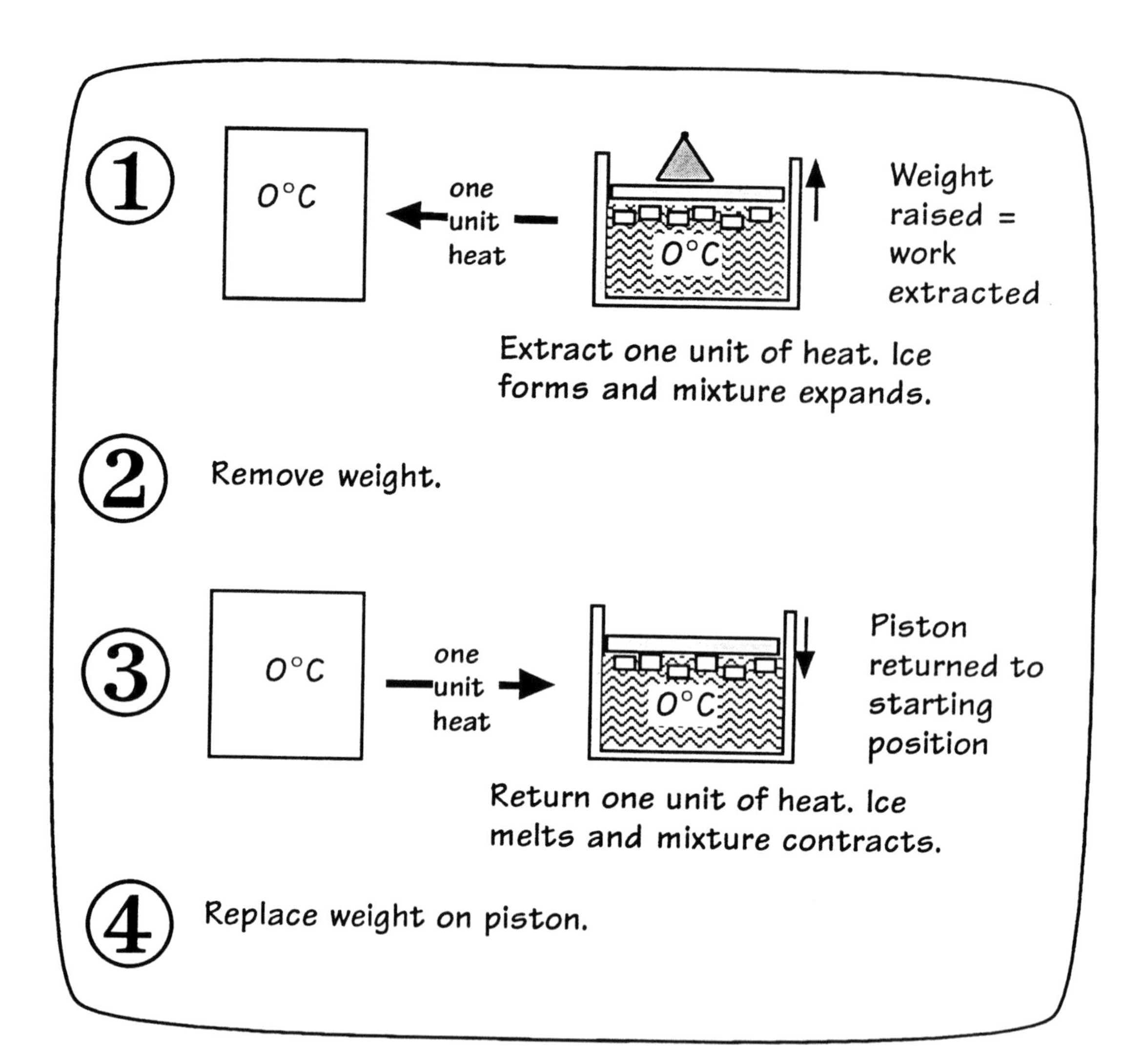

1
0°C
one
unit
heat
0°C
Weight
raised =
work
extracted
Extract one unit of heat. Ice
forms and mixture expands.
2
Remove weight.
3
0°C
one
unit
heat
0°C
Piston
returned to
starting
position
Return one unit of heat. Ice
melts and mixture contracts.
4
Replace weight on piston.

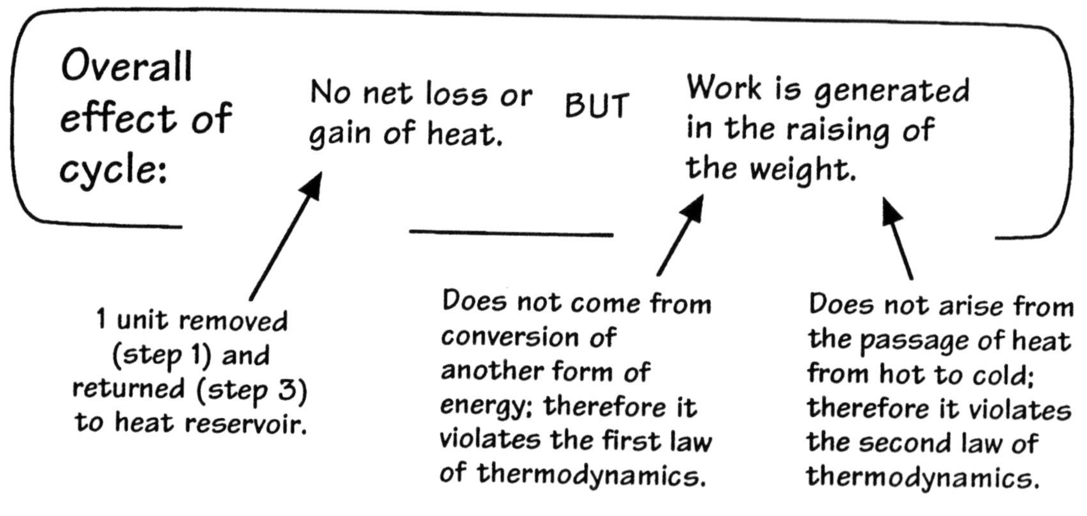

Overall
effect of
cycle:
No net loss or
gain of heat.
BUT
Work is generated
in the raising of
the weight.
1 unit removed
(step 1) and
returned (step 3)
to heat reservoir.
Does not come from
conversion of
another form of
energy; therefore it
violates the first law
of thermodynamics.
Does not arise from
the passage of heat
from hot to cold;
therefore it violates
the second law of
thermodynamics.

The engine operates with the four steps shown opposite. First a unit of heat is extracted reversibly from the ice/water mixture. This will freeze a small amount of the ice, expand the mixture and raise the piston. Work is generated in the raising of the weight. The remaining steps are designed to complete the cycle by bringing the engine back to its state at the start of the first step. The raised weight is removed. Then one unit of heat is returned reversibly to the cylinder. This one unit of heat will melt just that amount of ice formed in the first step, so the ice/water mixture is brought back to the proportions it had at the start of step 1. A new weight is placed on the piston and the cycle is complete.

The overall effect of this cycle is a thermodynamic catastrophe. First notice that there is no net gain or loss of heat in the cycle. The unit of heat removed from the reservoir in step 3 is replaced in step 1. yet an amount of work is produced in the cycle through the raising of the weight. The production of this work violates the first law of thermodynamics. That law states that energy in any form cannot be created from nothing; it must always result from the conversion of another form of energy. The work energy that raises the weight does not result from the conversion of any other form of energy. The means of production of the work violates the second law. The second law requires that any generation of work by a heat engine must be accompanied by some discharge of waste heat to a cold place. In this engine there is no discharge of waste heat to a cold place--or, more precisely, a place colder than the reservoir at $0^{o}C$ that supplies the heat for the engine.

James Thomson's solution

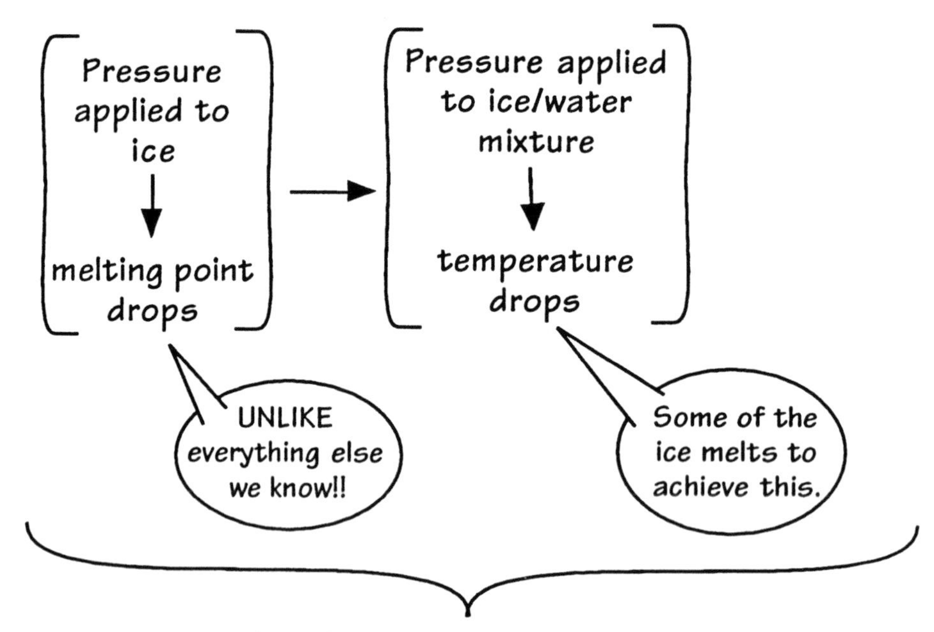

Therefore, engine will fail, since:

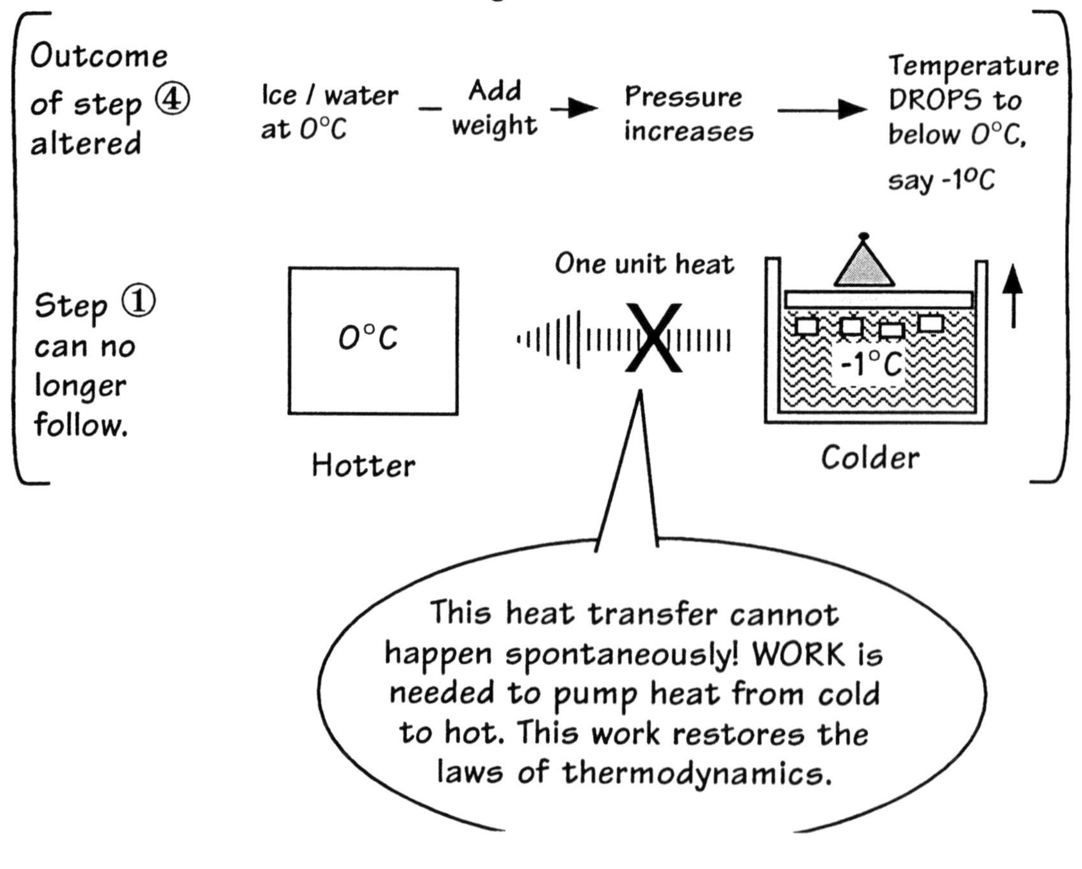

James Thomson saw that this thermodynamic catastrophe could be avoided with one simple hypothesis. This was the proposal that the melting point of ice depends on the pressure applied; in particular, as the pressure increases, the melting point drops.

The idea that the melting point should depend on the pressure applied is not so implausible. We are used to the boiling point of liquids varying with the pressure applied. That is why pressure cookers cook faster; the higher pressure due to the trapped steam increases the boiling point of the water inside the vessel. And that is why eggs take longer to hard boil on a mountain top; the lower air pressure reduces the boiling point of the water in the pot. If these effects did not happen for boiling water, similar thermodynamic catastrophes would result!

If the melting point of ice drops when pressure is applied, then the effect of increased pressure on an ice/water mixture is to leave the mixture at a higher temperature than the melting point of the ice. consider a block of ice at its melting point, 0^oC at 1 atmosphere pressure. Apply more pressure to it and the melting point point of the ice will drop to below 0^oC. But the ice temperature will stay at 0^oC, which is now above the ice's melting point. The ice will be like ice put into a warm drink. The ice is now above its melting point and it will start to melt. In melting it will cool the mixture and the temperature will drop.

This new effect is all that is needed to prevent the ice/water heat engine from operating and to rescue the laws of thermodynamics. The effect will prevent step 4 of the cycle proceeding as described before. If the ice/water mixture starts at 0^oC, the addition of the weight will increase the pressure on the mixture. Therefore its melting point will drop. Some of the ice will melt, cooling the temperature of the ice water mixture to the new lower melting point. For concreteness, let us say it drops to -1^oC. So the ice/water mixture at the start of step 1 will be below the temperature of the heat reservoir at 0^oC. Therefore heat cannot pass spontaneously from the mixture to the reservoir in step 1. The cycle stops and there is no violation of the laws of thermodynamics.

What happens now in Step ①

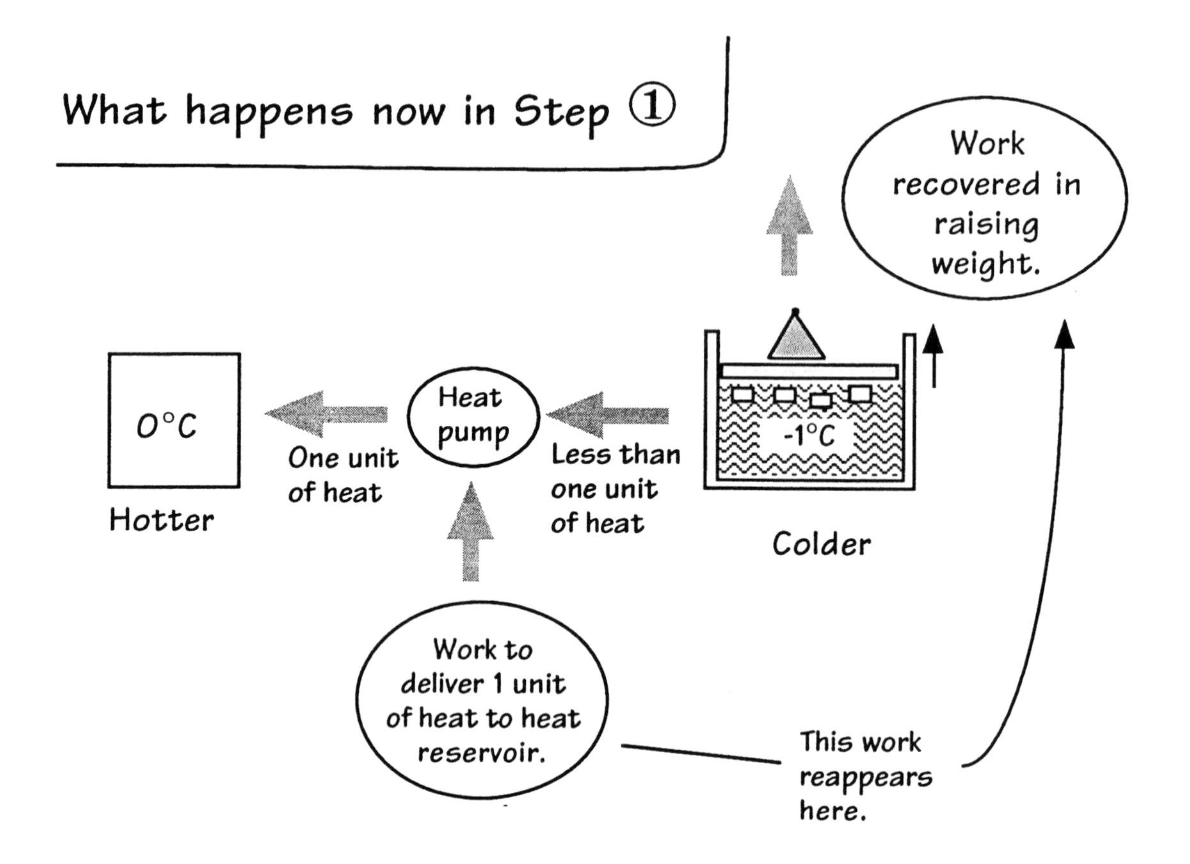

Overall effect of complete cycle in steps ①- ④

- Heat reservoir and cylinder are returned to their original states. There is no net gain or loss energy in each.
- No creation or destruction of work energy if we assume: Work consumed to deliver heat in step ① = Work recovered in raising weight in step ①

No violation of second law since no net work is produced.

First law saved since energy is now conserved.

To see more clearly how the two laws of thermodynamics are saved, we can add the extra components needed to make the cycle operate. If step 1 is to proceed, a little heat pump would be needed to extract the heat from the colder mixture at -1°C and deliver it to the reservoir at 0°C. This heat pump would consume work energy. Thus the step that produces work will also require work. The remaining steps will stay essentially the same. The other change is that the temperature of the mixture will return to 0°C when the weight is removed at the end of step 2. This means that the transfer of heat from the reservoir to the cylinder in step 3 can happen reversibly and spontaneously since both will be at the same temperature.

If we assume that the work required to move the heat in step 1 exactly equals the work recovered in the raising of the weight, then both laws of thermodynamics are saved. In the course of one complete cycle, both the cylinder and the reservoir are returned to their original states. Moreover the system as a whole neither absorbs nor generates any heat energy or work energy. This is what the first law requires. There is no violation of the second law. For the second law requires that the production of work be accompanied by a discharge of waste heat to a colder place. But in the amended cycle, we have no generation of work.

Notice that the addition of the heat pump is not the only way to amend the cycle so that it can operate. Another way would be to use two heat reservoirs. One heat reservoir at 0°C would supply the heat needed for step 3. A second heat reservoir at -1°C would receive the heat discharged in step 1. This discharge could be done reversibly without a heat pump since both reservoir and mixture would be at the same temperature. If the cycle were amended in this way, it would produce work. But this would not violate the second law of thermodynamics since the device would simply be operating as a heat engine. The hot place is the reservoir at 0°C; the cold place the reservoir at -1°C; and the engine operates by transferring heat from the hot place to the cold place. The first law would also not be violated since it would turn out that slightly less heat energy is discharged to the cooler reservoir in step 1 than is taken up in step 3. The difference is the work energy produced.

How William Thomson tested the escape in an experiment.

Condition: Energy is exactly conserved.

Work consumed in step 1 = Work recovered in step 1

—Re-expressed by James and William Thomson →

Condition on how much an increase in pressure decreases the freezing point of water.

Reduction of freezing point in °F = 0.0135 x Increase in pressure measured in atmospheres

William Thomson tested this experimentally (1850):

Pressure applied to ice/water mixture	Observed depression of temperature	Depression expected by theory
8.1 atmospheres	0.106 °F	0.109°F
16.8 atmospheres	0.232°F	0.227°F

The agreement is so good that Thomson was surprised!

"In this very remarkable speculation, an entirely novel physical phenomenon was predicted in anticipation of any direct experiments on the subject, and the actual observation of the phenomenon was pointed out as a highly interesting object for experimental research."

W. Thomson (1850)

James Thomson's resolution depends on the exact equality of the work consumed in step 1 to drive the heat pump and the work produced in step 1 in lifting the weight. for only when these are equal is energy conserved and the second law saved. It is natural to suppose that these two are equal since that is what the laws of thermodynamics require. But they can only be equal if the decrease in the melting point of water under pressure is precisely tuned to allow the two quantities to balance. The Thomsons turned this consideration around. Assuming they do balance, they inferred what the relationship must be between the applied pressure and the decrease in melting point. That result is given opposite. Each atmosphere increase in pressure yields a 0.0135^{o}F reduction in temperature.

This is a relationship that can be tested by a real experiment. William Thomson did this. He procured a large vessel filled with ice and water. He carefully measured the temperature within as he varied its pressure. The results are given in the table. The decrease in temperature observed matches the decrease predicted so closely that Thomson was surprised. He did not expect his results to be this close!

William Thomson portrayed the episode as a great triumph. He proclaimed that the effect was not known before and praised the theory for its ability to anticipate it. We may well wonder if the melting of ice under pressure was not suspected prior to 1850. How else could one explain why ice is so slippery to stand on? Even if the effect was known prior to 1850, what could not have been known was the connection between the two odd properties of water: it expands on freezing and its melting point decreases with pressure. Thermodynamics shows the connection to us. It is no accident that water happens to have both these odd properties. For the laws of thermodynamics require that once water has one, it must have the other.

In a sense what all this tells us is that the alien story is correct in virtually all detail--excepting the part that says that the adept at thermodynamics was a visiting alien!

Morals of Our Excursion into Thermodynamics

- We have seen how a branch of science can be built on laws of great simplicity: the first and second laws of thermodynamics.

- While thermodynamics looks like abstract theory remote from practical concerns, it actually has enormous practical use. Thermodynamics maps out the boundary of what is possible in many technical and engineering areas--and these boundaries turn out to exclude far more than we would expect.

Our excursion into thermodynamics has illustrated two complementary aspects of science and scientific theorizing, both of great importance.

First we have seen how a scientific theory can be put together. Thermodynamics is built upon the foundation of a few simple laws. Once we accept the truth of these laws, the remainder of the construction of the theory is a matter of careful inference. The content of the theory becomes an unavoidable consequence. We saw this, for example, in the context of the conservation of energy. We start with a few simple assertions about machines; for example, perpetual motion machines are impossible. From these we infer to the existence of energy and its properties. Similarly, the supposition of the second law of thermodynamics leads to many consequences. Once the law is accepted, we cannot deny the remarkable results that all reversible heat engines have the same efficiency and that no other engines perform better.

Second, we see how the content of the theory has a very practical application in the real world. The law of conservation of energy places profound constraints on many processes. Why is it that we persist in our use of gasoline fueled cars, when they are the target of such bitter complaints? The answer is that gasoline is almost unique in its energy carrying and delivering abilities. The challenge of finding a replacement automobile engine is the challenge of finding competing ways of storing energy. The second law is similarly revealing. It tells us why steam locomotives have passed into history to be replaced by the diesel-electric; why air-conditioning is likely to be an expensive affair; and why traditional nuclear or coal-fired power plants discharge as much energy as waste heat as they deliver in electricity.

These two aspects support the immense importance of successful sciences such as thermodynamics. On the one hand they supply us a huge body of organized results that cannot be denied; on the other these results penetrate deeply into matters of immediate concern in our practical lives.

Assignment 8: Applications of the Theory of Heat Engines

1 Many people notice just how much heat seems to be wasted by heat engines. For example, a coal fired powerstation will exhaust roughly 2 units of heat to the air or a cooling river for every unit of electrical energy generated. That is energy that came from burning coal and it cost the power company real money. In a gasoline car engine, there is a similar waste of heat in the exhaust gases and in the heat given off by the radiator. All this heat comes from burning expensive gasoline. One may well wonder if it is not worth the effort of trying to recover this heat and convert it into usable work. The theory of heat engines explains why such efforts will not succeed: because of the low temperature of the waste heat, it is impossible to convert it to work with an efficiency that warrants the fuss and machinery.

To see this, you will compute the efficiency of a reversible heat engine that tries to utilize the heat energy given off by a car's radiator. Assume that the heat is supplied at the temperature of $220^oF=105^oC=378^oK$ and that the heat can be exhausted to the air at $68^oC=20^oC=293^oK$,

(a) What is the efficiency of a reversible engine operating between these temperatures?

(b) From a 100 kcal of heat energy given off by the radiator, how many kcal are converted into work and how many discharged as waste heat?

Notice that the efficiency of a real engine operating under these conditions would be even less.

2 The simplest way to use electrical energy to heat a house is to put it into an electric radiator. That way one gets one unit of heat energy in the house for each unit of electrical energy consumed. It is hard to imagine that one could do better than this 100% complete conversion. Over a hundred years ago, William Thomson realized that one could do a lot better. He foreshadowed the use of what we now call "heat pumps" for heating houses. An air conditioner uses electrical work to move heat from a cooler house to the warmer outside in order to keep the house cool. A heat pump does the reverse. It uses machinery very similar to an air conditioner to move heat from the cooler outside in winter to the warmer indoors. The device is simply a heat engine running backwards. It uses Work to drive heat from a cold place at T_{cold} (the outside) to deliver it to a hot place at T_{hot} (the house).

Assume we wish to deliver warm air to the house at $120^oF=49^oC=322^oK$ and that the outside air is at $32^oF=0^oC=273^oK$.

(a) What is the ratio $\dfrac{\text{Work}}{\text{Heat delivered to house}}$ for such an engine assuming it is reversible?

Use the following form of Result III: $\dfrac{\text{Work}}{Q_{hot}} = 1 - \dfrac{T_{cold}}{T_{hot}}$ where Q_{hot}=Heat delivered to house.

(b) How much work (in kcal) is needed to deliver 100kcal of heat to the house?

(c) How much heat (in kcal) is delivered for each kcal of work?

Notice that the performance of a real heat pump will be less due to unavoidable inefficiencies in the pump. In practice, heat pumps are only practical when the temperature difference bewteen house and outdoors is not too great. When that becomes great, such as in the Northern U.S., then more electrical energy is required to pump the heat and the use of heat pumps tends to be less advantageous.

Chapter 9

Chance and Probability Theory I

Introductory Notions

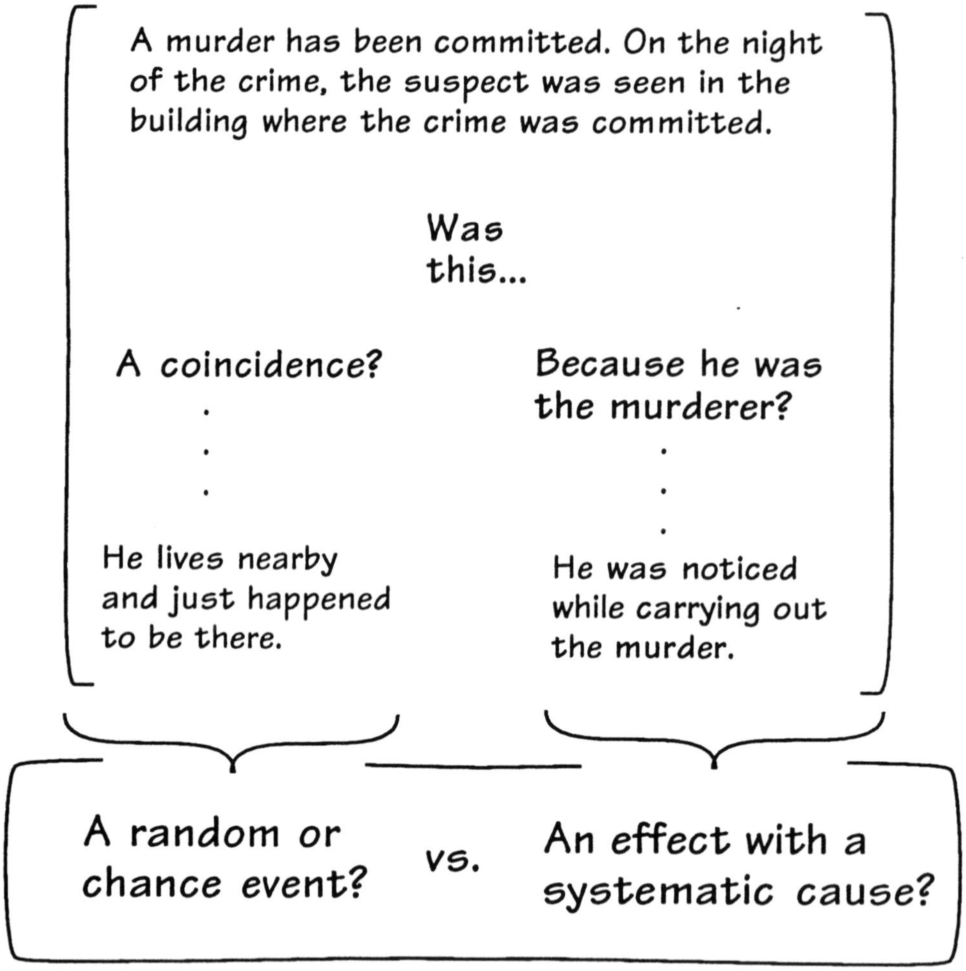

A murder has been committed. On the night of the crime, the suspect was seen in the building where the crime was committed.

Was this...

A coincidence?

He lives nearby and just happened to be there.

Because he was the murderer?

He was noticed while carrying out the murder.

A random or chance event? vs. An effect with a systematic cause?

Our excursion into thermodynamics showed us the content of science by looking in some detail at a particular scientific theory and its applications. We now turn to look at the methods of science. There are many methods and techniques used in science. Some of the best developed are the methods for dealing with randomness or chance effects. These are contained within the theories of probability and statistics.

We tend to think that a random process, such as flipping a coin, is necessarily inscrutable. This is not so. While we may not know whether the coin will come up heads or tails on the next throw, there will be a very well defined order in the long-term sequence of heads and tail thrown. Probability theory and statistics allows us to see that order and exploit it. We have already seen one example of the immense power of these methods. When Cyril Burt reported the results of his experiments on the inheritance of intelligence, we could be sure that something was wrong exactly because his results did not display the right sort of order expected amongst the randomness. A good statistician can wrestle huge amounts of information from quite slender sets of data.

A great deal of our efforts in the chapters to come will be spent in investigating whether a particular outcome is a purely random effect or the result of a systematic cause. This important distinction is quite familiar to us from everyday life. We see it in the mini-murder mystery opposite. If the effect has a systematic cause, then we can point to a definite reason for why it happened. The suspect was in the building because he was the murderer--he *had* to be there. If the effect happens by chance or randomly, then we will be unable to point to a single reason for its occurrence. To be sure, if the suspect were not the murderer, there would be all sorts of reasons for the suspect being in the building. But it was just a chance combination of these reasons that conspired to have him there. He happened to take the day off because of a mild illness. He was visiting his doctor whose offices just happen the be in the building where the murder was committed. All of this explains why he was there. But they offer no systematic explanation since they *do not* guarantee that he *had* to be there when the murder was committed.

Why do we need a theory to help us deal with randomness?

Fact 1: It is often very important to decide if an effect:

-happened randomly, by chance
or
-had a systematic cause.

* Sept. 8, 1994 — USAir Flight 427 crashes approaching Pittsburgh International Airport. All 132 passengers killed.

USAir has 5 catastrophic accidents in 5 years

Sept. 8, 94	USAir 427	Pittsburgh	132	killed
July 2, 94	USAir 1016	Charlotte	37	killed
March 22, 94	USAir 405	New York	27	killed
Feb. 1, 91	USAir 1493	Los Angeles	34	killed
Sept. 30, 89	USAir 5050	New York	2	killed

Is USAir as safe as other airlines and these are just due to chance? or Is USAir a more dangerous airline?

Let us look at the question a little more closely. Why do we need a theoretical tool to help us negotiate randomness? Two facts make it a pressing need.

First, many important decisions reduce to determining whether some outcome is simply a random occurrence or whether it is due to a systematic cause. Here are some examples.

On September 8, 1994, shortly after 7p.m., USAir flight 427 was nearing its destination of Pittsburgh International Airport. The plane suddenly rolled left and, after a 23 second dive, crashed into the ground, killing all 132 people on board. The nation was shocked by the catastrophe. Press coverage, especially in Pittsburgh, was intense. The following day the *Pittsburgh Post-Gazette* began to delve into the safety record of USAir. The headline of the article (p.A-5, Sept. 9) read "USAir crash its fifth since 1989" and a subheading provided readers what seemed to be a telling expert quote: "USAir 'has had a lot more crashes lately per thousand flights than other major U.S. airlines.'--Arnold I. Barnett" No doubt many reader who just scanned these headlines formed the impression that the record of five crashes in five years was strong evidence for some sort of unreliability in the airline compared to others. It was evidence for a systematic cause of the record of crashes. But such readers would have prejudged the issue. In the body of the article, Barnett, an M.I.T. professor in operations research and statistics, reported that analyses of *earlier* USAir crashes "found that the airline's performance could plausibly be attributed to bad luck" and that "he did not know whether that analysis could apply to yesterday's accident." This uncertainty persisted. Days later, statisticians were still debating the significance of USAir's record ("Statisticians differ on USAir safety record," *Pittsburgh Post-Gazette*, Sept. 13, 1994, p.B-11). An extensive safety audit eventually exonerated USAir ("Audit deems USAir safe, revises procedures," *Pittsburgh Post-Gazette*, Mar. 18, 1995, p.A-1).

Does USAir's record of five crashes in five years reveal an intrinsically unsafe airline? Or is it just a bout of bad luck, such as could afflict any airline? These questions are of paramount concern. They amount to a decision between a random effect or one with a systematic cause.

* You live in a town with 5,000 people. The national average for deaths by all forms of cancer is 2 per 1,000 each year.

You expect 10 cancer deaths each year on average. — 2 per 1,000 scaled up to 5,000

But you have:

... 10, 11, 12, 13, 14, 15, 16, 17, 18, 19, 20 ...

Small deviations from 10 cases expected compatible with CHANCE fluctuations.

Where do we divide?

With larger deviations, you would suspect a special cause, e.g. factory spilling carcinogens.

* The summer of 1988 was exceptionally hot. ⇨ Chance fluctuation or evidence of global warming ?

The winter of 1993/94 was exceptionally cold. ⇨ Chance fluctuation or evidence of a coming ice age ?

The national average for deaths from cancer each year is 2 in 1,000. So, in your town of 5,000 people, you expect approximately 2x5=10 deaths from cancer each year. Imagine that in some year there are 10 cancer deaths. That would not surprise you. Of course you may not get exactly ten cancer deaths. It may be 8 or 9; or it may be 11 or 12. These, you would probably say, are just chance fluctuations around the 10 expected--just as a fair coin tossed 20 times will not always show exactly 10 heads. But what if the death rate were 15 or 16? Or if it were 20 or 25? At some point you will want to say that these higher numbers could not arise by chance fluctuations. At some point you would want to insist that the cases of cancer in your town are due in part to an extra cause that does not operate nation-wide. (You may suspect the factory upstream that spills carcinogens into the water.) Your crucial decision in a vital matter of public health is when to sound the alarm--at 15 deaths? At 20 deaths? When? That decision amounts to distinguishing a random effect from one with a systematic cause.

The summer of 1988 was unforgettable for anyone who lived through it. Many cities suffered the highest temperatures ever recorded, such as the 103°F visited upon Pittsburgh in July. There was much talk that these high temperatures were a result of global warming, a conjectured slow increase in the earth's surface temperature that would eventually bring environmental disaster. But was this a fair assessment? Every year summers vary in temperature. Was this high temperature just a part of such random variation? Or should we seek a systematic cause in global warming and take it as a warning of impending disaster? As if to test our statistical acumen, 1994-95 supplied us a fiercely cold winter with the temperature in Pittsburgh in January reaching the lowest ever recorded, -22°F. Was this another random fluctuation? Or was it evidence of a coming ice age?

In scientific studies, deciding between the two types can be just as important. Imagine a controlled study of some drug. If just slightly more people who were treated with the drug recovered than those who took a placebo, is the higher number just a random fluctuation? Or is it evidence of the effectiveness of the drug? We decide the effectiveness of the drug by deciding whether the study outcome exhibits a random fluctuation or the result of a systematic cause at work.

Fact II Our intuitive judgments of what happens by chance are notoriously unreliable.

* Shared Birthdays:

There are 365 days in a year. What is the chance that at least 2 people in a group have the same birthday if the group is:

	Chance	— "High, Low" "1 in 2" "1 in 10" "0.01 etc...
Size of this class		
25 people		
50 people		
100 people		
200 people		
367 people		

↑ Write your guess here.

There were two facts that made it important to have a theoretical tool to aid us in analyzing randomness. The second fact is about us. We normally evaluate chance phenomena by impressions, hunches, intuition and guesses. It turns out that our evaluations are notoriously unreliable. Because judging randomness well can be very important, we are taking a great risk if we let our hunches decide. We can get a clear sense of this unreliability if we look at a few very simple systems involving randomness. Even though they are simple and the nature of the randomness involved is quite clear, virtually everyone misappraises them.

Consider two randomly chosen people. What is the chance that they have the same birthday? That is easy to estimate. The first person will have a birthday on one of the 365 days of the year, say January 20. The other person will also have a birthday on one of the 365 days of the year. But only one of these 365 will coincide with the birthday of the first person. So there is a 1 in 365 chance that the two share a birthday. This is a rather small chance--less than 0.3%

Now imagine that a third person joins the group. What is the chance that at least two of these three people share a birthday. Even if the first two do not, then the third's birthday may coincide with one of the first two. Of the 365 days available for the third person's birthday, 2 would now lead to a shared birthday in the group. A 2 in 365 chance is still pretty small.

Imagine that we keep adding people to the group. Eventually we can be sure that at least two people share a birthday. In the worst case possible, we could somehow assemble 366 people none of whom share a birthday. (Note that 366 is needed to allow for February 29!) But then, when we add the 367th person, no dates are left. That person's birthday must coincide with another's in the group.

So, somewhere between groups of 2 and of 367, the chance of at least two people sharing a birthday rises from rather unlikely to certain. The question for you to ponder is this: at what point does it start to become rather likely that at least two people will share a birthday? e.g. at what point do the chances hit 1 in 2 so that it is as likely as not? Fill in the squares opposite as best you can, assessing the probability in whatever scale you feel most comfortable, e.g. "high probability," "about 1 in 2," etc.

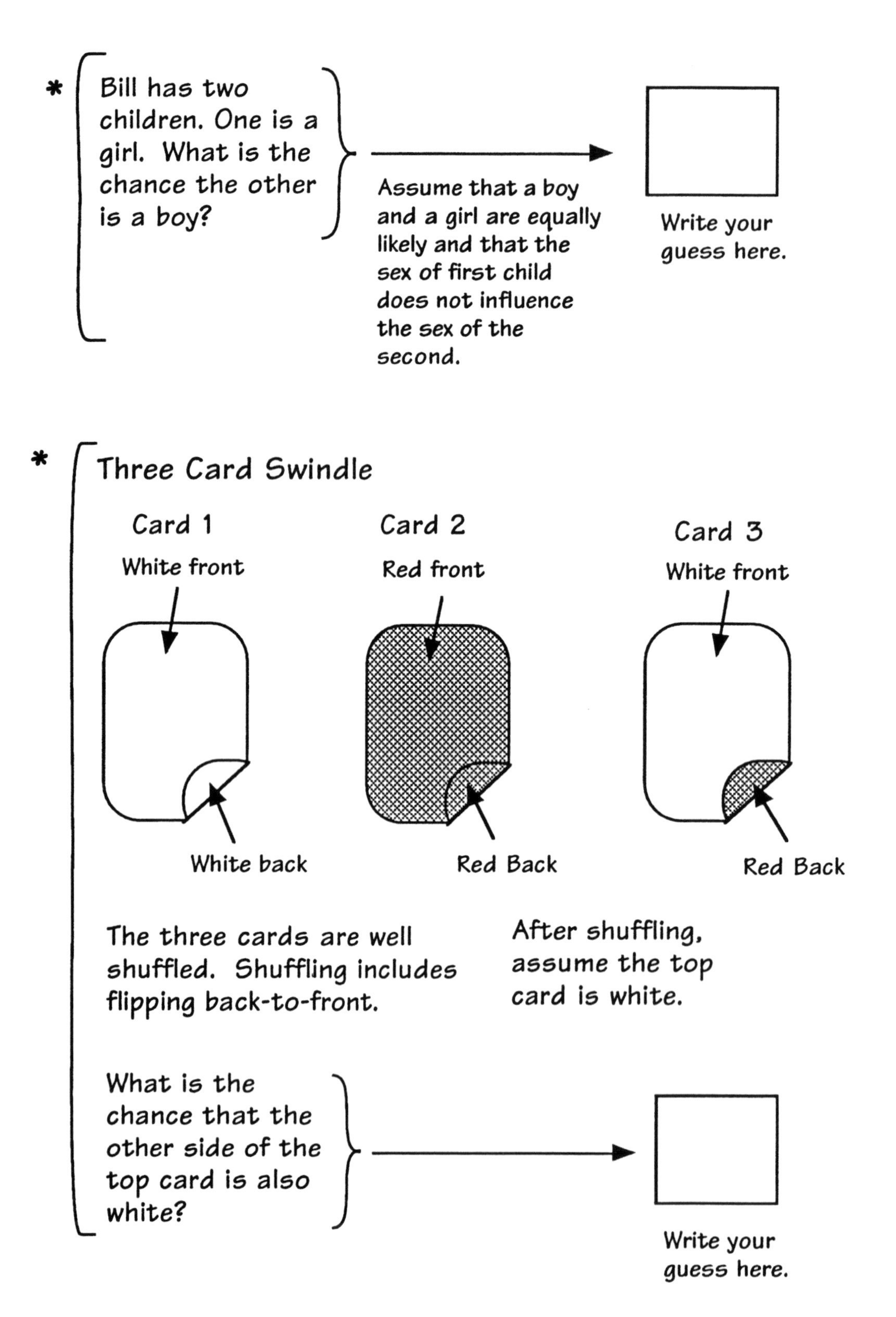
*
Bill has two children. One is a girl. What is the chance the other is a boy?
Assume that a boy and a girl are equally likely and that the sex of first child does not influence the sex of the second.
Write your guess here.
*
Three Card Swindle
Card 1
White front
White back
Card 2
Red front
Red Back
Card 3
White front
Red Back
The three cards are well shuffled. Shuffling includes flipping back-to-front.
After shuffling, assume the top card is white.
What is the chance that the other side of the top card is also white?
Write your guess here.

In this diabolical puzzle, you are told that Bill has two children and that one of the them is a *girl*. You are asked, what is the chance that the other child is a *boy*? You are allowed to make the obvious assumptions: you can assume that a boy child and a girl child are equally likely and that the sexes of the two children are independent--that is, whatever may be the sex of the first child does not influence the sex of the second. (e.g. if the first is a boy, then that does not make it any more or any less likely that the second will be a boy).

The three card swindle is a fine way to win money off your friends--although they may not remain friends if they find out the trick! In the swindle, one has three cards as illustrated: one is white on both sides; one is red on both sides; and the last one is white on one side and red on the other. The three cards are shuffled well. The shuffling includes turning the cards over many times individually so that either face of each card may be on top with equal chance when the shuffling is over. When the shuffling is complete, the squared deck is place on the table. The top face showing will either be white or red. Let us assume that it happens to be white. What is the chance that the other side of this top card is also white? (The answer is not what most people expect. So, once you know the right answer, you can easily induce them to accept unfavorable bets.)

Basic Notions of Probability Theory

Outcome space = set of all possible outcomes

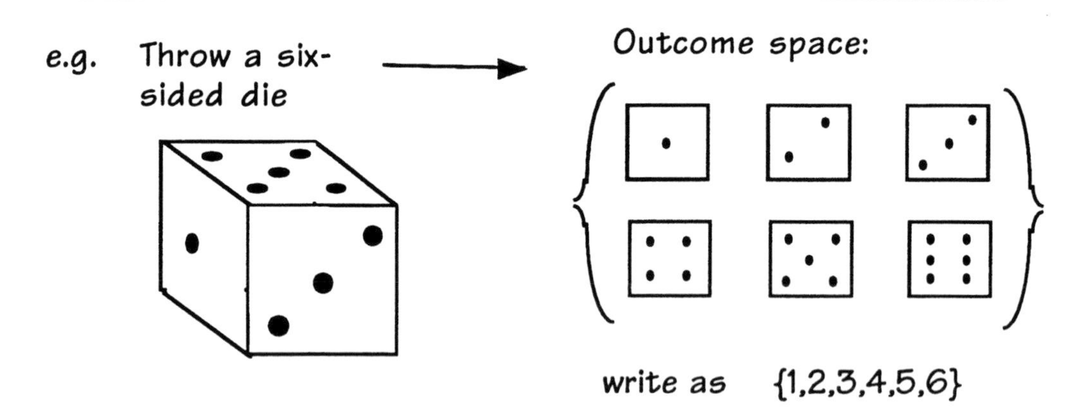

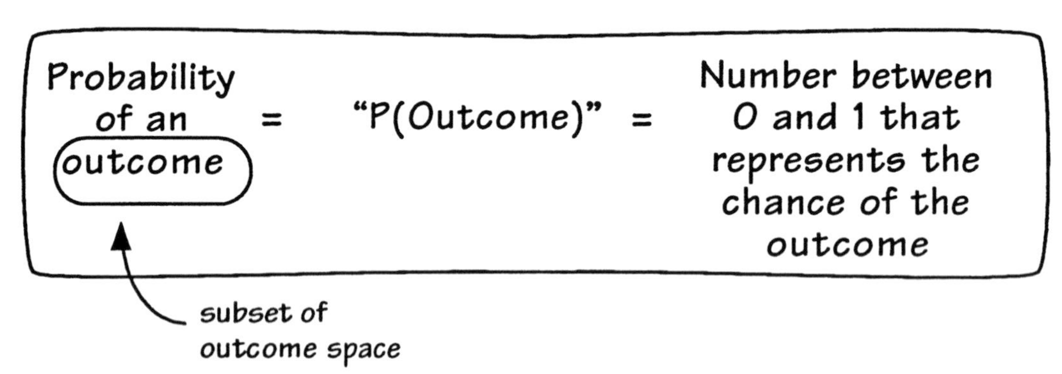

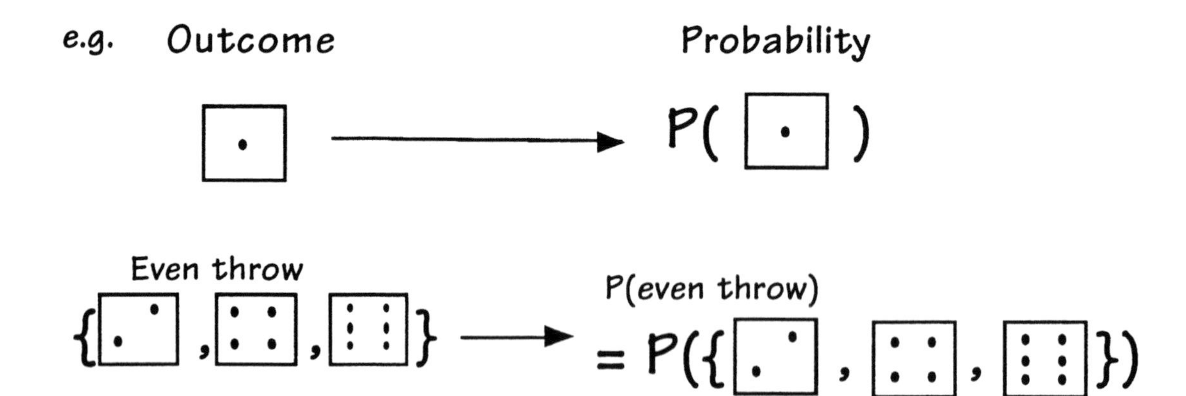

Probability theory is the theory of chance; it provides the framework within which modern statistical analysis proceeds. We begin our study of the theory with its central notions. The first is the notion of an "outcome space". This is just the set of all possible outcomes for some random process. For example, our random process may be the throwing of a six-sided die. The possible outcomes are that a 1, 2, 3, 4, 5 or 6 may show. This is the outcome space.

Note that there is a special way of writing the outcome set. You do NOT write it as 1 2 3 4 5 6. The outcome space is a set and sets are written using braces ("curly brackets"), which look like { }. The outcome set is always written as {1, 2, 3, 4, 5, 6} with the members of the set separated by commas. Writing the outcome set correctly is simply part of good grammar in probability theory. You will get away with writing it as 1 2 3 4 5 6. But it will not be correct. It would be similar to an error of English grammar. You intend to say "The cat sat on the mat." If, instead you say "Cat sat mat." you will be understood, but your listeners will wonder about your command of English!

An outcome is a subset of the outcome space. For example, the die may show a 1 after it is thrown. This outcome is a subset by definition and therefore also a set, so it should be written as {1}. In this special case in which there is only one member, the set is often written in the abbreviated form 1, that is, without the { }, although this is not strictly correct. (Correspondingly in English we may abbreviate "I am" to "I'm".) There are other more complicated outcomes. For example, the die may show an even face, that is, one of 2, 4 or 6--but we may not care or know which. This outcome is written as {2, 4, 6}.

The theory assigns a number to each outcome, called its "probability." This number will lie in the range of 0 to 1 (including 0 and 1). It measures the chance of the outcome happening. Notice again that there is a correct grammar to writing down the probability of an outcome. It is written as P(outcome), that is an upper-case P followed by the outcome enclosed in parentheses. So the outcome of an even throw is written as P({2,4,6}). Note that we have both ( ) and { }. The ( ) belong to the P and the { } belong to the outcome. As a shortcut we may choose to label the outcome {2, 4, 6} by "even throw". Then a correct, alternative way of writing the probability is P(even throw).

P (outcome)	=	1	Certain
P (outcome)	=	0.99	Near certainty
P (outcome)	= 0.5 =	1/2	As likely as not
P (outcome)	=	0.01	Near impossible
P (outcome)	=	0	Impossible

Sharpen vagueness of "near certainty" *etc.*

Connection to frequencies

P(outcome of die throw is 3) = 1/6 → If die thrown many times, frequency of ⚂ is around 1/6.

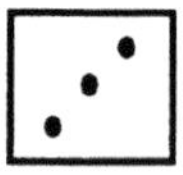

Outcome of many repeated trials = 4 ,1, 1, 3, 6, 5, 4, 2, ...

In the long run, 3 appears with frequency 1/6 = .1666

The probability of an outcome is a number between 0 and 1 inclusive that represents the chance of the outcome happening. The table opposite gives some idea of what the various values represent. The extreme value 1 represents the case in which the outcome is certain to happen. The other extreme value of 0 represents cases that are impossible. The mid point, 0.5 = 1/2, of the scale represents outcomes that are as likely to happen as not. For example, the probability of tossing a head with a fair coin is 0.5. You can interpolate to get a sense of what the other values mean. As the probability drops from 1, we have cases of high likelihood. A probability of 0.99 is nearly certain; a probability of 0.95 is still very likely, but not quite so. Correspondingly, probabilities near zero represent unlikely outcomes.

Notice the correct grammar for assigning these numbers. If the probability of a head is 0.5, you do NOT write "head = 0.5" You would probably be understood; but it looks as bad as saying "Cat sat mat." The correct way of writing this is P(head) = 0.5, or P(head) is 0.5.

So far, there is a messy vagueness in our explanation of the meaning of the numerical values. If a probability of 0.99 represents near certainty, what exactly do we mean by "near certainty"? This vagueness is eliminated by the connecting the notions of probability and frequency.

If the probability of throwing a 3 with a die is 1/6, then, if the die were thrown many, many times, then we would see the 3 in roughly one in six of the outcomes, that is, we would see it with a frequency of 1/6 = 0.1666. So, the probability of an outcome fixes its long term frequency. Thus, if an outcome has a probability of 0.99, then, if we run many, many trials, it would occur in 99 out of 100 trials in the long run, that is, with a frequency of 0.99. This gives a clearer meaning to "near certainty" and lets us distinguish it from the lesser certainty of a probability of 0.95.

Beware! These are the *long term* frequencies. In any short number of trials, the frequency of the outcome may differ from the probability. In six trials, we may get the 3 showing once as we expect; or it may not appear at all; or it may appear twice. In 60 trials, we may just get the ten 3s we expect; or we may get only 8 or 9 or 11 or 12. As the number of trials gets larger, however, the more closely the frequency of 3s will approach the 1/6 that we expect.

The case of equally probable ("equiprobable") outcomes

All the individual outcomes are equally probable. ⇒ then

$$P(\text{individual outcome}) = \frac{1}{\text{number of individual outcomes}}$$

Throw a six-sided die:
Outcome space = {⚀, ⚁, ⚂, ⚃, ⚄, ⚅}

6 equiprobable outcomes.

⇒ P(⚀) = P(⚁) = ... = P(⚅) = $\frac{1}{6}$

Toss a fair coin:
Outcome space = {head, tail}

2 equiprobable outcomes.

⇒ P(head) = P(tail) = $\frac{1}{2}$

BUT

Predict tomorrow's weather?
Outcome space = {rain, no rain}

NOT necessarily equiprobable.

⇒ P(rain), P(no rain) could be anything.

e.g. P(rain) = 0.3
P(no rain) = 0.7

In practice it is not easy to determine the probabilities of outcomes. Often much effort is put into the job and the results remain estimates. Consider, for example, the effort that has gone into determining the probability that you will contract various disease if you are or are not a smoker. We have a strong idea of trends and know roughly the ranges of the probabilities, but the exact values remain elusive.

There is one special case in which the probabilities are easy to determine. This is the case in which all the individual outcomes are equally probable or, in the common and useful abbreviated form, "equiprobable." In that case the probability of each individual outcome is 1/(number of individual outcomes).

We have already seen one example of this. There is an equal probability that you will get any of the six outcome 1, 2, 3, 4, 5 or 6 when a fair die is thrown. Therefore the probability of each of these outcomes is 1/6. Similarly, if you toss a fair coin, you may get a head or a tail. These two outcomes are equiprobable. Therefore the probability of a head is 1/2; and the probability of a tail is 1/2.

The real work in determining these probabilities comes in deciding that the individual outcomes are equiprobable. In the two cases above, we are able to say that they are because we know a lot about the physical circumstances that lead to the outcome. Indeed those circumstance--the energetic roll of the die or flip of the coin--are especially designed to guarantee equiprobable outcomes. Without these special circumstances, we could not judge the outcomes equiprobable. If I tell you there is a coin on my table and I ask you the probability that it has heads uppermost, you cannot say. The coin may have arrived there by a randomizing flip in which case the probability is 1/2; or I may have an obsession with keeping all coins on my table heads up, in which case the probability will be very close to 1.

Thus, that there are only a few outcomes does not make them equiprobable. For example, if you consider tomorrow's weather. The outcomes are that it might rain or it might not. We have no grounds for expecting these outcomes to be equiprobable. Thus we cannot expect them to 1/2. Indeed, in general, they are not. Anyone who listens to weather forecasts knows they can vary throughout the range of 0 to 1 (or, as they say in the forecasts, from 0% to 100%).

OR Rule

What is probability of throwing "1 OR 2"?

Outcomes of many repeated trials: = 4, 1, 1, 3, 6, 3, 2, 5, 6, 3, 3, 2, ...

frequency of 1 = 1/6 | frequency of 2 = 1/6 | frequency of 1 OR 2 = 1/6 + 1/6 = 1/3

P(1 OR 2) = P(1) + P(2)

generalize

$$P(\text{Outcome A OR Outcome B}) = P(\text{Outcome A}) + P(\text{Outcome B})$$

for *mutually exclusive* outcomes A and B.

This means that AT MOST ONE of outcomes A and B can happen. } IMPORTANT !

Examples

P(even throw) = P(2 OR 4 OR 6) = P(2) + P(4) + P(6)
= 1/6 + 1/6 + 1/6 = 3/6 = 1/2

Use OR rule since 2, 4, 6 are mutually exclusive.

P(5 OR even throw) = P(5) + P(even throw)
= 1/6 + 1/2 = 1/6 + 3/6 = 4/6 = 2/3

Use OR rule since 5, even throw mutually exclusive.

BUT

P(2 OR even throw) = ?

We cannot use the OR rule, since 2, even throw are NOT mutually exclusive. Both can happen if 2 is thrown.

A great deal of work with probabilities involves computing new probabilities from others we already know. This is done using a number of rules. The first we shall look at is the OR rule. If we know the probabilities of two outcomes A and B, it lets us compute the probability of a new outcome A OR B--that is the outcome in which one of A or B happen. The rule is very simple: the probability of the outcome A OR B is just the sum of the probabilities of two outcomes A, B separately.

The examples show how to use the rule. Consider throwing an even number with a six-sided die. This outcome is really the outcome 2 OR 4 OR 6. We know the probabilities of throwing a 2, 4 or 6 separately, they are 1/6 in each case. So the probability of an even throw is just the sum of these three probabilities 1/6 + 1/6 + 1/6 = 1/2. Another example is the probability of throwing a 5 or an even number. The probabilities for each separate outcome are 1/6 and 1/2. We just add these up to get the probability of 5 OR even, 1/6 + 1/2 = 2/3.

It is very important to remember that the OR rule can only be used in a special case. *The two outcomes A and B have to be mutually exclusive.* That means that at most one of them can happen. This condition is satisfied by the two examples we have looked at. In the case of 2 OR 4 OR 6, at most one of these can happen. If you throw a 2, then you cannot also have thrown a 4 or a 6! Thus the OR rule can be applied. Similarly, the outcomes 5 and even are mutually exclusive: if you throw a 5 then the throw cannot also be even, and *vice versa*. So the OR rule can be applied.

But mutual exclusivity fails in the case of 2 OR even. For you can throw both a 2 and even number at the same time. That happens when you throw a 2! The OR cannot be applied to P(2 OR even).

It is easy to see why the OR rule works if we recall the connection between frequencies and probability. If a die is thrown many times, we will see 1s and 2s each with a frequency of about 1/6. Since the outcome of a 1 and 2 are mutually exclusive, we will see an outcome that is either 1 OR 2 with a frequency of 2/6 = 1/3. Since these frequencies approach the corresponding probabilities in the long run, it follows that we can find P(1 OR 2) by adding P(1) and P(2), just as we added the corresponding frequencies. If we generalize this result to other cases, we have the OR rule.

Everything Rule

Entire outcome space of die throw

P(1 OR 2 OR 3 OR 4 OR 5 OR 6)
= P(1) + P(2) + P(3) + P(4) + P(5) + P (6)
= 1/6 + 1/6 + 1/6 + 1/6 + 1/6 + 1/6 = 6/6 = 1

generalize

$$P(\text{entire outcome space}) = 1$$

NOT Rule

P(NOT-6)= P(1 OR 2 OR 3 OR 4 OR 5)
= P(1) + P(2) + P(3) + P(4) + P(5)
= 1/6 + 1/6 + 1/6 + 1/6 + 1/6 = 5/6

same as 1 - 1/6 = 1 - P(6)

generalize

$$P(\text{NOT-outcome A}) = 1 - P(\text{outcome A})$$

Example

Given: P(rain) = 0.3

Use NOT rule to compute:
P(NOT-rain) = 1 - P(rain) = 1 - 0.3 = 0.7

Check:
P(entire outcome space)
= P(rain OR NOT-rain) = P(rain) + P(NOT-rain) = 0.3 + 0.7 = 1

As expected.

Use OR rule since rain and NOT-rain are mutually exclusive.

The next rule lets us compute the probability of the entire outcome space. In the case of a die throw, the entire outcome space corresponds to the outcome 1 OR 2 OR 3 OR 4 OR 5 OR 6. Its probability is 1/6+1/6+1/6+1/6+1/6+1/6 = 1. So the probability of the entire outcome space is 1. This holds generally. Whenever we set up an outcome space, the probability of the entire outcome space will be 1. I have called this the Everything Rule, because the outcome space contains everything that can happen, as far as the probabilistic calculation is concerned.

For every outcome, there is another that says it does not happen. When we toss a coin, it may come up heads. The outcome NOT-heads, says it will not come up heads. In this case NOT-heads is just the same as tails. With a die, the outcome NOT-6 is the outcome in which 6 does not happen. This corresponds to the outcome 1 OR 2 OR 3 OR 4 OR 5. Applying the OR rule we find that P(NOT-6) = 1 - P(6). This simple connection between the probabilities of outcome 6 and NOT-6 holds generally. It is the NOT rule shown opposite.

An example shows us how we can tie these rules together. Let us say that we know that the probability of rain tomorrow is 0.3. That is, P(rain) = 0.3. We use the NOT rule to compute the probability of no rain, that is the outcome NOT-rain and find P(NOT-rain) = 0.7. We can now check that these two probabilities are are compatible with the other two rules, the OR rule and the Everything rule. The Everything rule tells us that the probability of the entire outcome space is 1. In this case, the outcome space is {rain, NOT-rain}, so the Everything rule applied to P(rain OR NOT-rain). We can apply the OR rule to rain OR NOT-rain, since the two outcomes are mutually exclusive. If there is rain, we do not count it as a no-rain day; if there is no rain, we do not count it as a rain day. Applying the OR rule, we compute P(rain OR NOT-rain), using the two values P(rain) = 0.3 and P(NOT-rain) = 0.7, and find the final value 1, which is what the Everything rule told us to expect.

AND Rule

Throw 2 dice.

Outcome space = { ⚀⚀ , ⚀⚁ , ... , ⚄⚅ , ⚅⚅ }

(first die, second die)

P (⚀⚀) = P (⚀ on first die AND ⚀ on second die)

= 1/6 x 1/6 = 1/36

(1/6 = P(⚀ on 1st die); 1/6 = P(⚀ on 2nd die))

generalize ⇩

$$P(\text{Outcome A AND Outcome B}) = P(\text{Outcome A}) \times P(\text{Outcome A})$$

for *independent* outcomes A and B

This means that whether A happens does not affect the probability of B (and *vice versa*). } IMPORTANT!

Examples

Throw two dice:

P(even throw first die AND even throw second die) = P(even throw first die) x P(even throw second die)

= 1/2 x 1/2 = 1/4

We can use the AND rule since the outcomes are independent. The outcome of the first die throw does not affect the probabilities for the second die throw.

BUT

Throw one die:

P(2 on die AND 3 on same die) = ?

AND rule cannot be used since the outcomes are not independent. Throwing a 2 reduces the probability of 3 from 1/6 to 0!

The AND rule allows us to compute the probability of an outcome formed by ANDing together two others. The rule simply says that we find the probability of outcome A AND outcome B by multiplying the separate probabilities of outcome A and outcome B.

Generally, it is hard to apply to simple outcome spaces like those associated with a single die throw. We need more complicated cases, such as the throwing of two dice. In this case, the example shows its use. What is the probability of getting an even throw on the first die AND and even throw on the second? The probability of an even throw on the first is 1/2; the probability of an even throw on the second is 1/2; therefore the probability of an even throw on the first AND and even throw on the second is 1/2x1/2 = 1/4.

Notice that the outcome space for two throws is a great deal more complicated. With one die we need only consider the six faces that may show. With two dice, we need to consider the six faces on each and all possible combinations. So the outcome space is built up of a big collection of individual outcomes: 1-on-the-first-die-AND-1-on-the-second-die, 1-on-the-first-die-AND-2-on-the-second-die and so on through all 6x6=36 combinations.

It is very important to remember that the AND rule can only be used in a special case. *The two outcomes A and B have to be independent.* That means that whatever happens with the first outcome must not affect the probabilities of the second and *vice versa.* This restriction is illustrated in the examples opposite. With two dice being thrown, whatever the first die may show will not affect the probabilities of the outcomes on the second die. If the first die shows an even number or if it fails to, in both cases the probability of an even number on the second die is 1/2. The outcomes are independent; the AND rule may be used. This is not so if we throw just one die. Consider getting a 2 AND getting a 3 on the same die. If we succeed in getting a 2, then this outcome forces a change in the probability of getting a 3. It is now impossible to get a 3; its probability drops from 1/6 to 0. The outcomes are not independent; the AND rule cannot be used.

It can be helpful to think of the AND rule in terms of frequencies. If we throw two dice, what is the probability of a 1 on both dice? 1/6th of the time the first die shows a 1; 1/6th of the time the second also shows a 1; so 1/6x1/6=1/36th of the time they both show a 1.

Further Optional Material

Generalized OR Rule

For outcomes that may not be mutually exclusive

P (Outcome A OR Outcome B) =
P (Outcome A) + P(Outcome B)
- P(Outcome A AND Outcome B)

Example
One die throw

even = (2,4,6}

P(2 OR even) = P(2) + P(even) - P(2 AND even)

(where P(2) = 1/6, P(even) = 1/2, P(2 AND even) same as P(2)= 1/6)

= 1/6 + 1/2 - 1/6 = 1/2

This is just what we expect since the outcome (2 OR even) is just the same as even and P(even) = 1/2.

Conditional Probability

P(Outcome A | Outcome B) = Probability of Outcome A given that Outcome B occurs

Example
One die throw P(even throw | low throw) = 1/3

even = {2,4,6}
low = {1,2,3}

Generalized AND Rule

For outcomes that may not be independent

P(Outcome A AND Outcome B)
= P(Outcome A | Outcome B) x P(Outcome B)

Example
One die throw

P(even AND low)
= P(even | low) x P(low) =1/3 x 1/2 =1/6

This is what we expect since the outcome (even AND low) is just the same as the outcome 2 and P(2) = 1/6.

The rules we have seen so far are sufficient for all the calculations of the chapters that follow. But they are not sufficient for many other problems. The OR and AND rules generalize to the more powerful rules shown opposite. These general rules will not be needed later.

In cases in which two outcomes, A and B, are not mutually exclusive, the OR rule can be used in the generalized form shown, in which an extra term, P(Outcome A AND Outcome B) is subtracted from the right hand side. This extra term accounts for any lack of mutual exclusivity in the outcomes. The example shows its use for the case of (2 OR even) for a die throw. These outcomes are not mutually exclusive. One can throw a 2 and still have an even throw. We expect the probability of (2 OR even) to be the same as the probability of even, since the outcome (2 OR even) is the same as the outcome even--and that is what happens.

In applying the generalized OR rule, we need to know P(2 AND even). In the example, it was found by noting that outcome (2 AND even) is just the same as outcome 2, so the probability P(2 AND even) is the same as P(2)=1/6. We could not use the AND rule to compute this probability since the outcomes 2 and even are not independent. Whether 2 happens does affect P(even). For example, if 2 does happen, then the probability of P(even) jumps from 1/2 to 1 (certainty), since 2 is an even outcome.

To formulate a generalized AND rule which can tolerate outcomes that are not independent, we need the notion of a conditional probability as shown opposite. In the example, P(even | low) is the probability that the outcome is even given that we know it is low. Now the outcome is low if it is in {1,2,3}. Only one of these three equiprobable outcomes is an even number. So we can see that the probability of an even outcome, given that it is low, is 1/3. We write: P(even | low) = 1/3.

The generalized AND rule replaces the P(Outcome A) on the right hand side by the conditional probability P(Outcome A | Outcome B). This rule can be applied to P(even AND low), whose two outcomes are not independent. The result, as shown opposite, is compatible with our expectation that this probability is the same as P(2).

Of course, the generalized OR rule works for mutually exclusive outcomes and the generalized AND rule for independent outcomes.

Answers

Shared Birthdays	
Number of people in group	Probability that at least 2 share birthday
2	.0027
10	.117
23	.507
25	.569
36	.832
40	.891
50	.970
57	.990
70	.9991
80	.99991
89	.999992
90 and over	Very close to 1
367	1

See table of the Presidents of the U.S. at end of chapter.

Among 36 death dates: Adams, Monroe, Jefferson share July 4. (36: .832)

Among 40 birthdays: Polk, Harding share Nov.2 (40: .891)

How big must a group of people be before we are pretty sure that at least two will share a birthday? We know the answer lies somewhere between 2 and 367. With 2 people the chance is pretty small--1/365 or 1 in 365.25 if we allow for leap years with February 29. By the time we have 367 in the group, we are sure that at least two must share a birthday, even allowing for the February 29 of leap years. That is, the probability has risen to 1.

Most people guess that we need to have a group with one or two hundred people before there is a 50% chance (i.e. probability of 0.5) that at least two people will share a birthday. This seems to rely on the fact that one or two hundred is about half way between 1 and 367. The actual number is much lower as the table shows. By the time there are merely 23 people in the group, the probability has just exceeded 0.5. By the time there are fifty people in the group, the probability has risen to 0.97. That is a high probability. Think about what it means. If you have a group of 50 people thrown together randomly, 97 times out of 100, at least two of those people will share a birthday. In other words, you will find at least two people with a shared birthday almost always.

By the time the size of the group has reached 100, the probability that at least two share a birthday is so close to 1--certainty--that the difference is hard to discern.

As a test of these results, consider the birth and death days of the first 40 presidents. For 40 people, there is a 0.891 chance that at least two share a birthday, so the outcome is quite likely. And it did happen: Polk and Harding share November 2. For the 36 death dates in the table, there is a probability of 0.832 that at least two share the same death date; again reasonably likely. And it did happen, with *three* presidents: Adams, Monroe and Jefferson all share July 4.

Answers

Calculation of probabilities for shared birthdays.

Take the case of the presidents.
The first president (Washington) was born February 22.

$$P\left(\text{2nd president does not share 1st president's birthday}\right) = \frac{364}{365}$$

The 2nd president can be born on any of the remaining 364 days. Assume that each day is equally probable.

$$P\left(\text{3rd president does not share 1st \& 2nd president's birthday} \;\middle|\; \text{2nd president does not share 1st president's birthday}\right) = \frac{363}{365}$$

The 3rd president can be born on any of the remaining 363 days.

⋮

$$P\left(\text{23rd president does not share 1st - 22nd president's birthday} \;\middle|\; \text{1st - 22nd president does not share 1st president's birthday}\right) = \frac{365 - 22}{365} = \frac{343}{365}$$

Combine using generalized AND rule.

$$P\left(\text{1st - 23rd president does not share 1st president's birthday}\right) = \frac{364}{365} \times \frac{363}{365} \times \ldots \times \frac{343}{365} = 0.493$$

Recover the probability we want using the NOT rule.

$$P\left(\text{some of presidents 1st - 23rd share birthday}\right) = 1 - P\left(\text{1st -23rd president do not share birthday}\right) = 1 - 0.493 = 0.507$$

The other values in the table are computed similarly.

Why does the probability of a shared birthday rise so quickly as we add people to the group? The simple answer is that this is what we find when we compute the probabilities using standard methods. The calculation is shown opposite. It is given for the case of 23 presidents and shows the probability is 0.507. (Other entries in the earlier table are computed similarly.) This calculation requires the use of the conditional probabilities and the generalized AND rule, so review it only if you care to. We will not need this material later.

In any case, the calculation gives the result but does not really explain why the probability rises so quickly. *That* can be explained more easily. As the group grows in size, what matters is not the number of people, but the number of *pairs* of people that can be formed. Each pair represents one chance for a shared birthday and there is a probability of 1/365 that a pair will share a birthday. As the number of people in the group increases, the number of pairs increases and it does so very quickly. If there are two people, there is only one pair, ①-②. If there are three people, there are three pairs: ①-②, ②-③ and ①-③. By the time there are 5 people, there are 10 pairs; by the time there are 23 people, there are 253 pairs.

Number of people	①–②	①, ②, ③ (all joined)	①, ②, ③, ④ (all joined)	①, ②, ③, ④, ⑤ (all joined)	23	50
Number of pairs	2	3	6	10	253	1225
Probability that at least two share birthday	.0027	.0082	.0164	.117	.507	.970

If each pair represents a 1/365 chance of a shared birthday, then the 253 pairs of a group of 23 people represents many opportunities for a shared birthday. It is no longer so surprising that the probability of at least one shared birthday passes 0.5 with a group of this size. By the time the group has 50 people, there are 1225 pairs, each representing a 1/365 chance of a shared birthday. So it is not at all surprising that the probability of at least one shared birthday in the group has risen to a very high level. It is not surprising and this review helps us understand how the probabilities arise--but this kind of rough analysis cannot replace the full calculation shown opposite.

Answers

Bill's two children

ONE child is a girl. ⇨ The probability that the OTHER is a boy is 2/3.

FIRST child is a girl. ⇨ The probability that the SECOND is a boy is 1/2.

Three card swindle

The probability that the other side is also white is 2/3.

The cards:

	Card 1	Card 2	Card 3
Front	white	red	white
Back	white	red	red

Outcome Space = { Card 1 Front up (white),
Card 1 Back up (white),
Card 3 Front up (white),
Card 3 Back up (red),
Card 2 Front up (red),
Card 2 Back up (red) }

(The first three: These three equally probable outcomes correspond to a white face showing.)

In TWO of THREE cases, the other side is white too. Therefore, P(other side white)= 2/3

In the case of Bill's children, the probability that the other child is a boy is 2/3. Most people expect it to be 1/2. Why it is 2/3 is seen most clearly in a short calculation of the probabilities. That is set as an exercise at the end of the chapter. However, we can get an idea of why the result is 2/3 and not 1/2 if we notice that there are two distinct cases. If we knew the first child was a girl, then there would be two equally probable options left; the second child is a girl or the second child is a boy. This leaves a probability of 1/2 that the second is a boy. In our case, we know only that one child is a girl. We do not know whether it is the first or the second child. This leaves more possibilities and they lead to the probability of 2/3 that the other child is a boy.

In the case of the three card swindle, if the top card shows a white face, most people suppose there is a probability of 1/2 that its hidden face is also white. The correct answer is 2/3. This difference between the real and generally supposed probability allows the swindle to operate. The swindler bets $10 on a white hidden face; the sucker bets $10 on a red hidden face. The sucker thinks it a fair bet since the sucker mistakenly believes there is an equal chance of a white or red hidden face. But 2/3rd of the time the swindler will win and 1/3rd of the time lose--an average gain of $3.33 with each bet.

It is not hard to see why the probability of a hidden white face is 2/3. Consider the outcome space as shown opposite. Altogether there are six equally probable outcomes. Three of these correspond to the case that interests us: a white face showing. Now, in two of these three outcomes, the hidden face is white. Since these three outcomes are equiprobable, it follows that the probability of a hidden white face is 2/3.

In its crudest form, this is the trick. There is only *one* way that the hidden face can be red: the top card is card 3, front face up. But there are *two* ways the hidden face can be white: the top card is card 1 with the front face up and the top card is card 1 with the back face up. In two of three equally likely cases, we have a white hidden face.

PRESIDENTS OF THE U.S.

No.	NAME	BORN	DIED
1	George Washington	1732, Feb 22	1799, Dec. 14
2	John Adams	1735, Oct 30	1826, July 4
3	Thomas Jefferson	1743, Apr 13	1826, July 4
4	James Madison	1751, Mar 16	1836, June 28
5	James Monroe	1758, Apr 28	1831, July 4
6	John Quincy Adams	1767, July 11	1848, Feb 23
7	Andrew Jackson	1767, Mar 15	1845, June 8
8	Martin Van Buren	1782, Dec 5	1862, July 24
9	William Henry Harrison	1773, Feb 9	1841, Apr 4
10	John Tyler	1790, Mar 29	1862, Jan 18
11	James Knox Polk	1795, Nov 2	1849, June 15
12	Zachary Taylor	1784, Nov 24	1850, July 9
13	Millard Fillmore	1800, Jan 7	1874, Mar 8
14	Franklin Pierce	1804, Nov 23	1869, Oct 8
15	James Buchanan	1791, Apr 23	1868, June 1
16	Abraham Lincoln	1809, Feb 12	1865, Apr 15
17	Andrew Johnson	1808, Dec 29	1875, July 31
18	Ulysses Simpson Grant	1822, Apr 27	1885, July 23
19	Rutherford Birchard Hayes	1822, Oct 4	1893, Jan 17
20	James Abram Garfield	1831, Nov 19	1881, Sept. 19
21	Chester Alan Arthur	1829, Oct 5	1886, Nov 18
22	Grover Cleveland	1837, Mar 18	1908, June 24
23	Benjamin Harrison	1833 Aug 20	1901, Mar 13
24	Grover Cleveland	1837, Mar 18	1908, June 24
25	William McKinley	1843, Jan 29	1901, Sept 14
26	Theodore Roosevelt	1858, Oct 27	1919, Jan 6
27	William Howard Taft	1857, Sept 15	1930, Mar 8
28	Woodrow Wilson	1856, Dec 28	1924, Feb 3
29	Warren Gamaliel Harding	1865, Nov 2	1923, Aug 2
30	Calvin Coolidge	1872, July 4	1933, Jan 5
31	Herbert Clark Hoover	1874, Aug 10	1964, Oct 20
32	Franklin Delano Roosevelt	1882, Jan 30	1945, Apr 12
33	Harry S. Truman	1884, May 8	1972, Dec 26
34	Dwight David Eisenhower	1890, Oct 14	1969, Mar 28
35	John Fitzgerald Kennedy	1917, May 29	1963, Nov 22
36	Lyndon Baines Johnson	1908, Aug 27	1973, Jan 22
37	Richard Milhous Nixon	1913, Jan 9	1994, April 22
38	Gerald Rudolph Ford	1913, July 14	
39	Jimmy (James Earl) Carter	1924, Oct 1	
40	Ronald Reagan	1911, Feb 6	
41	George Bush	1924, Jun 12	
42	Bill Clinton	1946, Aug 19	

Assignment 9: Chance and Probability I

STOP! Before going any further, fill in this table according to the instructions.

We are the subject of this experiment. The purpose is to compare our intuitive notions of randomness with true randomness. In order to do this, **imagine** that you have an unbiased six sided die and that you will throw it 60 times. The result will be a sequence of 60 numbers drawn from 1-6, for example: 4, 3, 5, 1, ... What might the resulting 60 outcomes look like? **Imagine** that you have done the experiment.

In the table, write in a set of 60 numbers that might have come from such an experiment.

1	2	3	4	5	6	7	8	9	10
11	12	13	14	15	16	17	18	19	20
21	22	23	24	25	26	27	28	29	30
31	32	33	34	35	36	37	38	39	40
41	42	43	44	45	46	47	48	49	50
51	52	53	54	55	56	57	58	59	60

1.(a)__________

1.(b)__________

OVER

Assignment 9: Chance and Probability I

STOP! Before going any further, fill in this table according to the instructions.

We are the subject of this experiment. The purpose is to compare our intuitive notions of randomness with true randomness. In order to do this, **imagine** that you have an unbiased six sided die and that you will throw it 60 times. The result will be a sequence of 60 numbers drawn from 1-6, for example: 4, 3, 5, 1, ... What might the resulting 60 outcomes look like? **Imagine** that you have done the experiment.

In the table, write in a set of 60 numbers that might have come from such an experiment.

1	2	3	4	5	6	7	8	9	10
11	12	13	14	15	16	17	18	19	20
21	22	23	24	25	26	27	28	29	30
31	32	33	34	35	36	37	38	39	40
41	42	43	44	45	46	47	48	49	50
51	52	53	54	55	56	57	58	59	60

1.(a)__________

1.(b)__________

OVER

1 In order to assess whether the outcomes of your imaginary die throws are truly random, count up the following and write your answers on the previous page:

(a) How many 3s are there?

(b) How many doubles are there?

A double occurs whenever two numbers in the sequence are the same. For example:

...3,4,4,5,... --> one double since the second 4 is the same as the preceding 4.

...3,5,5,5,6,... --> **two** doubles since the second 5 is the same as the preceding (first) 5 and the third five is the same as the preceding (second) 5.

Be sure to count doubles that extend over the end of a line. For example, a 5 in square 10 and a 5 in square 11 counts as a double.

2 (a) Since your imagined die is unbiased, the probability of throwing a 3 is 1/6. On average, how many 3s do you expect to get in 60 throws?

(b) The outcomes of two throws of your imagined die are independent. Therefore there is a 1/6 chance of that any outcome will match the one before it. Therefore there is a 1/6 chance that any pair of outcomes will be a double. Given that there are 59 pairs in the 60 throws of your experiment, how many doubles do you expect to get?

It is unlikely that the actual numbers of 3s and doubles counted in 1. agree exactly with the expectations of 2. But, if the imagined die is to behave like a real die, they ought to be close. How close is close enough? The answer will come in the next lecture!

In questions 3, 4 and 5

assume that, in a family, the probability that a child is a boy ("B") or a girl ("G") is 1/2=0.5 and that the sex of one child in the family does not affect this probability for the other children. (e.g. if the first child is known to be a boy, that does not make it any more or less likely that the second is a boy).

3 Mary has just one child. We are interested in the probability that it is a boy or a girl.

(a) What is the outcome space? Write "B" for boy, "G" for girl.

(b) What is P(B)? What is P(G)?

(c) Are outcome B and outcome G mutually exclusive? Explain.

(d) What is P(B OR G)?

(e) Outcome B is the same as NOT-G. Use the NOT rule to compute P(B) from P(G).

(f) Explain why outcome B and G are not independent.
(Hint: the B and G apply to the same child. If the outcome is B, what does that do the probability that it is also G?)

(g) Does this mean we cannot use the simple AND rule to compute P(B AND G)?

4 Bill's family has two children. We are interested in the probabilities of various combinations of boys and girls.

(a) What is the outcome space. Write "BG" for first child boy, second child girl.
(Hint: there are four outcomes in the space)

(b) Is the outcome that the first child is a boy independent from the outcome that the second child is a girl? Explain.

(c) What is P(BG) = P(First child boy AND Second child girl)?

(d) What are the three probabilities of the three remaining outcomes of the outcome space.

5 (a) In Bill's family, if we know that the **first** child is a girl, then the outcome space of (a) is reduced. What is the new outcome space?

(b) Which outcomes correspond to one girl and one boy?

(c) Each outcome in (a) is equally probable. Therefore, given that we know that the **first** child is a girl, what is the probability that the other is a boy?

6 (a) In Bill's family, if we know that **one** (and possibly more) of the children is a girl, then the outcome space of (a) is reduced. What is the new outcome space?

(b) Which outcomes correspond to one girl and one boy?

(c) Each outcome in (a) is equally probably. Therefore, given that we know that **one** of the children is a girl, what is the probability that the other is a boy?

7 In a game of darts, the probability of a player hitting the bull ("H") is 1/3 and the probability of missing ("M") is 2/3. Assume that a player's success or failure on one throw does not affect the probability of success or failure on the other throws.

(a) Consider four throws. Are the four outcomes independent?

(b) In four throws, what is the probability that all are misses? That is, use the AND rule to compute P(MMMM) where we understand

"MMMM" = M on 1st AND M on 2nd AND M on 3rd AND M on 4th.

(c) What is the probability that the **first** throw is a hit and the rest misses, i.e. P(HMMM)?

(d) What is the probability that the **second** throw is a hit and the rest misses, i.e. P(MHMM)?

(e) What is the probability that the **third** throw is a hit and the rest misses, i.e. P(MMHM)?

(f) What is the probability that the **fourth** throw is a hit and the rest misses, i.e. P(MMMH)?

(g) What is the probability that exactly one of the four throws is a hit?
That is, what is P(HMMM or MHMM or MMHM or MMMH)?

Chapter 10

Chance and Probability Theory II

We Meet the Bell Curve

The sort of problem we want to be able to solve easily:

An unbiased die is thrown 60 times.

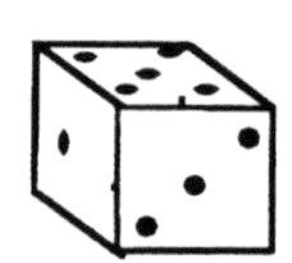

P(⚂) =1/6

Expect *about* 10 ⚂ in the 60 throws.

But typically we will not get *exactly* 10.
We may get:

mean = the average number we get

0, 1, 2, 3, 4, 5, 6, 7, 8, 9, 10, 11, 12, 13, 14, 15, 16, 17, ...

Zero ⚂'s are not likely in 60 throws. If there are no ⚂ we should suspect something is wrong!

If everything is normal, we expect the number of ⚂'s to be in this range.

These larger numbers of ⚂'s start to look suspicious again. Too many ⚂'s.

How many ⚂'s will we have?

To handle this problem precisely:

Compute the probability of throwing 0, 1, 2 ... ⚂'s

So far we have learned how to analyze fairly simple problems in probability theory. Real life is rarely so kind as to offer us simple problems. If probability theory is to be of use to us in real life, we must develop methods for handling tougher problems. For example, recall the problem of comparing the actual death rates from cancer in some town with those expected. We want to be able to calculate if the actual rate is far enough from the expected for us to be concerned. It turns out that many real problems like this have the same mathematical structure as the problem shown opposite concerning dice throws. So a technique for handling this problem will have far wider use.

If we throw an unbiased die 60 times, how many 3s will there be? Of course we cannot say exactly. Just as we do not know if a 3 will show on one die throw, we cannot say exactly how many times a 3 will show on 60 throws. But we can narrow things down a little. The probability of throwing a 3 is 1/6. So, on average, 1 in 6 of the throws will be 3s. There are sixty throws. So, on average we expect 10 3s, although it could be more or less.

To use the technical language, this means that the *mean* number of 3s thrown will be 10. The mean is the average number in the long run.

The actual number of 3s could differ from this. We may have fewer. Perhaps we may have only 9 or 8 or 7; or perhaps only 6 or 5. As the number gets smaller, it seems less and less possible that it could be that low. Certainly we would be very surprised if there were no 3s at all. Imagine it: 60 throws of the die without a 3 ever thrown! Somewhere between 10 3s and no 3s, the outcome moves from likely to very unlikely. The same thing happens as we consider higher values. 11, 12 or 13 3s seems quite possible. But higher values seem less and less possible. Could we have 20? Or 30? How many 3s will we have?

The most complete and most precise answer we can give to this question is probabilities of the different numbers of 3s that may be thrown: the probability of no 3s, of one 3, or two 3s, of three 3s, and so on.

What is the probability of 0, 1, 2, ... ⚁ 's?

On any one throw,

P(⚁) = 1/6

P(NOT- ⚁) = 5/6

Begin computation.

NOT rule:
P(NOT- ⚁) = 1 - P(⚁) = 1 - 1/6)

P (No ⚁ in 60 throws) = P(0)

= P (NOT- ⚁ on 1st
AND NOT- ⚁ on 2nd
AND ... AND NOT- ⚁ on 60th)

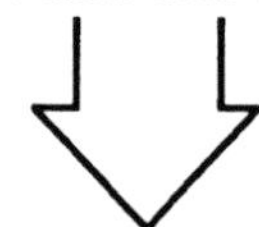

These outcomes are independent, so we can use the AND rule.

= P (NOT- ⚁ on 1st)
x P (NOT- ⚁ on 2nd)
x ... x P (NOT- ⚁ on 60th)

= 5/6 x 5/6 x ... x 5/6

(60 times)

= 0.0000177479

This is a tiny probability! It is very unlikely that we have no ⚁'s.

Let us start computing the probability of no, 1, 2, ... 3s in sixty throws of the die. All we need to know to compute these probabilities is that the probability of a 3 on any one throw is 1/6 and the probability of something other than a 3--a NOT-3--is 5/6. (We get the 5/6 from an application of the NOT rule as shown.)

First consider the probability of having no 3s at all in sixty throws. We will write this as "P(0)". The outcome has very many parts. No 3s at all is really: NOT-3 on the first throw AND NOT-3 on the second throw AND NOT-3 on the third throw and so on, all the way through to 60 throws.

This huge outcome is 60 outcomes ANDed together. Fortunately each of these smaller outcomes are independent of each other. Whether we throw a 3 or not on the first throw will not affect the probability of throwing a 3 on the second throw, the third throw, and so on. Therefore we can use the AND rule and replace the ANDs by multiplication of probabilities. The probability of the huge outcome is equal to 60 probabilities of smaller outcomes multiplied together. The probability of each of these smaller outcomes is 5/6. So our answer is just 5/6 x 5/6 x 5/6 ...(60 times)... x 5/6, which turns out to be 0.0000177479.

P (One ⚁ only in 60 throws) = P(1)

= P (⚁ on 1st AND NOT on any others OR ⚁ on 2nd AND NOT on any others OR ... OR ⚁ on 60th AND NOT on any others)

These outcomes are mutually exclusive. Therefore use OR rule.

= P (⚁ on 1st AND NOT on any others) + P (⚁ on 2nd AND NOT on any others) + ... + P (⚁ on 60th AND NOT on any others)

= .000003549 + .000003549 + ... + .000003549

= .000212964

These are calculated in a computation that is essentially the same in each case. Here it is for the second term.

P (⚁ on 2nd AND NOT on any others) = P (NOT-⚁ on 1st AND ⚁ on 2nd AND NOT-⚁ on 3rd AND ... AND NOT-⚁ on 60th)

The outcomes are independent. Therefore use AND rule.

= P (NOT-⚁ on 1st) x P (⚁ on 2nd) x P (NOT-⚁ on 3rd) x ... x P (NOT-⚁ on 60th)

= 5/6 x 1/6 x 5/6 x ... x 5/6

= .000003549

What is the probability that we have exactly one 3 in sixty throws; that is, what is P(1)? Now the calculation becomes more complicated. There are many ways that we could have just one 3. We may have a 3 on the first throw and none on the rest; or we may have a 3 on the second throw and none on the rest; or we may have a 3 on the third throw and none on the rest; and so on for all sixty cases. So the probability of having exactly one 3 is the probability that one of these outcomes happens; that it, it is the probability of these sixty outcomes ORed together, as shown opposite.

These sixty outcomes are mutually exclusive. If we have a 3 on the first and none of the other throws, then each of the other 59 outcomes is precluded. In addition to it, we cannot also have a 3 on the second throw and none on the other throws; and so on. Therefore we can use the OR rule and replace each OR by an addition of probabilities as shown. We now evaluate each of these 60 probabilities. As we will see in a moment, each turns out to be equal to 0.000003549. Sum 60 of these and we have our final result, 0.000212964.

A second computation was needed to find the probabilities of each of the sixty probabilities just summed. The computation for one is shown opposite. The other 59 are essentially the same, differing slightly in the order of the terms, but giving the same final result.

What is the probability of a 3 on the second throw and none on the rest? This is the probability of a huge outcome: NOT-3 on the first throw AND 3 on the second AND NOT-3 on the third AND ... AND NOT-3 on the sixtieth. Since these outcomes are independent, the AND rule can be applied. This gives us sixty individual probabilities multiplied together. Putting in the values for each we end up with the probability 0.000003549 mentioned above.

P(Two ⚂ only in 60 throws) = P(2)

= P (⚂ on 1st AND ⚂ on 2nd AND NOT on any others OR ⚂ on 1st AND ⚂ on 3rd AND NOT on any others OR ...)

= AARGGH!!

This is very tedious for us to calculate, but it is easy if you have a computer to do all the sums!

P(0) =	0.000017747	P(11) =	0.124557
P(1) =	0.000212964	P(12) =	0.101722
P(2) =	0.00125649	P(13) =	0.0751177
P(3) =	0.00485842	P(14) =	0.0504362
P(4) =	0.0138465	P(15) =	0.0309342
P(5) =	0.0310162	P(16) =	0.0174005
P(6) =	0.056863	P(17) =	0.0090073
P(7) =	0.0877314	P(18) =	0.00430349
P(8) =	0.116244	P(19) =	0.0019026
P(9) =	0.134327	P(20) =	0.000780064
P(10) =	0.137013		

"P(10)" is the probability of 10 ⚂'s.

Now let us consider the outcome in which we have exactly two 3s in sixty throws. What is its probability? As before, it can arise in many ways. We may have 3s on the first and on the second throws, but on no others; we may have 3s on the first and third throws, but on no others; and so on for all of what will prove to be 1770 different combinations. The probability of exactly two 3s is the probability that one of these 1770 individual outcomes happens; that is, it is the probability of these 1770 individual outcomes ORed together.

The calculation has started to become rather tedious! It will get no simpler when it comes to the remaining cases of exactly three, four five, ... 3s in sixty throws. There are ways to shorten the calculation. We need not go into them here. The whole problem can be fed into a computer which will oblige us by computing the probability of each of the cases that interests us. The table opposite shows the results of these calculation. It gives us the probability in 60 throws of having a set number of 3s. For example P(10) = 0.137013 is the probability of ten 3s. The table could be continued past P(20) to P(21), P(22),..., but the probability of these higher number of 3s has dropped so low that we will have no interest in them.

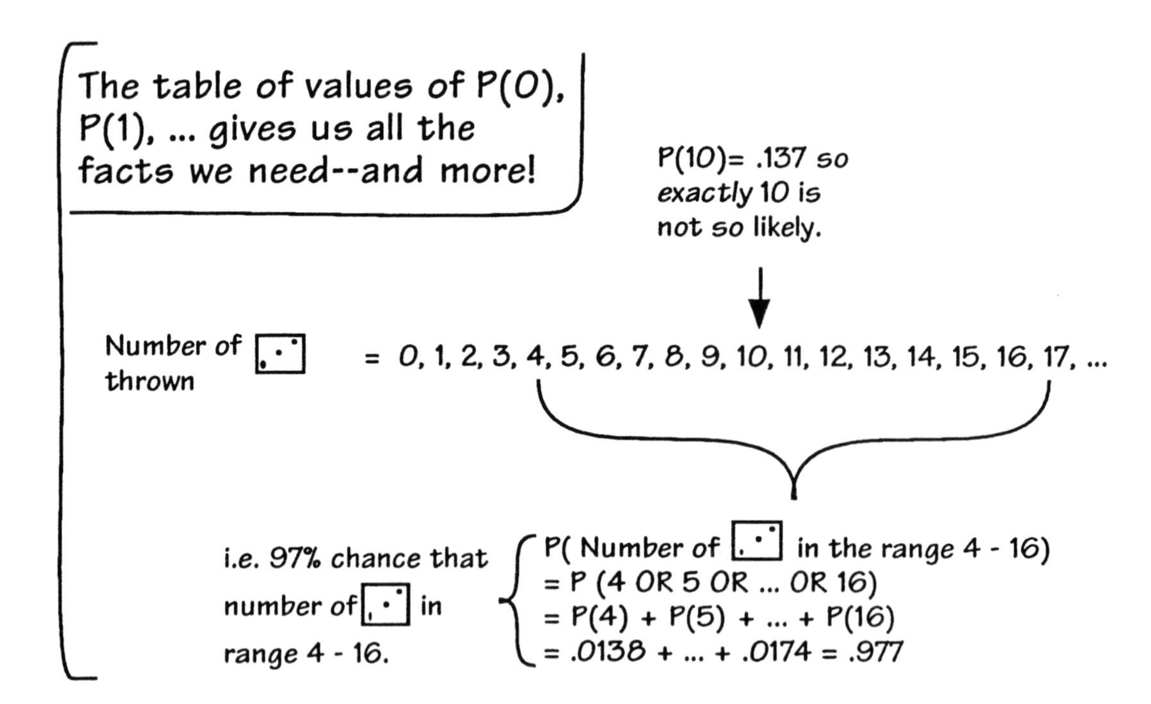

Now we need to make life easier!

Problem	Problem
The results of this computation are very hard to absorb.	The calculation is too much work!

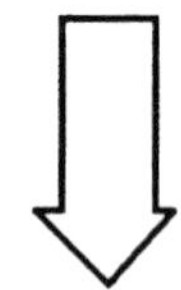

Solution	Solution
Use a graphical method to present the results so we see them in a picture.	Use a numerical shortcut to compute very rapidly only the part of the results we really need.

The previous table gives us all the facts we need to assess the possibility of the various outcomes. To begin, notice that the probability of any particular number of 3s is quite small. There is a 1/6 chance of throwing a 3; so in 60 throws we expect *on average* that there will be 1/6x60=10 3s, the mean. But the probability that we have *exactly* ten 3s is quite small, just 0.137. That is the highest probability for any number of 3s. Other numbers are less probable. So the table tells us not to expect any particular number of 3s. Any particular number selected in advance is quite unlikely. What is likely, however, is that the number of 3s will be close to the mean of 10. Simple addition gives us the probabilities of various ranges. For example, we can compute the probability that there will be from 4 to 16 3s thrown. As shown opposite, that turns out to be 0.977. This is a very high probability. It tells us that we can be very confident that the number of 3s thrown will be in this range. Anything outside of this range will be very unlikely.

This last result completes the analysis. But we are left with serious problems of a practical nature. The dice throwing experiment is similar mathematically to many others of practical importance. The methods used so far are just too difficult for practical application. The first problem is that the results are overwhelming and hard to absorb. The table of P(0), P(1), ... does give us all the facts we need, but these facts are not informative until we have stared at the table for a while and done some more calculations. Our solution will be to present the information in the table in a picture that tells us at a glance the most essential parts of the results.

The second problem is that there are far too many steps needed to arrive at the results in the table. Here we can simplify things by noticing that we do not actually need the values for each individual probability P(0), P(1), ... How does it help to know that P(8)=0.116? What actually matters to us are the probabilities of ranges of values, such as 4 to 16. It turns out that there is a very quick way of estimating the probabilities of outcomes in intervals like this. The method only gives estimates, so they may not be exactly correct. But they are usually close enough for practical purposes and the tremendous gain in simplicity far outweighs the loss in precision.

The remainder of the chapter will develop these methods for making our life simpler.

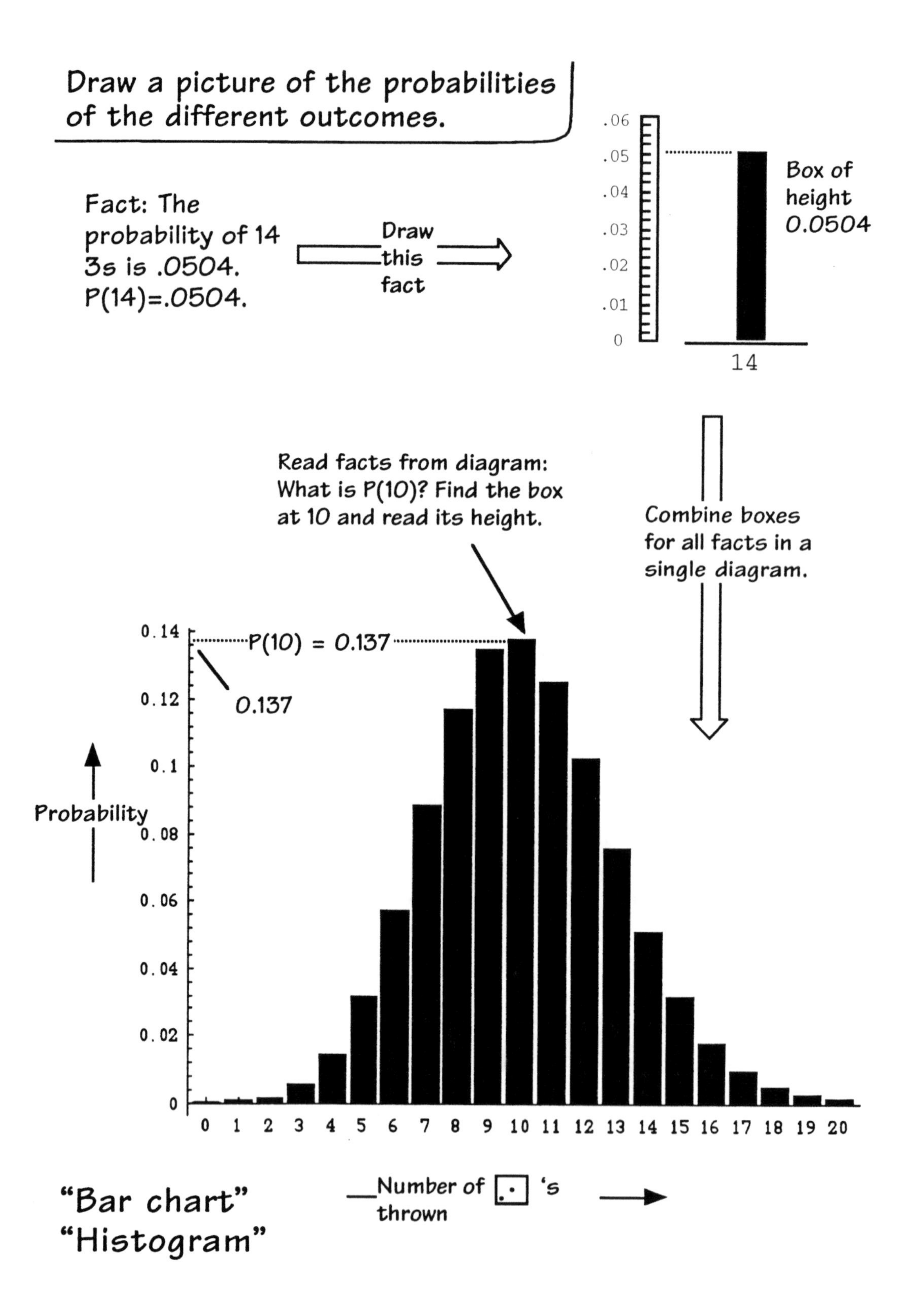
Draw a picture of the probabilities of the different outcomes.
Fact: The probability of 14 3s is .0504. P(14)=.0504.
Draw this fact
.06
.05
.04
.03
.02
.01
0
14
Box of height 0.0504
Combine boxes for all facts in a single diagram.
Read facts from diagram: What is P(10)? Find the box at 10 and read its height.
P(10) = 0.137
0.137
0.14
0.12
0.1
0.08
0.06
0.04
0.02
0
Probability
0 1 2 3 4 5 6 7 8 9 10 11 12 13 14 15 16 17 18 19 20
Number of 's thrown
"Bar chart"
"Histogram"

To draw a picture of the results, we start by drawing a picture of each fact in the table. For example, we have P(14)=0.0504. So we draw a thin box or bar whose height is 0.0504 units on some scale. So we know what the scale is, it is given by the ruler standing next to the box. The bottom of the box aligns with the 0 mark on the ruler; its height extends to 0.0504.

To draw the complete table, we draw a similar box for each of the facts P(0)=0.0000177, P(1)=0.000213, ... and we combine them into a single diagram. The scale across the bottom tells us which box is which; that is, it tells us which box corresponds to the case of 0, 1, 2, 3, ... 3s thrown. The scale on the left lets us read off how high each box is. Since the height of each box is the probability of the corresponding outcome, this scale is labeled "probability."

This diagram now contains all the information of the original table. We can read off whatever facts we want from it. For example, to find P(10), we locate the box numbered 10. We read off its height as 0.137 and thus recover P(10)=0.137. The only loss is that we cannot read the heights as accurately as the probabilities were given in the original table. The original table told us P(10)=0.137013, but we cannot read that many decimal places accurately from the scale. This will not worry us since we rarely need to know the probabilities to such accuracy.

This type of diagram is called a bar chart or histogram.

Now we can see what is likely and what is not merely by looking at the diagram.

How many should I expect ?

The likely range is where the bars are large.

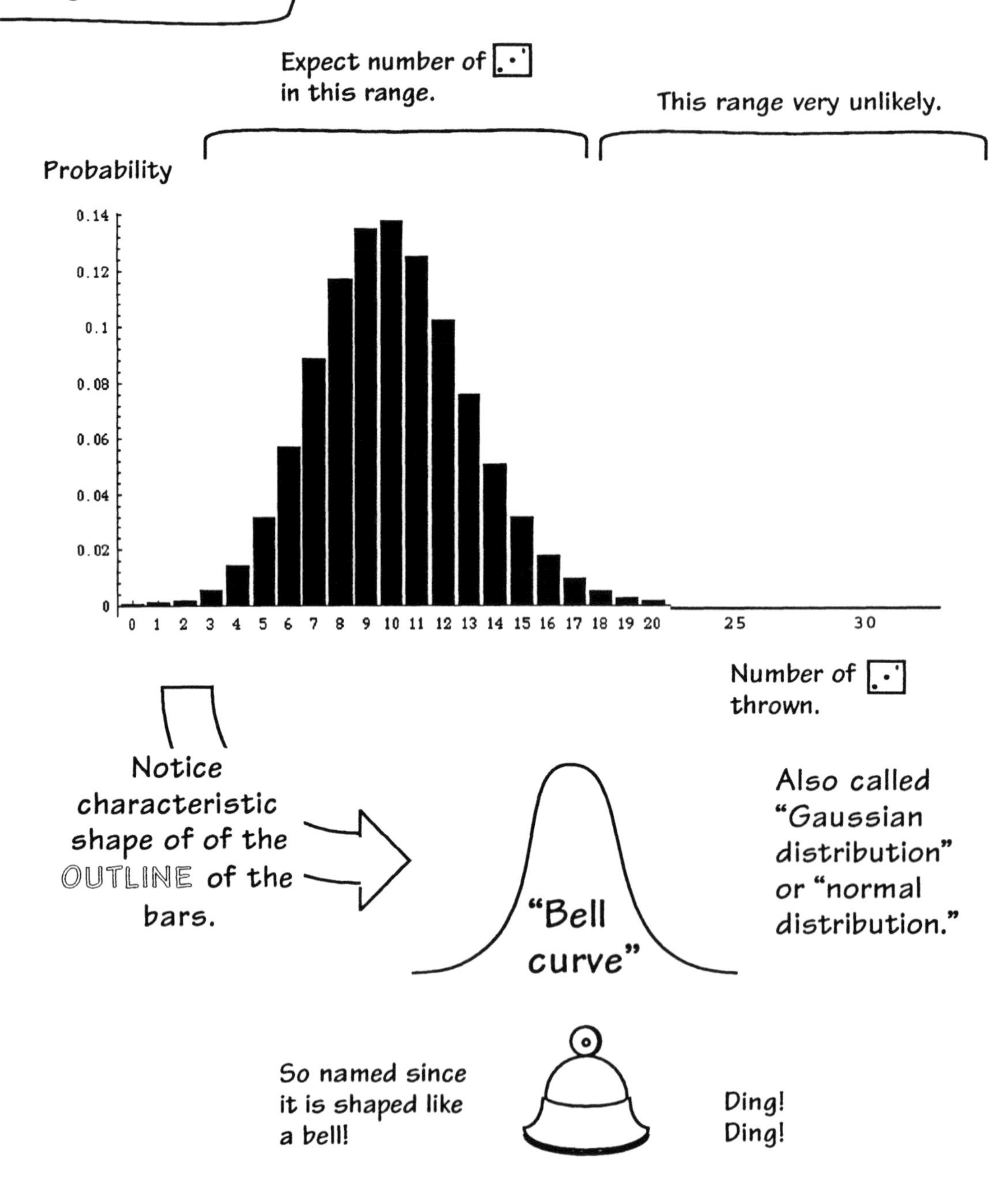

If we glance at the table of values of P(0), P(1), ... we cannot instantly see how the probabilities are distributed. If we glance at the bar chart, however, we see it immediately. The likely outcomes are where ever the bars are high. The range of outcomes stretches from 0 to 60. We see immediately from the bar chart that the likely outcomes are clustered around 10. The outcome is very unlikely to be far away from 10. We would not expect it in the range 20 to 30, for example.

10 itself has the highest probability--its bar is the highest. The outcomes close to it are almost as probable, with the probability dropping away as we move in either direction away from 10. Once we get far enough away from 10, the probability becomes negligible. The range in which the outcome is likely to be is just that range in which the tall bars are clustered. The tall bars are all clustered in the range of (roughly) 4 to 16. So we expect the number of 3s thrown to fall in that range, 4 to 16. (This fits with our earlier numerical analysis where we found the probability of the outcome in this range was 0.977.)

We can also see from the chart that setting the boundaries of this likely range is somewhat arbitrary. We could have chosen 3 to 17 and increased the chance that the range encompass possible outcomes--at the cost of letting in some, like 3 and 17, that are are less likely. We could have narrowed the range to 5 to 15, at the cost of eliminating the two least likely outcomes, 4 and 16.

In this bar chart we meet an important curve for the first time. The bars form a little hill or mound. Consider its outline. This outline approximates the characteristic bell shape--so called because it looks like a bell. We will devote a lot of attention to this curve. It will turn out to be one of the most important of all the curves considered in science. It will enable us to simplify our calculations very greatly. Because of the enormous significance this curve has in science, it is referred to under many names. Other terms that are used include "Gaussian distribution" and "normal distribution."

When will our die throwing experiment give us a bell curve?

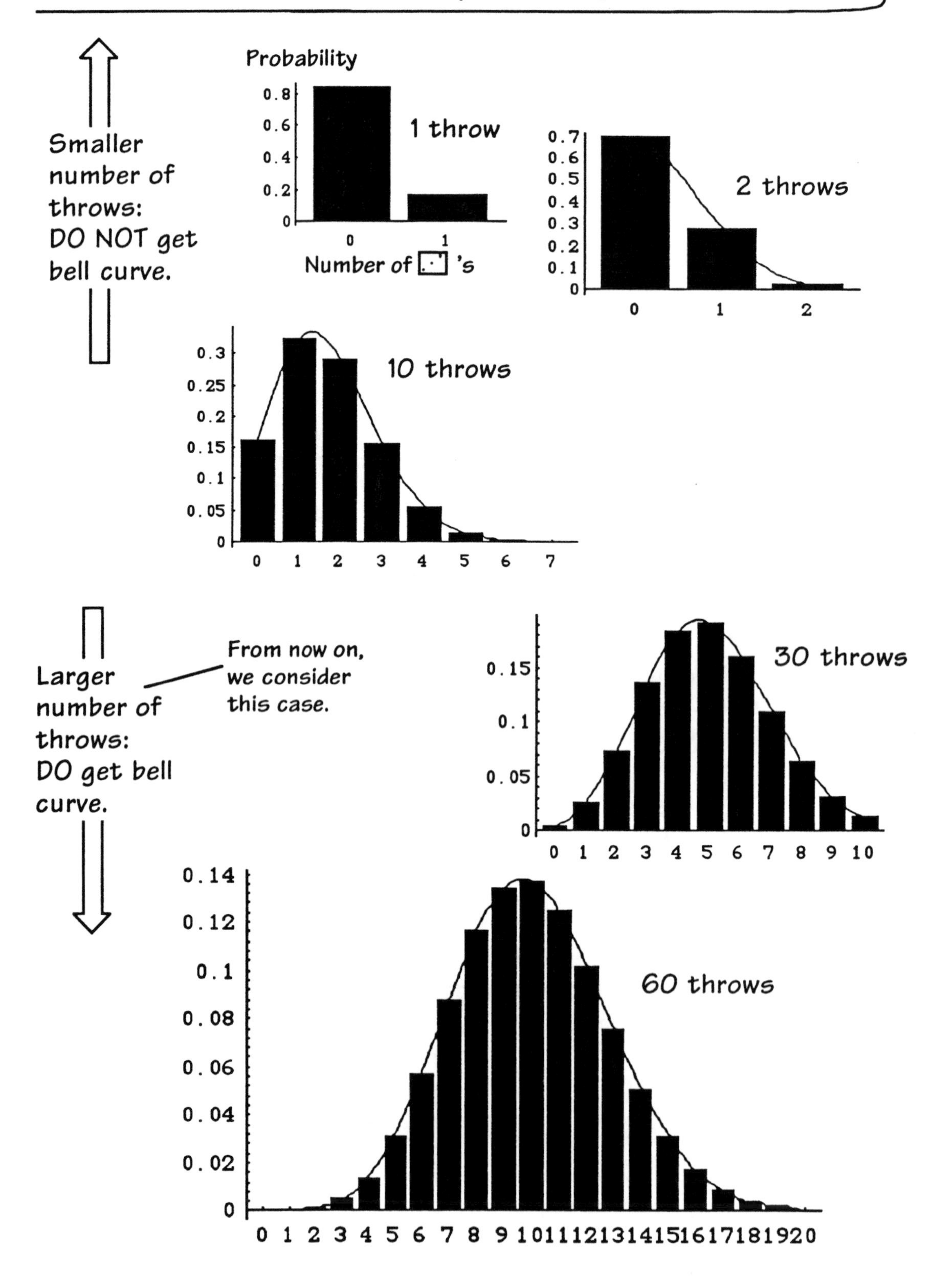

From now on, we will concern ourselves with systems exhibiting a bell curve. This is because such systems prove especially simple to analyze. Therefore we must start to develop a sense of when a system will exhibit a bell curve. In general a bell curve is exhibited by a chance outcome that results from the combination of very many parts. (We will see a more precise statement of this idea in Chapter 13.) The number of 3s thrown in a die throwing experiment results from combining the outcomes of the individual throws. These individual throws are the "parts" and we have the required "many parts" when we have many die throws.

The bar charts opposite illustrate what is needed for the die throwing experiment to yield a bell curve. Imagine that we do our die throwing experiment but instead of throwing the die 60 times, we throw it fewer times: say once, twice, 10 times or 30 times. In each case we ask after the probability of having different numbers of 3s. The bar charts represent these probabilities.

If the experiment involves smaller number of throws, then the bars do not exhibit a bell curve. In the smallest case, one throw, there are just two bars corresponding to the only possible outcomes, zero and one 3. With the case of 2 throws, there are three bars which decrease in size to the right--no bell curve. By the case of 10 throws, the outline of the bars form a hill. But it is not yet the type that fits the bell curve. The bell curve has the same shape on each side of its highest point; more precisely, each side is the mirror image of the other. In the case of 10 throws, the curve has a different shape on either side.

With 30 throws, the outline of the bars starts to look like a bell curve. The appearance is more encouraging than the actuality. If we compare the outline for 30 throws with a real bell curve, there are quite a few differences. If you look carefully you will see that the curve on the right hand side of the peak is still not the mirror image of the curve on the left hand side. By 60 throws, we have a curve that is close enough to a bell for practical purposes.

Notice again how informative these bar charts are. The five tables of numbers these charts represent would be overwhelming. But we can just look at the bar charts and recover almost everything we want to know with a glance.

How do we choose a bell curve to fit a given bar chart?

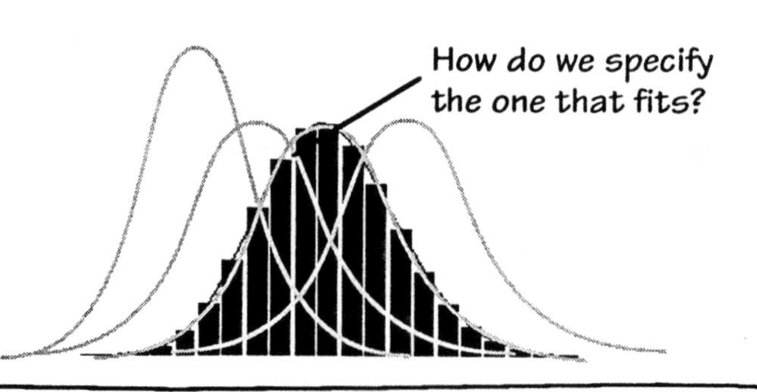

A bell curve is completely fixed if we specify:

- its MEAN
- its STANDARD DEVIATION

Location of Peak set by MEAN

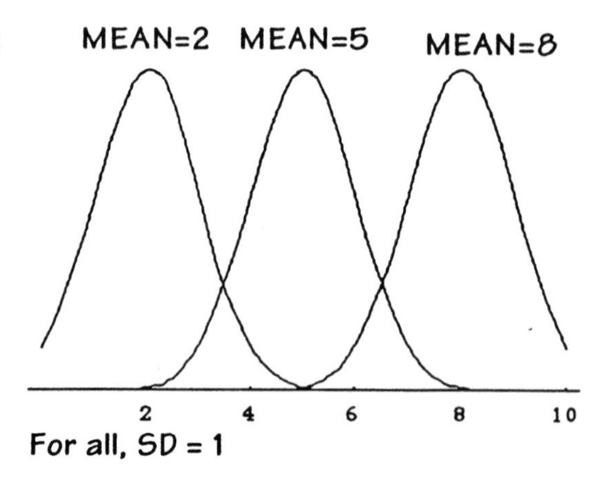

Spread of curve set by STANDARD DEVIATION

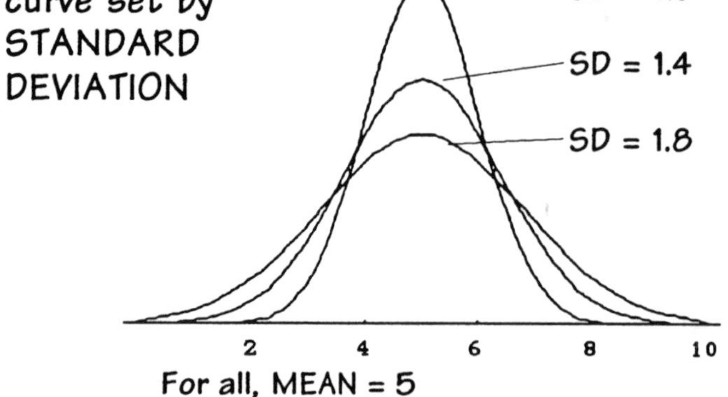

Analogy: fix a circle by specifying its center and radius.

Vary center

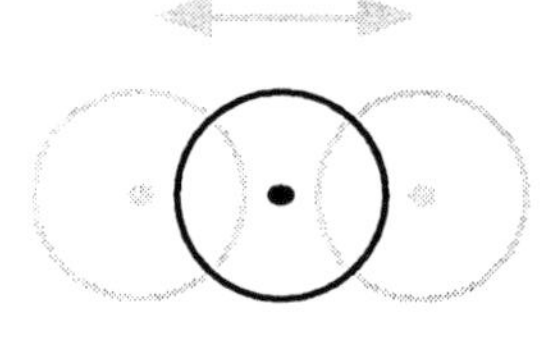

Vary radius

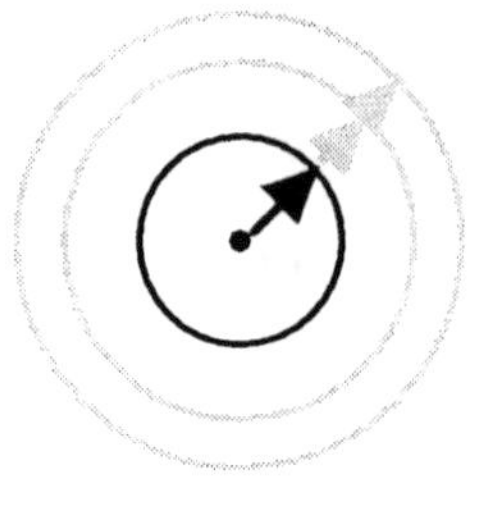

Once we have a bar chart that can be fitted by a bell curve, we need to know which bell curve is the one that fits. The situation is not so different from fitting shoes. Our feet have a standard shape. To find the foot that fits, we need only to find the size of the shoe. Once we have that number, life is greatly simplified. We can even order shoes through a catalog and still be reasonably assured that the shoes will fit.

In the case of the bell curve, instead of a single number for the shoe size, we need to specify two numbers, the MEAN and STANDARD DEVIATION in order to pick out a particular bell curve. Just as varying shoe size varies the size of the shoes bought, so varying the MEAN and STANDARD DEVIATION alters the shape of the bell curve.

The MEAN sets the location of the peak of the curve. Increasing or decreasing it slides the bell curve horizontally as shown opposite. The STANDARD DEVIATION determines how pointy the bell curve is. The smaller the STANDARD DEVIATION ("SD"), the more pointy; the larger SD, the more spread out, as shown opposite.

Adjusting these two parameters is akin to adjusting the center and radius of a circle. As we vary the center of the circle, we slide the circle over the page, while otherwise leaving it unchanged. As we vary the radius, it remains centered on the same spot but it grows or shrinks in size.

Thus, the task of fitting a bell curve to a bar chart reduces to finding the MEAN and STANDARD DEVIATION of the bell curve that fits.

Rules for fitting a bell curve to probabilities

that arise in the following type of experiment:

Repeat same trial many times.

e.g. throw die 60 times

Count the number of "Successes" = "outcome"

e.g. count number of [die face showing 2]

What are the probabilities of the different outcomes? Draw as a bar chart.

Probability

Numbers of successes

Fit a bell curve by finding its MEAN and STANDARD DEVIATION.

Rule (1)

MEAN = Expected number of successes = Probability of success on a single trial x Total number of trials

Example:
60 die throws;
"success" = [die face showing 2]

MEAN
= 1/6 x 60
= 10

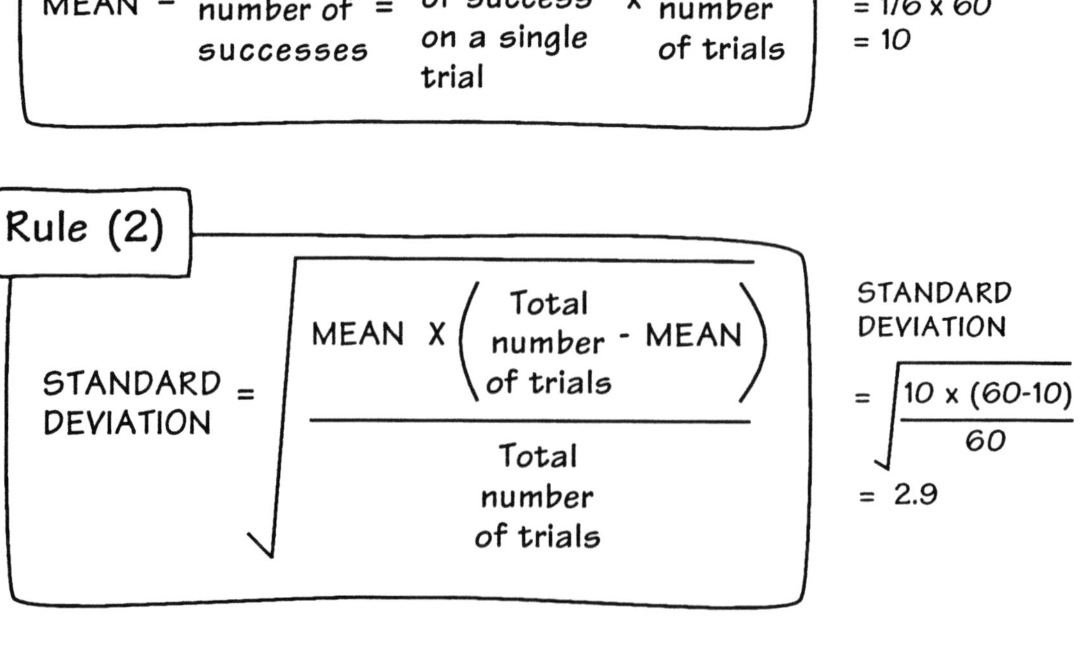

In the case of repeated die throws, it turns out to be especially easy to find the MEAN and STANDARD DEVIATION. This case is an example of a more general case in which we have simple rules for finding the MEAN and STANDARD DEVIATION. This general case is one in which we have a "trial" that is repeated some definite but large number of times. For example, a single trial may be the throw of a die, which we will repeat 60 times. In each trial, there will be an outcome that interests us and that we label as a "success." In the case of the die throws, we take throwing a 3 as a "success." Our major assumption is that the probability of success on each trial remains the same. The outcome of the experiment is just the total number of successes. In the case of the die throws, it is the number of 3s thrown.

Notice that the term "success" is used without any connotation of a happy eventuality. For example, our trials may be examining many different patients and "success" may be finding a cancer.

As long as there are many trials, the probabilities of the various outcomes will form a bell shaped bar chart; in other words, they will be "normally distributed." Rules (1) and (2) tell what the MEAN and theSTANDARD DEVIATION of the bell curve are.

The MEAN is just the expected number of successes. We have already seen that this is equal to the probability of success on an individual trial multiplied by the number of trials. For a single die throw the probability of throwing a 3 ("success") is 1/6 and there are 60 throws ("trials"), so the MEAN is 1/6x60=10.

The STANDARD DEVIATION is computed by inserting the values for the MEAN and total number of trials into the formula opposite. This calculation is carried out most conveniently with a hand calculator. Be sure it has a square root key! The result is 2.9.

Three very useful facts about the bell curve

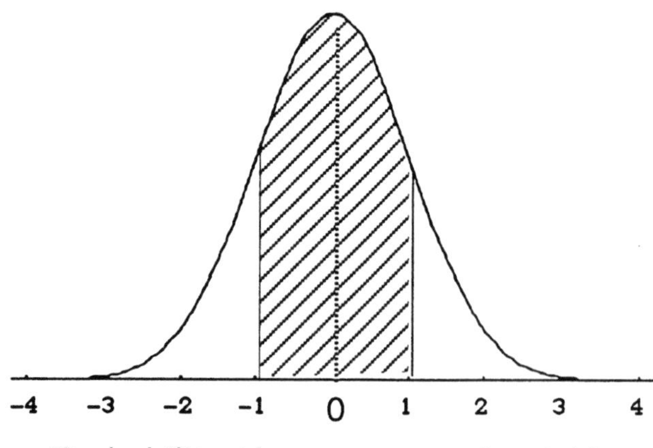

Probability that outcome is within ONE standard deviation of mean = .683

68.3 % of outcomes

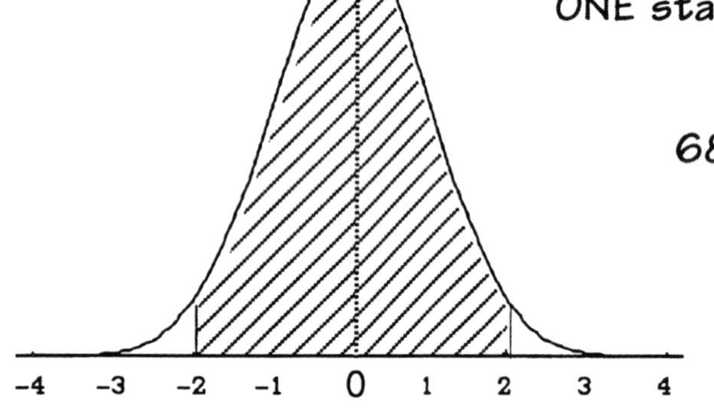

Probability that outcome is within TWO standard deviations of mean = .954

95.4 % of outcomes

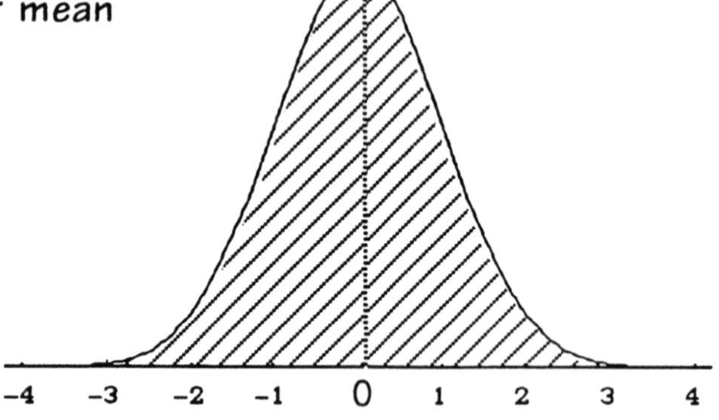

Probability that outcome is within THREE standard deviations of mean = .997

99.7 % of outcomes

These are the ±1SD, ±2SD, ±3SD confidence intervals.

To complete our treatment of the bell curve we need three very useful facts about the bell curve. These are the ones that will make the bell curve enormously useful to us in the chapters to follow. We have seen that the peaked part of the bell curve corresponds to the range of outcomes that are most likely. We can use the bell curve to tell us quickly how probable it is for the outcome to be in various ranges around the peak.

Start at the outcome that coincides with the mean of the bell curve, that is, with its peak. Opposite, that outcome is labeled "0".

If we move up or down by ONE standard deviation on the scale of outcomes, we have defined an interval of outcomes. There is a probability of 0.683 that the outcome will actually be in this interval. To put it another way, 68.3% of the time, the outcome will fall in this range. This interval is the "mean plus or minus ONE standard confidence interval."

If we move up or down by TWO standard deviations on the scale of outcomes, we have defined another interval of outcomes. There is a probability of 0.954 that the outcome will be in this interval; that is, 95.4% of the time, the outcome will fall in this range. This interval is the "mean plus or minus TWO standard deviations confidence interval."

If we move up or down by THREE standard deviations on the scale of outcomes, we have defined yet another interval of outcomes. There is a probability of 0.997 that the outcome will be in this interval; that is, 99.7% of the time, the outcome will fall in this range. This interval is the "mean plus or minus THREE standard deviations confidence interval.

We will see a lot of these three probabilities: 0.683, 0.954 and 0.997, corresponding to the three confidence intervals. Notice that the probability of the outcome falling in the three intervals increases as we move from the one to two to three SD confidence interval. A probability of 0.683 represents only a fairly modest probability--the outcome will be in that range slightly more than 2 in 3 times The probability of 0.997, however, represents a very high probability. The outcome will be in the relevant range in all but 0.3% of cases.

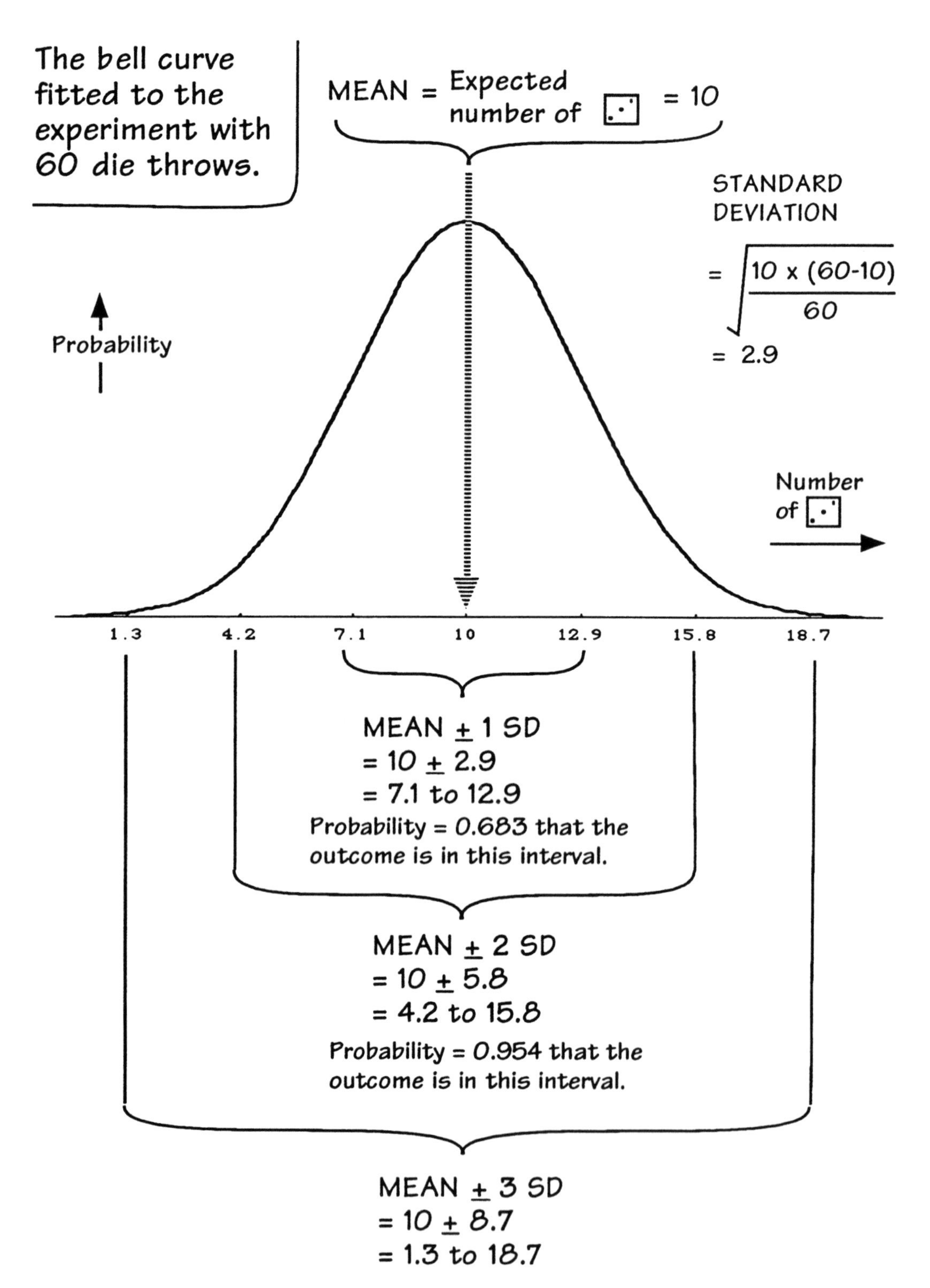
The bell curve fitted to the experiment with 60 die throws.
MEAN = Expected number of ⚁ = 10
STANDARD DEVIATION
$= \sqrt{\frac{10 \times (60-10)}{60}}$
= 2.9
Probability
Number of ⚁
1.3
4.2
7.1
10
12.9
15.8
18.7
MEAN ± 1 SD
= 10 ± 2.9
= 7.1 to 12.9
Probability = 0.683 that the outcome is in this interval.
MEAN ± 2 SD
= 10 ± 5.8
= 4.2 to 15.8
Probability = 0.954 that the outcome is in this interval.
MEAN ± 3 SD
= 10 ± 8.7
= 1.3 to 18.7
Probability = 0.997 that the outcome is in this interval.

10.24

Let us now see how these three confidence intervals can be used in the case of our 60 die throws. We have already determined that a bell curve can be fitted and that its MEAN is 10 and SD is 2.9. We now need to compute the three confidence intervals.

The mean plus or minus one confidence interval is found by starting at the mean 10 and going up by 2.9 and down by 2.9; that is to 10+2.9=12.9 and to 10-2.9=7.1. The interval is 7.1 to 12.9. Therefore, there is a probability of 0.683 that the number of 3s thrown will be in this interval. Of course the fractional part of the interval makes no sense; you cannot have 7.1 3s being thrown! These fractional parts arise because the bell curve is still only an approximation. So we round the values off at 7 and 13 to get meaningful results.

We repeat this calculation for the remaining confidence intervals. Two standard deviations are given by 2xSD = 2x2.9 = 5.8. So the confidence interval is 10 ± 5.8, that is, 4.2 to 15.8. We can round this to 4 to 16. Our rule tells us that there is a probability of 0.954 that the number of 3s thrown will be in this interval. Notice that this disagrees with the result of 0.977 calculated directly from the table of P(0), P(1),... This latter result of 0.977 is the correct one. It differs from the one computed here because the fit of the bell curve is still only an approximate one, so these small errors are to be expected. For larger numbers of throws, these errors become smaller.

We can now see how much the use of the bell curve has simplified our analysis of the die throwing experiment. In our original analysis, we needed to compute a large table of values of P(0), P(1), ..., something that was only really practical with a computer. If we use the bell curve, all the calculations we need fit on the page opposite. Once we are satisfied that a bell curve will fit, we need only to spend a moment calculating the MEAN and SD. The MEAN is calculated by multiplying 1/6x60; the SD comes from the short square root formula repeated opposite. Then we proceed directly to the three confidence intervals and the calculation is complete. The use of the bell curve has introduced some errors of approximation. But our gain is that a lengthy and burdensome calculation has been replaced by one that is very brief and easy.

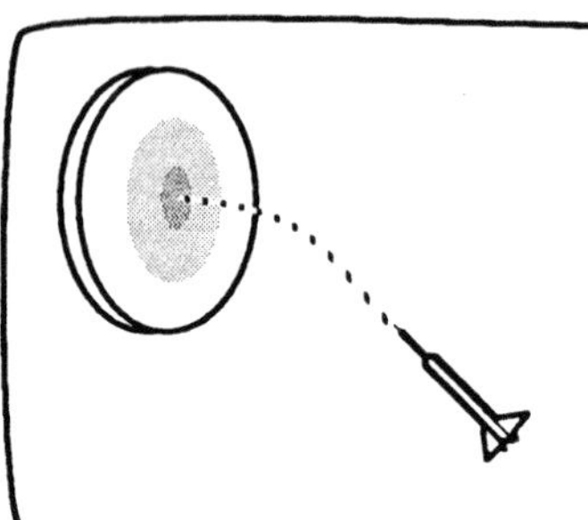

In a game of darts, the probability of a hit is 1/4. What range of hits do we expect in 100 throws?

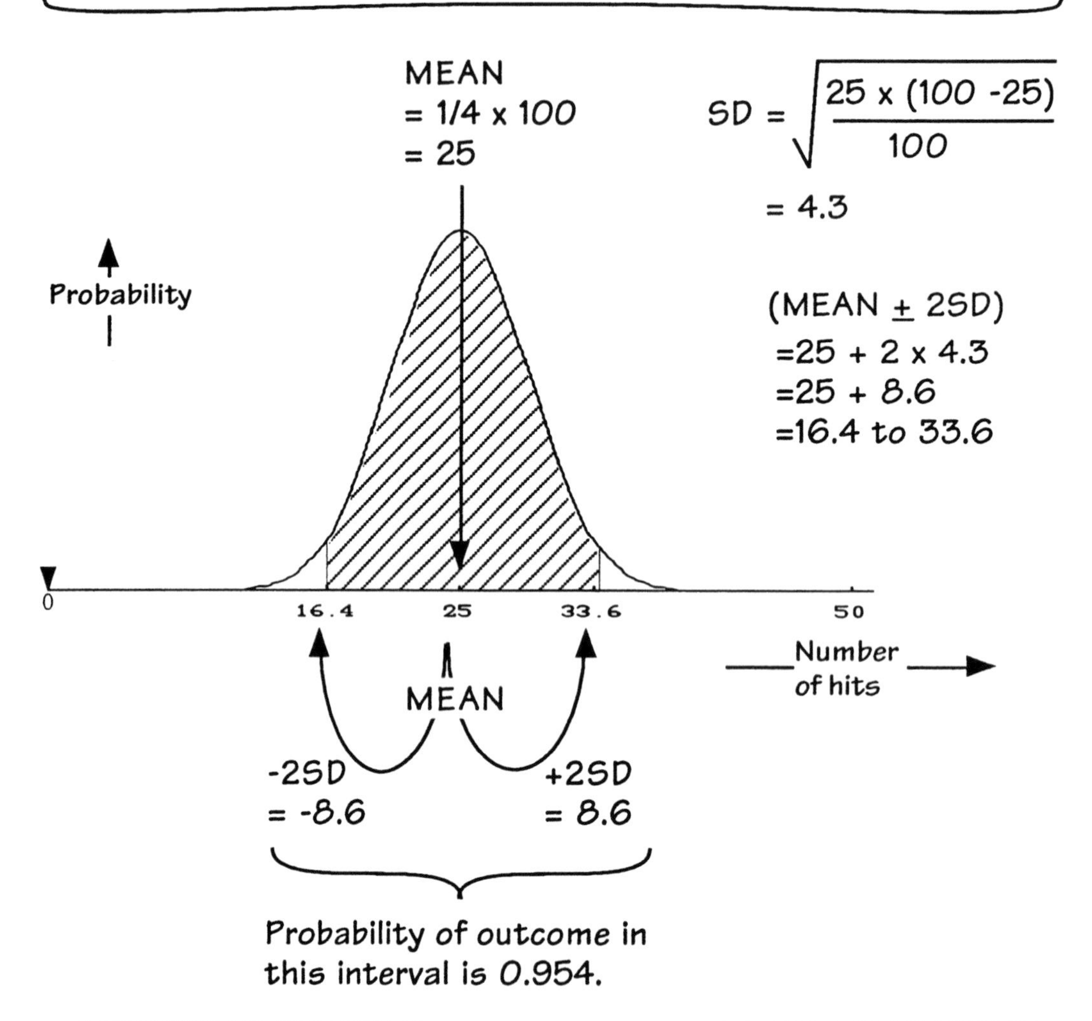

Probability of outcome in this interval is 0.954.

The probability of approximately 16 to 34 hits = 0.954.

Power of the bell curve approximation: Notice how little we had to compute to get this result!

We have learned a very simple and rapid technique for assessing the likely outcomes of experiments such as the repeated throw of a die. Here is the complete calculation for this method used to solve a related problem. When a dart is thrown in some dart game, there is a probability of 1/4 of a hit. In what range do we expect to find the number of hits in 100 throws?

Notice just how little actual calculation is needed to arrive at the result. To get a sense of the power of the technique, before you review the calculation, try to estimate the range in which the number of hits will lie. It will be distributed around 25, but how far away from 25 can we reasonably expect it to stray? 24 to 26? 20 to 25? 15 to 35? Without some assistance, it is impossible to say. The bell curve approximation will take only moments to tell us.

The possible outcomes will range from 0 to 100 but with varying probability. We expect that these probabilities will be normally distributed so that the bar chart of probabilities will conform to the bell curve.

We compute the MEAN of the curve as follows: The probability of success (a hit) is 1/4. The total number of trials (throws) is 100. Therefore MEAN = 1/4x100 = 25. Thus the peak of the curve lies at 25 hits and we expect the actual number of hits to be somewhere near to 25.

Computing the standard deviation tells us how near to 25 we can expect the number of hits to be. We substitute the values for the MEAN (25) and total number of trials (100) into the formula for the STANDARD DEVIATION. The answer is SD = 4.3.

We are free to choose which confidence interval to compute. Let us compute the two standard deviation interval. Two standard deviations are 2xSD = 2x4.3 = 8.6. Therefore the two standard deviation confidence interval is 25 $\pm$ 8.6 = 16.4 to 33.6. We will round this to 16 to 34. Thus there is a probability of 0.954 that the number of hits will lie in the interval 16 to 34. Thus 95% of the time--in roughly 19 of 20 times--the outcome of the experiment will lie in this interval.

Assignment 10: Meet the Bell Curve

An unbiased coin is tossed 20 times. ("Unbiased" means that the probabilities of heads H and tails T are the same and thus equal 0.5.) Using the techniques we have seen earlier, we can compute the probability of there being 0 heads, 1 head, 2 heads, etc. as the outcome of the experiment. The result of this computation is shown in the table. (Be glad you didn't have to compute it!)

Number of heads	Probability
0	0.000000953674
1	0.0000190735
2	0.000181198
3	0.00108719
4	0.00462055
5	0.0147858
6	0.0369644
7	0.0739288
8	0.120134
9	0.160179
10	0.176197
11	0.160179
12	0.120134
13	0.0739288
14	0.0369644
15	0.0147858
16	0.00462055
17	0.00108719
18	0.000181198
19	0.0000190735
20	0.000000953674

1. To get a sense of where the most likely outcomes are grouped, draw a histogram of the probabilities in this table. Notice that the probabilities for smaller numbers of heads (0,1,2,3...) and larger numbers (20, 19, 18, ...) are very small compared with the large values so that these will *look* like they have a zero probability on your chart.

2. To get a sense of which are the most likely ranges of numbers of heads, use the OR rule to compute from the table:

(a) the probability that there are 8 to 12 heads

= P(8 heads OR 9 heads OR 10 heads OR 11 heads OR 12 heads)

(b) the probability that there are 6 to14 heads

(c) the probability that there are 4 to 16 heads.

The rules for fitting a bell curve/normal distribution to the probabilities of outcomes generated by repeated trials are:

MEAN of bell curve = expected number of success
= probability of success of single trial x total number of trials

$$\text{SD} = \text{STANDARD DEVIATION} = \sqrt{\frac{\text{MEANx(Total number of trials-MEAN)}}{\text{Total number of trials}}}$$

If the bell curve fits the probabilities of the various outcomes, then

Probability that outcome is in range MEAN - SD to MEAN + SD = 0.683
Probability that outcome is in range MEAN - 2SD to MEAN + 2SD = 0.954
Probability that outcome is in range MEAN - 3SD to MEAN + 3SD = 0.997

3. Use this rule to fit a bell curve to the probabilities of the various outcomes in the 20 coin toss experiment.
(a) What is the MEAN?
(b) What is the STANDARD DEVIATION?
Round off your answer to 3(b) to 2. Then use the rules for probabilities of outcomes governed by a bell curve to compute:
(c) the probability that there are 8 to 12 heads
(d) the probability that there are 6 to14 heads
(e) the probability that there are 4 to 16 heads.

Notice that the answers to 3(c),(d),(e) do not agree exactly with those of 2(a),(b),(c). This is because the bell curve only *approximates* the exact probabilities of the table. As the number of trials gets larger, the approximation gets better and better.

4. To get a sense of how all the probabilities we have computed relate to real life, do the experiment with 20 coin tosses ten times. (The easiest way is to put 20 pennies in a jar in which they can flip freely. Shake them ***well*** , tip them out and count the number of heads. Do this ten times.)
(a) What are your outcomes for the ten experiments.
(b) On the basis of your answer to 2., how many of these outcomes ought to lie in 8 to 12 heads?
(c) How many did lie in 8 to 12 heads?
(d) Do your answers to (b) and (c) agree well enough?

OPTIONAL QUESTION. You should be sure that you know how to compute the STANDARD DEVIATION from the MEAN and number of trials. To get some practice, fill out this table:

Trials I MEAN=	10	20	30	40
50				
100				
500				
1200				

Chapter 11

Applications of the Bell Curve I

Testing Hypotheses

The Logic of Hypothesis Testing

We start with an HYPOTHESIS we are inclined to believe unless there is evidence against it.

We TEST it by comparing the actual outcome with range of outcomes allowed if the hypothesis is true

Agreement → Accept hypothesis

Disagreement → Reject hypothesis

Therefore hypothesis testing cannot be self-contained. We must *already* have strong preferences for the hypothesis.

Example

Cancer death rates in a town of 5,000.

You live in a town of 5,000 people. The national average for deaths from all forms of cancer is 2 per 1000 each year. You expect 10 deaths per year on average. How far from 10 deaths per year must we stray before we suspect that some out-of-the-ordinary factor is at work?

HYPOTHESIS under test: The chance of dying of cancer in our town is the same as the national average.

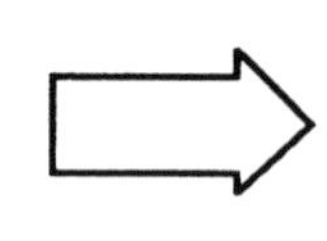

TEST by comparing: actual number of deaths with the range allowed if the hypothesis is true.

Hypothesis testing is one of the most common and most important forms of statistical analysis. Its logic is summarized opposite. We must begin with an hypothesis that we are inclined to believe unless we find evidence against it. This hypothesis should allow predictions about some outcome that we can check. We test the hypothesis by checking whether the actual outcome fits with the prediction. If it does, we accept the hypothesis; if it does not, we reject it.

Our earlier example of cancer death rates in a town illustrates hypothesis testing. We start with the default hypothesis that our town is just like other towns; that is, we assume that the average cancer death rate in our town will be the same as the national average, 2 deaths per thousand people each year. If our hypothesis is true, then we can predict the approximate number of cancer deaths that we should have in a year. In our town of 5,000 people, it will be around 10. If the actual number of deaths is close enough to 10, we find agreement and accept the hypothesis that our town is just like others. If the actual number is too far away to be compatible with an average of 10, we reject the hypothesis.

The first step in the logic of hypothesis testing is very important and easily overlooked. We must start with an hypothesis we are inclined believe unless we find evidence against it. Thus the failure to refute the hypothesis in the test leads us to accept it. This acceptance depends heavily on our prior commitment to the hypothesis. Without it, agreement of the actual outcome with the predicted outcome could not lead us to accept the hypothesis. Imagine the actual number of deaths is close to ten. All that really tells us is that the average death rate from cancer in our town per year is *around* 2 in 1000. It may also be compatible with rates of 1.95 in 1000 and 2.05 in 1,000. However we choose 2 in 1,000 and say we accept the hypothesis that our town is like others because we have a preference for that hypothesis.

If we find that the actual outcome disagrees with the predicted, then we reject that hypothesis. Our prior preference for the hypothesis does not come into play in this case.

Our preference for the hypothesis under test is typically not emphasized when hypothesis testing is conducted. We must not forget it. It can be very significant to the final result of the test.

How can we find the probabilities of the different numbers of cancer deaths possible in our town each year?

Hypothesis: Average cancer death rate in our town of 5,000 is 2 per 1,000 per year.

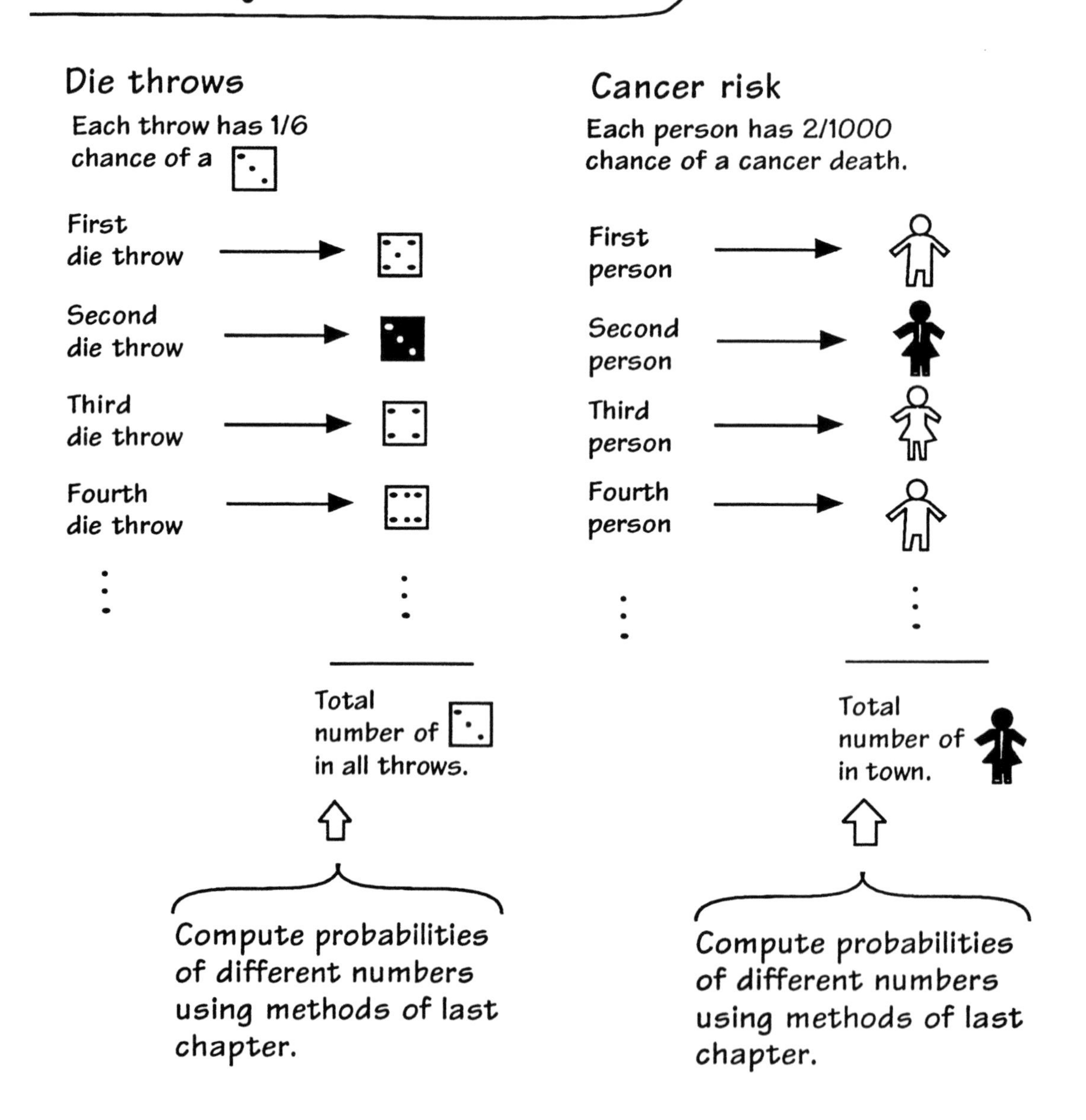

In our example of cancer death rates, we check whether the actual number of deaths is close enough to 10 to agree with the hypothesis that the average cancer death rate in our town is 2 per 1,000. But how do we know what is "close enough"? We need to know whether the actual number of death lies in an interval around 10 that is probable if the hypothesis is true. We will see in a moment that, most probably, the number of cancer deaths in our town will lie in the interval 4 to 16 if the average cancer death rate is 2 per 1,000 each year. But how do we know that 4 to 16 is this most probable interval?

It turns out that the techniques of the last chapter are just the ones we need to compute this interval. The problem of estimating the number of 3s on repeated die throws is essentially the same as the problem of estimating the number of cancer deaths once the average rate is assumed. The two problems map exactly onto each other.

The core of the die throwing experiment is the throw of the die and the check of whether it yields a 3. These individual throws correspond to taking each of the 5,000 people in town and checking which are cancer deaths. In the die throwing experiment, we expect on average 1 in 6 throws to yield a 3. That is, more precisely, with each of the 60 throws there is a probability of 1/6 that we throw a 3. In our town, we assume that on average there are 2 cancer deaths per 1,000 in each year. That is, more precisely, with each of the 5,000 people, there is a probability of 2/1000 of a death from cancer each year.

In the case of repeated die throws, we expect that about 1/6th of the 60 throws (i.e. 10) will be 3s, but the number may be more or less. We saw how to calculate the probabilities for throwing ... , 7, 8, 9, ... 3s on the die and that they will conform to a bell curve. So we could use the properties of the bell curve as a shortcut for finding the intervals in which the actual number of 3s will most probably lie.

In the case of our town, we expect about 2/1000th of its 5,000 people (i.e. 10) to die of cancer, but the number may be more or less. In the same way we can calculate the probabilities of ... , 7, 8, 9, ... cancer deaths and we will find that they too conform to a bell curve. So again, we can use the properties of the bell curve for finding the interval in which the actual numbers of cancer deaths will probably lie.

Estimate range of deaths allowed if the hypothesis is true.

Hypothesis: Average cancer death rates in our town of 5,000 matches the national average of 2 per 1,000 per year.

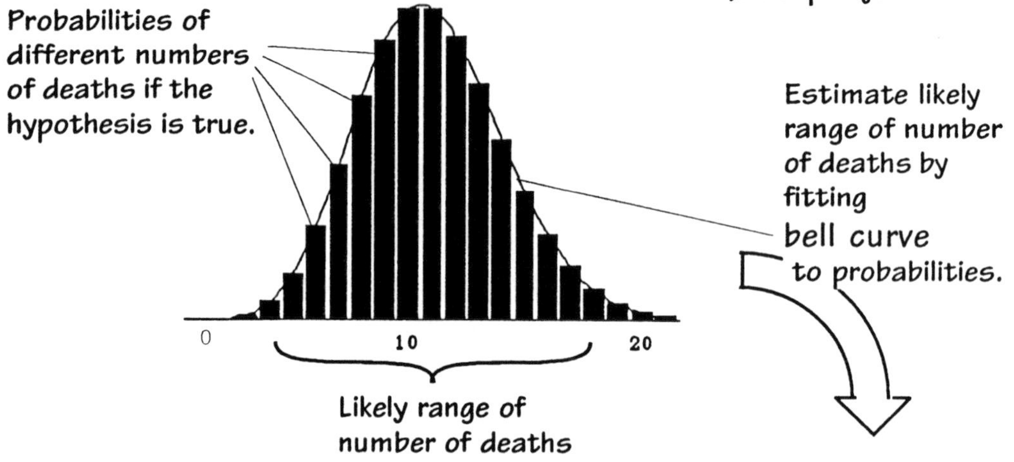

$$\text{MEAN} = \text{Expected number of deaths} = \frac{2}{1000} \times 5000 = 10$$

$$\text{STANDARD DEVIATION} = \sqrt{\frac{10 \times (5000-10)}{5000}} = 3.2$$

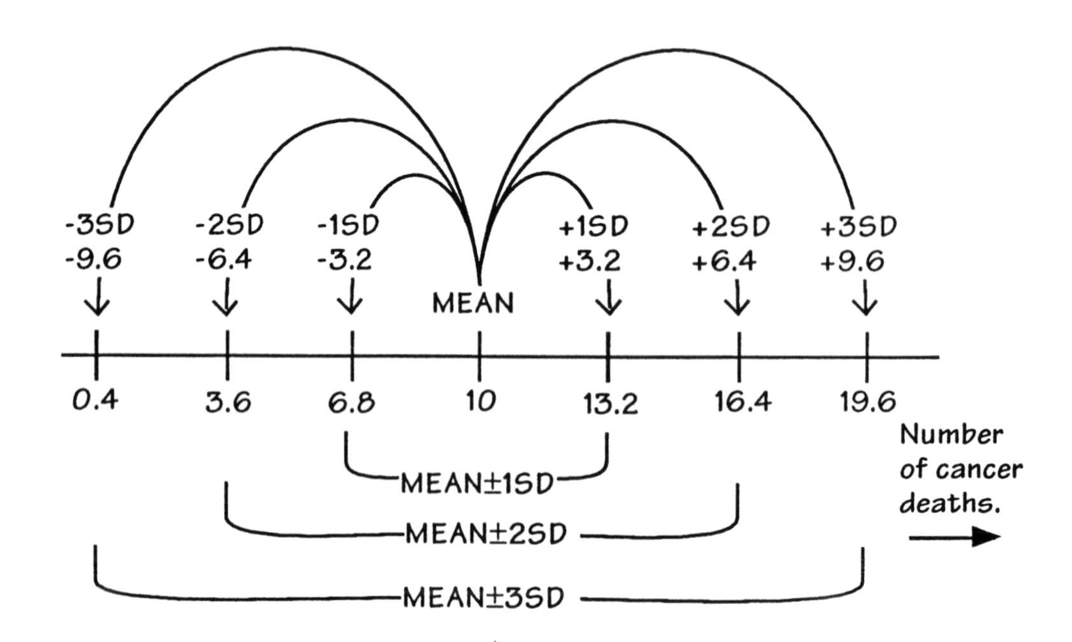

Consider the probabilities of the different numbers of cancer deaths in a year under the assumption that the hypothesis is true. The probabilities are drawn opposite in the bar chart. We are not so interested in the probabilities of the different numbers individually. What we really need to know is the interval around 10 deaths per year in which the actual number number of deaths will most probably lie. That we can find without a tedious calculation of the probabilities of the individual numbers of deaths. A shortcut is possible since these probabilities conform to a bell curve.

Proceeding as in the last chapter, we fit the bell curve to these probabilities by determining the values of the MEAN and STANDARD DEVIATION. The MEAN is just the expected number of deaths, 10. It also happens to be the location of the peak of the curve. The STANDARD DEVIATION is computed from the formula of the last chapter. That formula requires a MEAN and "Total number of trials". The "Total number of trials" is the number that corresponds to the total number of die throws. In this case it is 5,000, the number of people in our town. Just as each die throw is a trial in which a 3 may result, so each person in town is a "trial" in which a cancer death may result.

The results are a MEAN of 10 and a STANDARD DEVIATION of 3.2. We now form the confidence intervals as before. Starting at the MEAN of 10, we go down and up by one STANDARD DEVIATION to arrive at the interval 6.8 to 13.2. This is the MEAN plus or minus one STANDARD DEVIATION confidence interval. Two STANDARD DEVIATIONS are 2x3.2=6.4. Going down and up by two STANDARD DEVIATIONS from the MEAN gives us the MEAN plus or minus two STANDARD DEVIATION confidence interval of 3.6 to 16.4. A similar calculation gives us the MEAN plus or minus three STANDARD DEVIATION confidence interval of 0.4 to 19.6.

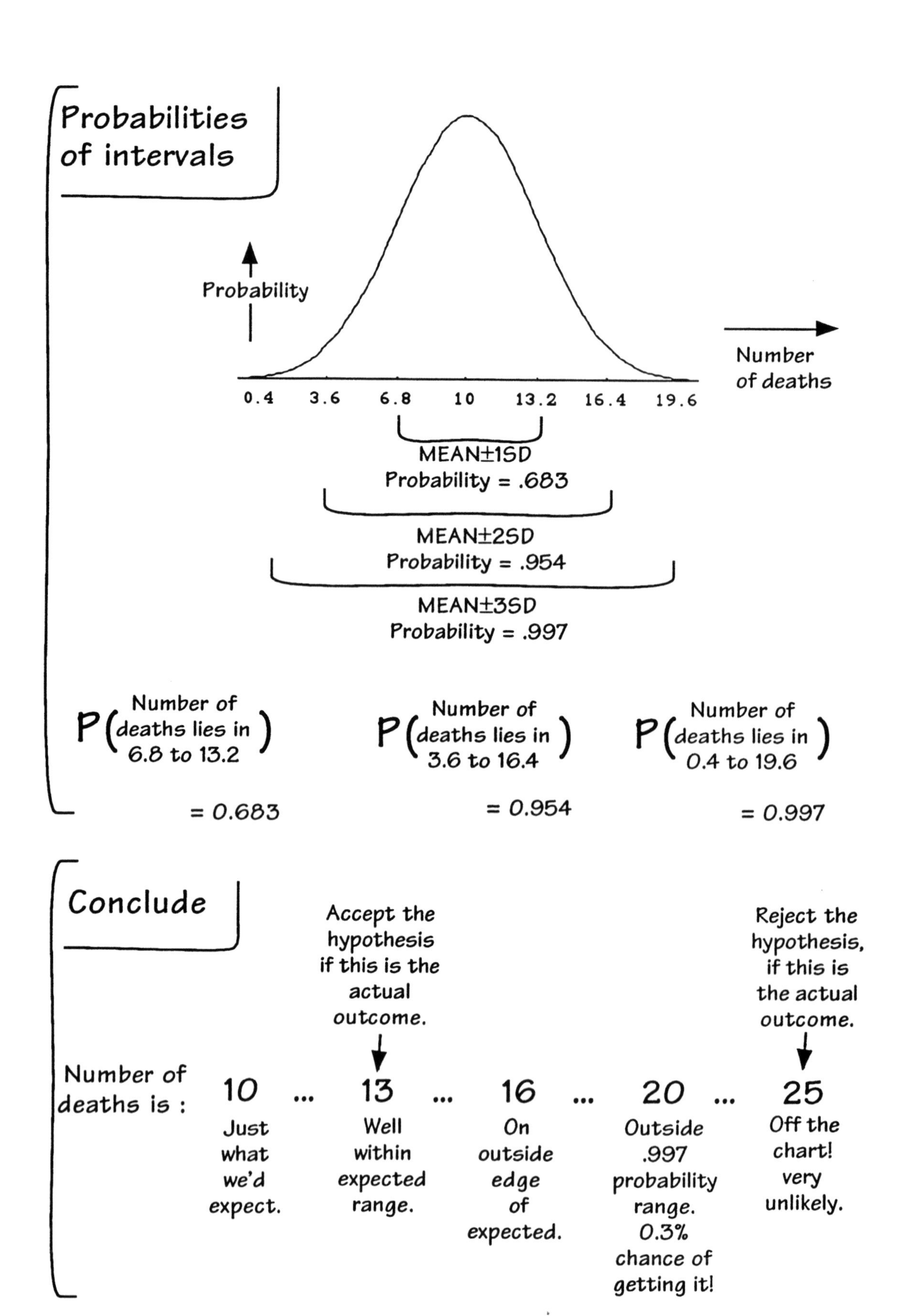
Probabilities of intervals
Probability
Number of deaths
0.4
3.6
6.8
10
13.2
16.4
19.6
MEAN±1SD
Probability = .683
MEAN±2SD
Probability = .954
MEAN±3SD
Probability = .997
P(Number of deaths lies in 6.8 to 13.2) = 0.683
P(Number of deaths lies in 3.6 to 16.4) = 0.954
P(Number of deaths lies in 0.4 to 19.6) = 0.997
Conclude
Accept the hypothesis if this is the actual outcome.
Reject the hypothesis, if this is the actual outcome.
Number of deaths is :
10 ... 13 ... 16 ... 20 ... 25
Just what we'd expect.
Well within expected range.
On outside edge of expected.
Outside .997 probability range. 0.3% chance of getting it!
Off the chart! very unlikely.

We now have the one, two and three SD confidence intervals calculated. For a bell curve, as we saw in the last chapter, the probabilities that an outcome lies in each of these three intervals is 0.683, 0.954 and 0.997 respectively. The fractional numbers of deaths on the interval boundaries are physically unrealistic, so we may choose to round these intervals to whole numbers. The second would be the interval 4 to 16. This interval was mentioned earlier as the one in which the actual number of deaths would most probably lie.

We can now use the actual number of deaths recorded from cancer in the year to determine whether we accept or reject the hypothesis that our town is typical. If the number is 10 or 13, it lies well within the range we would expect if the hypothesis were true. With these as actual outcomes we would accept the hypothesis. 25 deaths is highly improbable if the hypothesis is true. So if the actual number is 25 we would reject the hypothesis. We would strongly suspect that there is something atypical about our town that leads to higher than normal cancer death rates. We must be careful not to read too much into such a negative result. The test does not tell us why the rate is higher. People tend to presume sinister causes. For example they presume that a carcinogenic pollutant is to blame. The real cause may be quite benign. It may just be that our town has a larger elderly population than most. Cancer death rates among the elderly are far greater than in the population as a whole. This alone may explain why our death rate was so much higher.

Intermediate values, such as 16 and 20, pose a problem. They lie on the edge of what we would expect if the hypothesis were true. The outcome of 16 is at the edge of the MEAN±2SD interval. Deaths outside this interval can only arise with probability 1-0.954=.046 (about 5% or 1 in 20 chance) if the hypothesis is true. The outcome of 16 is just possible if the hypothesis is true, so we cannot feel secure in rejecting the hypothesis. But since it is so close to the impossible range, we cannot comfortably accept the hypothesis either! In real life, this result would raise a warning flag and we would want to investigate further. Similar remarks apply to an outcome of 20, which is just outside the MEAN±3SD interval. Results outside this range can only happen with probability 1-0.997=0.003 (i.e. 0.3% or 3 in 1,000 chance) if the hypothesis is true. In this case we would be more inclined to doubt the hypothesis, since the actual outcome is so much further into the range that is improbable if the hypothesis is true.

The language of statistics: "Significance Tests"

We are testing the hypothesis that the average cancer death rate in our town of 5,000 matches the national average of 2 per 1,000 each year.

Ordinary language	The outcome of 25 cancer deaths per year is far enough away from the range expected for us to reject the hypothesis.	"Far enough away" means that the outcome lies in a region that has a probability of 0.003 of containing the outcome if the hypothesis is true.
⇩	⇩	⇩
The language of statistical hypothesis testing	The outcome of 25 cancer deaths per year is statistically significant.	The level of significance is 0.003.

! Beware! we are asking if a result is "significant". In ordinary talk all results are significant--they mean something! The usage in statistics is much narrower. Many highly informative results are not "significant". For example, we would judge an actual number of cancer deaths of 10 as "not significant," even though it provides much useful information. **!**

We have seen how we conduct a test of an hypothesis in statistics. We now need to learn the special *language* statisticians use to describe these test and their results. We continue with the example of cancer rates in a town of 5,000. Let us imagine that the actual number of cancer deaths is 25. That is so far away from the range expected that we can confidently reject the hypothesis. In the special language of statistical hypothesis testing, we say that the result is "*statistically significant*". Correspondingly, if the outcome were 13, in ordinary language we would say that the outcome is compatible with the hypothesis. In the language of statistical hypothesis testing, we would say that the outcome was "*not statistically significant*".

Beware. The term "significant" is being used in a special, technical way. In ordinary usage, something is "significant" if it has special meaning to us. So we would say that an outcome of 13 is significant (ordinary language!) because it reveals something important about cancer deaths in our town. This ordinary meaning is *not* the one used in statistical hypothesis testing.

The outcome of 25 is far enough away from the expected range for us to doubt the hypothesis. How do we decide what is "far enough away"? Our first inclination might be to demand that the outcome is far enough away if its probability is 0 when the hypothesis is true. This is far too strict a demand. If the hypothesis is true, all outcomes are possible--up to and including the death of all 5,000 people in the town. These higher numbers are possible--but they are vastly improbable. A more realistic standard is that an outcome is "far enough away" if it lies in a region of very low probability. 25 cancer deaths lies outside the MEAN±3SD confidence interval. If the hypothesis is true, it is very unlikely that the actual outcome will lie in this region. To be precise, there is a probability 1-0.997=0.003 that the outcome can lie in this region if the hypothesis is true.

In the language of statistical hypothesis testing, we convey our standard of "far enough away" by specifying the "*level of significance*". In this case, the level of significance is 0.003.

With this talk of "significance", we have introduced no new content. We are just learning a slightly different way of speaking. Unfortunately it is rather clumsy until you get used to it. The advantage is that it is somewhat more concise.

0.05 level of significance

"Far enough away" is outside ±2 SD confidence interval.

Most common choice.

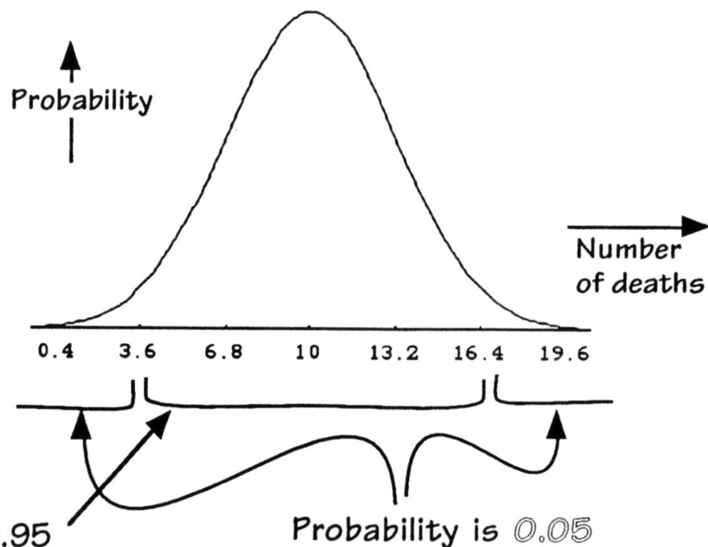

Probability is 0.95 that the outcome is in this interval.

Probability is 0.05 that the outcome is in the remaining range.

In sum:	An outcome is statistically significant at the 0.05 level.	=	The outcome can happen with probability 0.05 if the hypothesis is true.

Examples

Number of deaths = 15
Number of deaths = 8

Outcome lies within the range 3.6 to 16.4.

"Not statistically significant at the 0.05 significance level"

No reason to doubt the hypothesis.

Number of deaths = 18
Number of deaths = 2

Outcome lies outside the range 3.6 to 16.4

"Statistically significant at the 0.05 significance level"

Good reason to doubt the hypothesis.

To understand the notion of "level of significance" more clearly, let us take the specific case of 0.05. This also happens to be one of the most common choices. This level of significance arises when we choose the MEAN±2SD confidence interval as the interval in which we expect to find the actual outcome. If the outcome lies inside this interval, we will accept the hypothesis; if it lies outside, we will reject it. From our earlier calculations, we know that this interval extends from 3.6 to 16.4 deaths in a year.

If the hypothesis is true, the probability of an outcome inside this range of 3.6 to 16.4 is 0.954. It is customary to round this probability to 0.95. The probability of an outcome outside this range is 1-0.95 = 0.05. This is the 0.05 named in choosing the level of significance.

If the actual outcome lies in the MEAN±2SD interval of 3.6 to 16.4, we say that the outcome is *"not statistically significant at the 0.05 level"*. Examples are actual numbers of cancer deaths of 15 or of 8. They would give us no reason to doubt the hypothesis.

If the actual outcome lies outside the MEAN±2SD interval of 3.6 to 16.4 interval, we say that the outcome is *"statistically significant at the 0.05 level"*. Examples are actual numbers of cancer deaths of 18 or 21. They would give us reason to doubt the hypothesis.

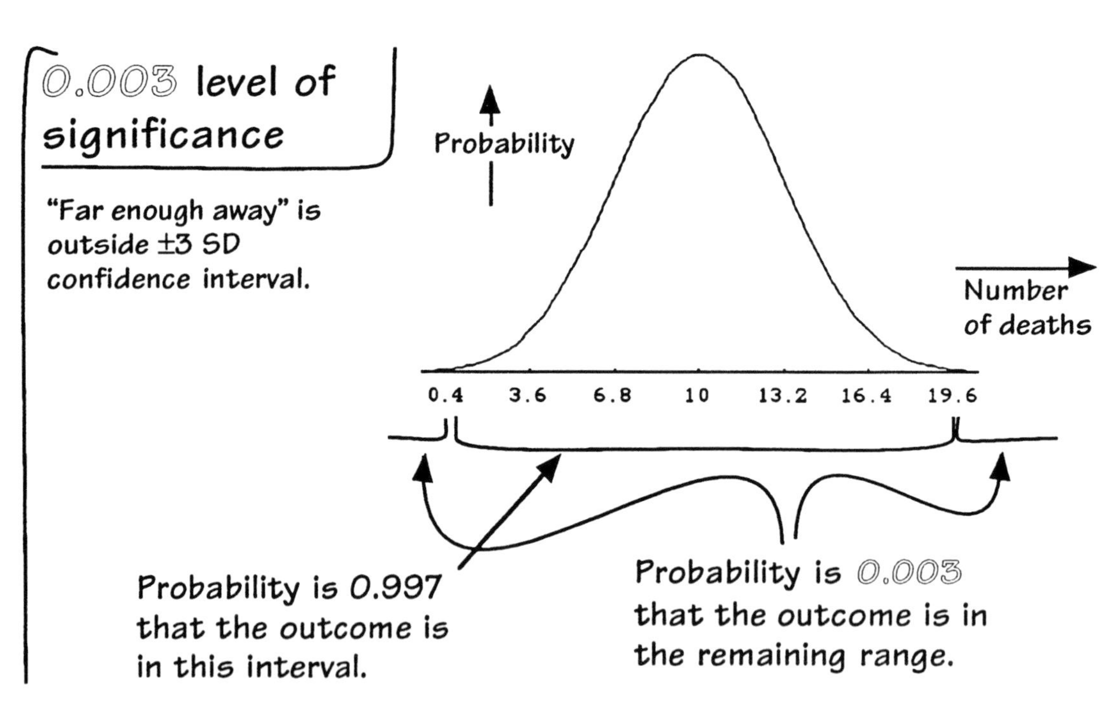

In sum: An outcome is statistically significant at the 0.003 level. = The outcome can happen with probability 0.003 if the hypothesis is true.

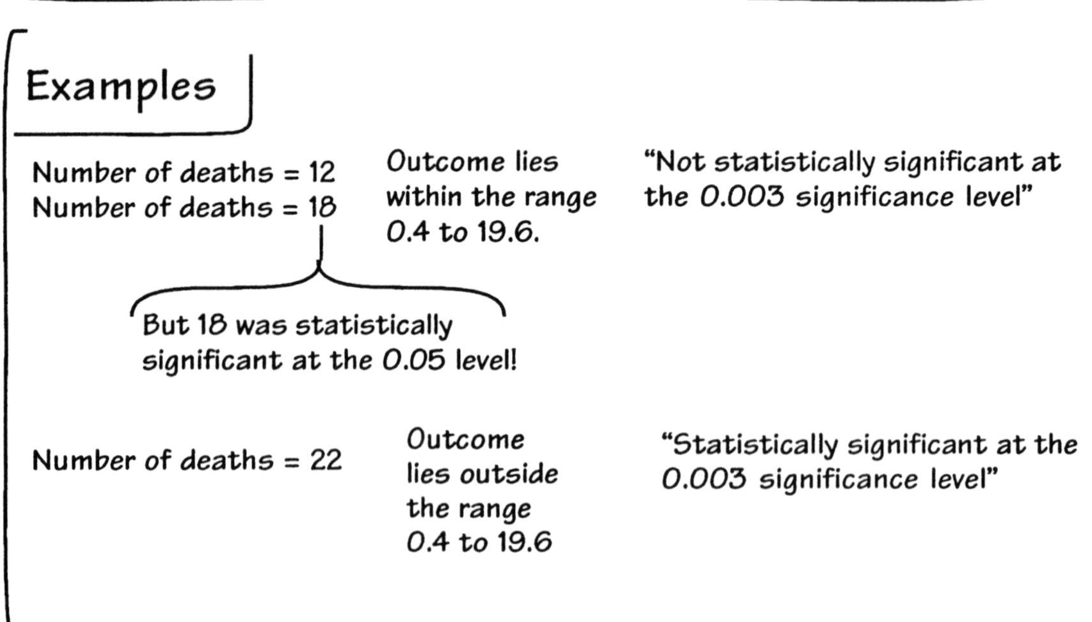

We are free to choose the level of significance. For example, we may pick 0.003. This level of significance is associated with the MEAN±3SD confidence interval. In this case it extends from 0.4 to 19.6 deaths. If we choose this level of significance, we will accept the hypothesis if the outcome lies within this interval and reject it if the outcome lies outside. If the hypothesis is true, the probability that the outcome lies within this interval is 0.997; the probability that it lies outside is 1-0.997=0.003.

Outcomes such as 12 and 18 lie within the interval. These we describe as "not statistically significant at the 0.003 level". Outcomes such as 22 and 25 lie outside the interval. These we describe as "statistically significant at the 0.003 level".

While we are free to choose the level of significance, it is very important to state our choice when we report results. Simply stating that a result is not statistically significant is incomplete and potentially very misleading. Your reader will have to guess the level of significance and therefore also guess the meaning of the result. If we choose different levels of significance, then the significance of an outcome can change. Such a change is illustrated in the examples opposite. When we worked at a level of significance of 0.05, the outcome of 18 deaths *was* statistically significant. But now, when we move to a level of significance of 0.003, the outcome is deemed *not* statistically significant. This shift of status would happen if we had any outcome that lay within 2 and 3 standard deviations of the mean.

The choice of level of significance can profoundly alter the interpretation of the outcomes. Therefore it is essential to state the level chosen in reporting your results. You may have the misfortune of encountering results for which a level of significance is not reported. You will have to guess the level of significance. The best choice is 0.05 since that level is most commonly used--but there are no guarantees that it was used in this case. Note finally that the level of significance may be reported indirectly. For example, a 0.05 level may be indicated by a remark that the outcome could only happen by chance in 5% of cases.

Choice of level of significance: no easy answers

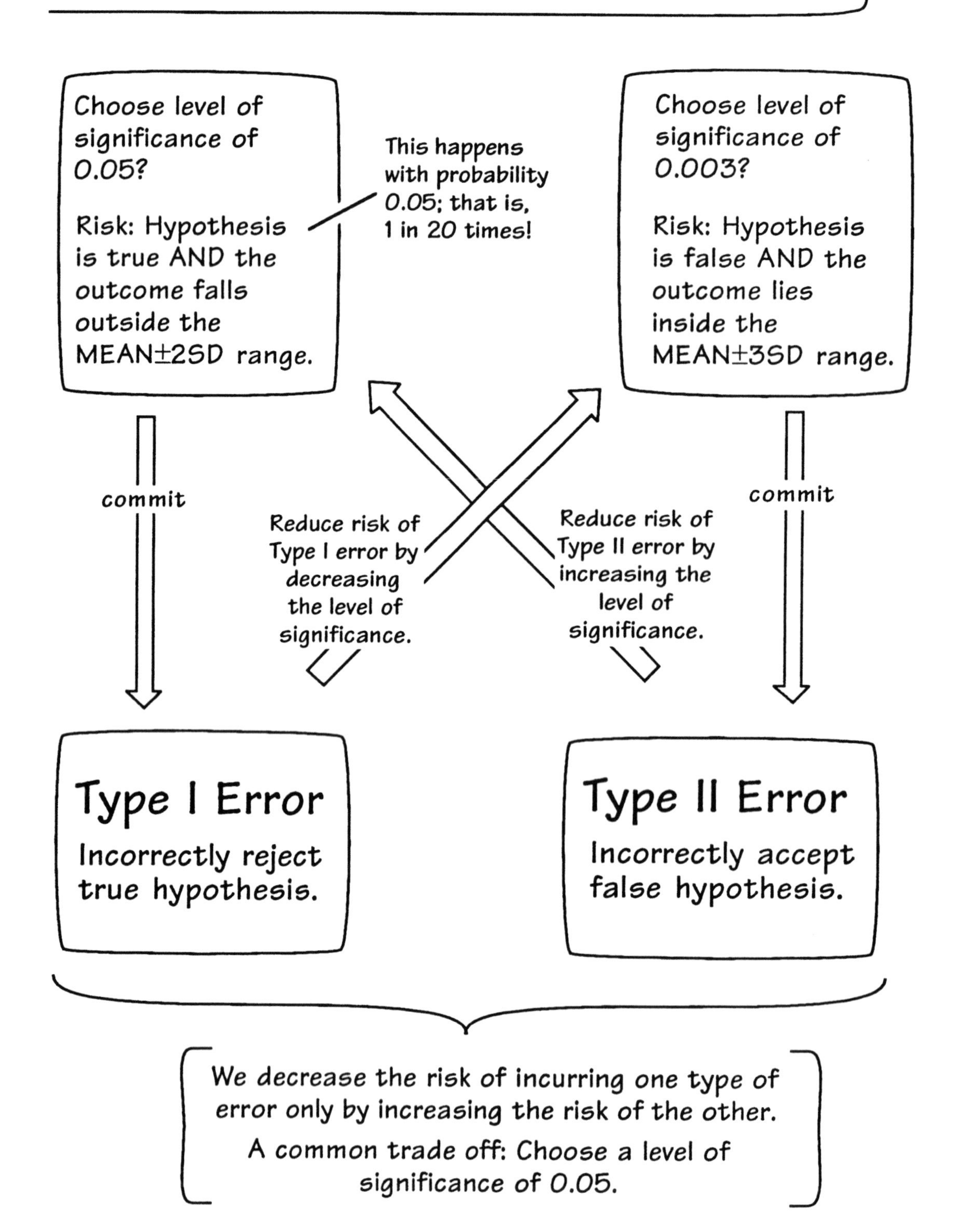

We decrease the risk of incurring one type of error only by increasing the risk of the other.

A common trade off: Choose a level of significance of 0.05.

We must choose the level of significance wisely, since this choice greatly affects the interpretation of our results. Unfortunately there are no easy answers to the question of what level of significance we should choose. In hypothesis testing, there are two types of errors we can make. We can alter the level of significance to reduce one type, but we do so at the cost of increasing risk of the other type of error.

To see how this unfortunate loop arises, imagine that we select a level of significance of 0.05. At this level, there is a probability of 0.05 that the hypothesis is true and the outcome lies outside the MEAN±2SD interval. We would label this outcome statistically significant (at the 0.05 level) and reject the hypothesis. This would be an instance of an error of Type I: incorrectly rejecting a true hypothesis. It would be a very unfortunate error. Yet it is one that we are sure to make eventually. With each hypothesis test, the probability that we make it is 0.05--a chance of 1 in 20. So, on average, we will commit this error 1 in 20 times! Oh no!

We can reduce the chance of this Type I error by decreasing the level of significance. Say we reduce it to 0.003. We commit a Type I error with probability 0.003--3 in 1,000 times. Now that is a great deal better! Errors of Type I have become infrequent. But there is a cost. In reducing the level of significance, we have greatly increased the size of the interval of outcomes that lead us to accept the hypothesis. It has grown from MEAN±2SD to MEAN±3SD, a growth of 50%. This 50% growth will allow us to detect only a few more cases of a true hypothesis: our success rate grows from 95% to 99.7%. But our exposure to a Type II error--incorrectly accepting a false hypothesis--is much greater. For if some other hypothesis is true, it may lead to outcomes in the new range we added. Just how serious our problem is depends on what other hypotheses may be true. In our simplest of analyses here, we have not considered other candidate hypotheses. In a more complete analysis that does consider them, we can quantify our exposure to Type II errors more precisely. Of course the remedy is to increase the level of significance, say, back up to 0.05. But now we must face the risk of Type I errors again. The loop is closed!

In practice, the choice of level of significance can be directed by our relative fear of Type I and Type II errors. We may choose a smaller level of significance if we especially fear Type I errors, for example. A common trade-off is the 0.05 level.

USAir Flight 427 , Sept. 8, 1994 132 deaths

Five crashes in five years. Do we have a real basis for worry about USAir's safety?

Question is too vague. We must specify more clearly what our worry is.

Candidates

* Is there a chance of dying when we fly USAir? ⇨ Yes, but too simple. Air travel involves some risk, but it is TINY! Chance of death per mile of regularly scheduled airline travel for one passenger: 1 in 5,000,000,000

* Is flying USAir more dangerous than other forms of transport? ⇨ Probably not. Safe low risk driver (age 40, wears seat belts, heavier than average car, safe roads -- e.g. rural interstates) Chance of death per mile traveled: 1 in 1,200,000,000

* Is travel on USAir more dangerous than on other airlines? ⇨ This gets closer to our worry. But we are not quite there since we are not interested in risks not specifically due to USAir.

- Chance of death far greater in smaller planes. → Chance of death per mile in general (=private) aviation = 1 in 220,000,000
- Chance of death far greater in take-off and landing. → Airline that has shorter flights will have greater chance of accidents *per mile flown* since it will have more take-offs/landings *per mile flown*.
- Local geography → Are some areas more prone to dangerous weather, more crowded, etc.?

Let us now consider another example. When USAir flight 427 crashed in September 1994 killing all 132 on board, there was considerable concern over the safety of the airline. It had amassed a record of five fatal crashes in five years. Does USAir's safety record give us good reason to doubt the airline's safety? This is a question of real concern to us and can only be answered by a statistical analysis.

So far we have emphasized the mathematical part of statistical analysis. In practice, it represents a smaller and easier part of the entire analysis. Before it can be done, we need to do two things that can be much harder. We must formulate our question appropriately and we must find data adequate to answer it. To see the problems of formulating the question, let us work through some candidates as shown opposite.

We are not worried merely by the risk of death on a USAir flight. All airline travel involves risks. We are not concerned to compare the risks of airline travel against other forms of travel. Airline travel seems to be safer than other forms of travel. In the figures shown opposite, the chance of a fatality if one makes a trip by road is 4 times greater than if the same trip was made by air. (These figures oversimplify a little since the dangers in air travel are distributed unevenly, being greater in takeoff and landing.)

Our worry is about USAir specifically. But it is not simply whether we have a greater chance of death flying on USAir. Many other factors can affect the safety of an airline. Fatal accidents are far more likely in take-off and landing. So an airline that flies mostly shorter flights can expect a greater accident rate *per mile flown*, since it has far more take-offs and landings per mile flown. Similarly, smaller airplanes are less safe than larger ones. So an airline that flies smaller planes can expect a poorer safety record even if its safety practices are as impeccable as an airline flying larger machines. Again, an airline whose routes lie in unfavorable geographic areas can expect a poorer record, all other things equal. In all these cases, the risks come from factors not directly under the airline's control. Should we penalize an airline since it supplies the shorter, more accident prone routes we choose to fly?

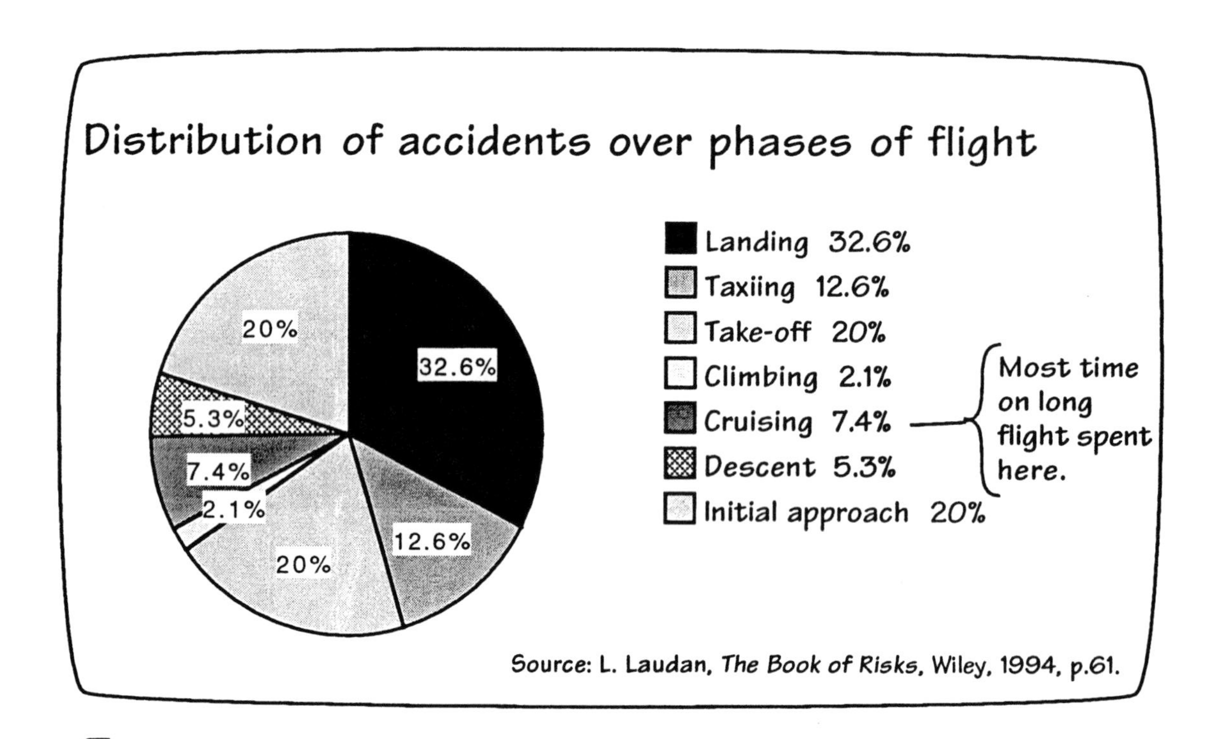

* Is USAir more risky compared to other airlines on the same routes flying the same equipment? ← This is what we really want to know--but can we find data on which to base an analysis?

Ideal test? USAir and other airlines fly the same routes with the equipment many times. Then compare accident rates.

Problems:

1. We cannot do the study prospectively.

2. The study cannot be blind.

3. The rate of serious accidents is very low--2 per million flights. Can we collect enough data to allow statistical trends to emerge?

The chart opposite shows the distribution of accidents over the different phases of flight. What is interesting is how uneven that distribution is. On long flights, most time is spent in cruising. But merely 7.4% of accidents happen during this phase of flight. The most dangerous phase with landing. Over 50% of accidents occur during the initial approach to landing and in the landing itself.

Data such as these indicate what we wish to know: imagine that we can hold fixed all factors such as differences in routes, flight durations and equipment, would USAir still be a riskier airline?

This question suggests a study. We would have each airline fly the same routes with the same equipment many times. We would then compare their accident rates. It does not take long to see that this is not a realistic proposal and that it may not tell us what we want to know even if we could do it.

First, the expense and risk make it unlikely that the airlines would agree to the test. To fly the same routes with the same equipment would require a huge dislocation to existing schedules. If the dislocation was associated with an accident, would we want to blame the study for this potentially deadly outcome?

Second, it would probably be impossible to hide the purpose of the test from the airlines. So they would know its purpose and we could not be sure that this knowledge would not consciously or unconsciously affect operating procedures.

Third, accident rates in the airline industry are, fortunately, very low--about 2 per million flights. Therefore the study would need a huge number of fights before any positive accident data could be amassed and then hugely more for trends to emerge. After a million flights we would expect there to be *about* 2 accidents. But it may be 0 or 1 or 3 or ... So, if one airline has had 2 accidents and another 1 in a million flights, we can hardly conclude that the first is twice as dangerous. We need many more flights to be flown for the true trends to be apparent. This would make the test vastly expensive.

Airline	Accidents	Total number of flights (millions)	Accidents per million flights
USAir	14	5.38	2.6
American	12	4.80	2.5
United	12	3.87	3.1
Delta	10	5.26	1.9
Continental	9	2.65	3.4
American West	5	1.16	4.3
Northwest	3	3.00	1
TWA	2	1.54	1.3
Southwest	2	2.22	0.9
Grand totals	69	29.89	2.31
Grand totals excluding USAir	55	24.50	2.24
Grand totals excluding American West	64	28.73	2.23

Data, reported in *US News & World Report* Sept. 19, 1994, p. 30. For 1989-July 7, 1994 except for USAir which runs through to Sept. 9, 1994.

"Accident" --"involve death or serious passenger injury or substantial damage to plane"

These data form a serviceable basis for analysis.

- Data presented as accidents per flight. → Tends to correct for greater risk of accident in take-off and landing.
- Data is for major carriers. → Covers larger planes on typical routes.
- Data is for "accidents" (=catastrophes) and near misses of catastrophes. → More incidents reported so trends can emerge.

These considerations show that our test will have to be retrospective; that is, it will have to rely on an analysis of the history of the airline industry. At least this history does contain the many millions of flights needed to allow trends to emerge.

The table of data opposite provides a serviceable, but not ideal, basis for this analysis. It covers the safety record of major carriers for the five years prior to the September 1995 crash. (The published data consisted just of the the number of accidents and the accidents per million flights. The remainder of the table was calculated from them.)

USAir records the most accidents of all carriers. But this is not so surprising since USAir recorded the most flights in accruing these accidents. If we compare the accident rates per million flights, USAir's record looks a lot better. The industry average is 2.3 accidents per million flights. USAir has 2.6 accidents per million flights--close enough for us to suspect that its performance is typical.

One deficiency in the data is the reporting period has been artificially extended in the case of USAir to so that the catastrophic crash of September 8, 1994, is *just* included in the period. This biases the data against USAir. Every airline would fare worse if we extended the reporting period so that its latest accident is just included. It guarantees recording of at least one extra accident for that airline.

This presentation of data does reduce or avoid other problems, however, Since the data is expressed as accidents *per flight*, there is some correction for the greater risks of take-off and landing. Reporting accidents per mile flown unfairly favors carriers who fly longer flights. They can dilute the dangers of take-off and landing over the greater miles. Since the data is for major carriers, there will be some uniformity in the type of equipment and routes flown, although carriers that specialize in one or other area may gain or suffer from special conditions in their areas. Finally, the data records more than catastrophic accidents. It also counts serious accidents that may not have lead to any deaths. There are more of these less catastrophic accidents, so that trends in accident distribution can emerge more reliably from the data. We would assume that a propensity for less catastrophic accidents goes with a propensity for more catastrophic accidents. So, if we measure the former, we have a good indication of the latter.

Hypothesis to be tested:

USAir is just as risky as other airlines.

The probability of an accident on one flight is the same for USAir as for other carriers
= 2.24 in 1,000,000
=0.00000224

We read this from the table. The accident rate for carriers *other than* USAir is 2.24 per million flights. This translates into a probability of 2.24 /1,000,000 for one flight.

"Experiment" to test hypothesis:

USAir flies 5.38 million flights 1989-94 — Analogous to 5.38 million die throws!

Outcome:

Actual number of accidents = 14 — Analogous to the number times we throw a ⚂

Is this actual number of accidents compatible with the interval expected if the hypothesis is true?

Fit a bell curve to the probabilities of the different outcomes.

$$\text{MEAN} = \text{Expected number of accidents} = \text{Probability of accident} \times \text{Total number of flights}$$

$$= 0.00000224 \times 5{,}380{,}000 = 12.05$$

$$\text{STANDARD DEVIATION} = \sqrt{\frac{\text{MEAN X} \left(\text{Total number of fights} - \text{MEAN}\right)}{\text{Total number of fights}}} = 3.47$$

To answer our question about USAir, we will conduct a test of the hypothesis that USAir is just as risky as other airlines. We will take this to mean that the probability of an accident on a USAir flight is the same as the probability of an accident on other carriers. Since the accident rate on other carriers is 2.24 accidents per million flights, the probability of an accident on just one flight is 2.24 per million.

Notice that the figure of 2.24 per million comes from the performance of carriers *excluding* USAir. This is essential. We cannot use the overall industry average of 2.31 accidents per million flights. USAir's own safety record is included in the data used to calculate this figure of 2.31 per million. If we used the 2.31 per million figure, we would be testing USAir partly against itself!

The test is a little more severe than it should be. The probability of an accident on a non-USAir flight is not 2.24 per million. The figure of 2.24 per million is only an *estimate* of that probability. It is the accident rate for 24.5 million flights in the 5 years prior to 1994. It is compatible with a probability of an accident that may also be slightly more or slightly less than 2.24 per million. A more complete statistical analysis would allow for this uncertainty, but its techniques extend beyond our methods. Fortunately it will turn out that the more complete analysis would not alter our results. This analysis would be slightly kinder to USAir. But USAir will not need it, since it is about to pass the test handily!

The test proceeds as before. USAir flies 5.38 million flights in the period for which we have records. This is the number of trials and corresponds to the number times we would throw a die in our original analysis. If the hypothesis is true, then there is a probability of 0.00000224 on each flight of an accident. From this we can compute the probabilities of various numbers of accidents: 0, 1, 2, ... The experiment has been done. USAir has flown its flights and there were 14 accidents. Is this outcome of 14 in the interval expected if the hypothesis is true?

To answer this question, we proceed to fit a bell curve to the probabilities of the different outcomes. The MEAN is just the expected number of accidents and is 12.05. To find the STANDARD DEVIATION from the customary formula, we substitute values for the MEAN, 12.05, and the total number of trials, 5,380,000. The result is 3.47.

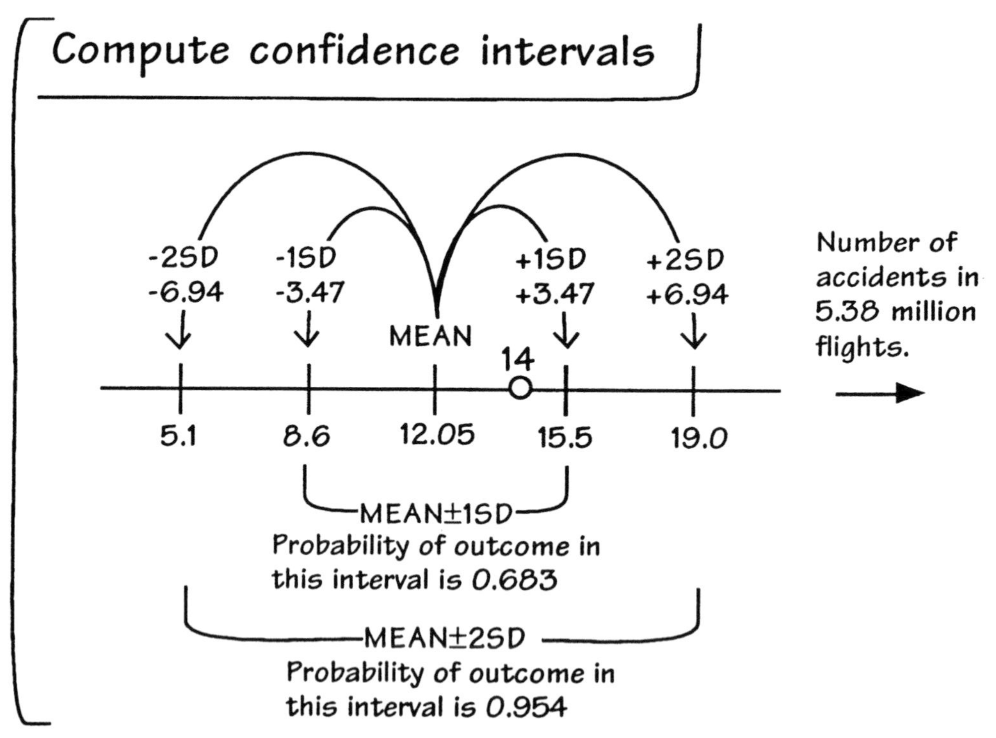
Compute confidence intervals
-2SD
-6.94
-1SD
-3.47
+1SD
+3.47
+2SD
+6.94
MEAN
14
Number of accidents in 5.38 million flights.
5.1
8.6
12.05
15.5
19.0
MEAN±1SD
Probability of outcome in this interval is 0.683
MEAN±2SD
Probability of outcome in this interval is 0.954

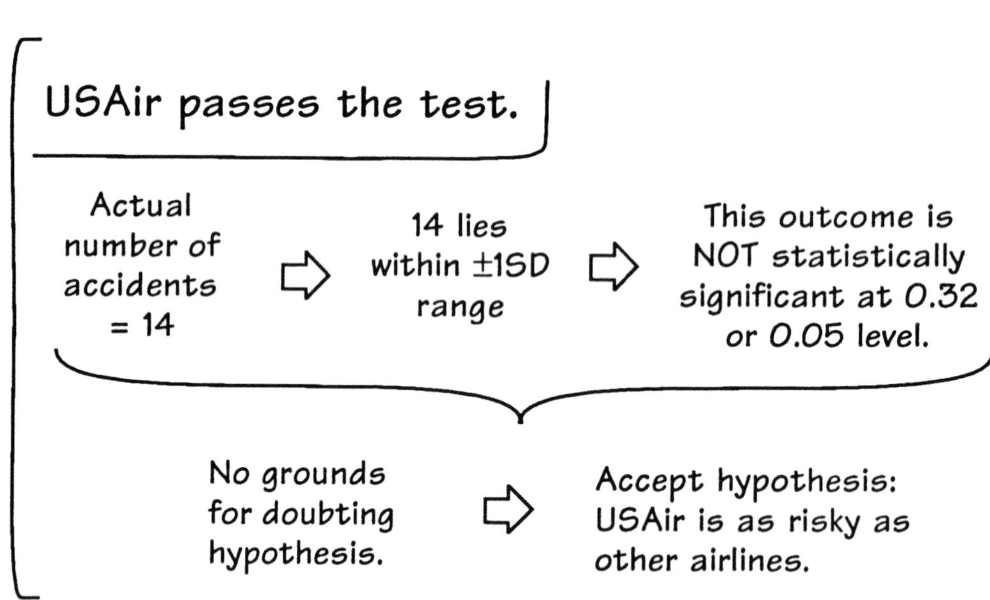
USAir passes the test.
Actual number of accidents = 14
14 lies within ±1SD range
This outcome is NOT statistically significant at 0.32 or 0.05 level.
No grounds for doubting hypothesis.
Accept hypothesis: USAir is as risky as other airlines.

We can now compute the MEAN$\pm$1SD and MEAN$\pm$2SD confidence intervals in the usual manner. They run from 8.6 to 15.5 and from 5.1 to 19.0, respectively. The actual outcome is 14. This lies within both confidence intervals. In other words, the actual number of accidents recorded by USAir lies right within the intervals expected if the hypothesis is true, that is, if the probability of an accident on a flight is 0.00000224. Thus the outcome is not statistically significant in either the 0.32 or 0.05 levels. We have no grounds for doubting the hypothesis.

On calculating the STANDARD DEVIATION

$$\text{STANDARD DEVIATION} = \sqrt{\frac{12.05 \times (5{,}380{,}000 - 12.05)}{5{,}380{,}000}} = 3.47$$

$$\frac{(5{,}380{,}000 - 12.05)}{5{,}380{,}000} \approx 1$$

(1) Compute the part in the box first to avoid overflow errors on your calculator.

(2) Because the part in the box is almost exactly equal to one, the expression for the STANDARD DEVIATION is greatly simplified:

$$\text{STANDARD DEVIATION} \approx \sqrt{\text{MEAN} \times 1} = \sqrt{\text{MEAN}}$$

$$\sqrt{12.05} = 3.47$$

This simplification will hold whenever:

MEAN	is very much less than	Total number of trials.
e.g. 12.05	is very much less than	5,380,000.

The calculation of the STANDARD DEVIATION in the last test raises some important practical and theoretical questions.

You must be a little cautious in computing the STANDARD DEVIATION on a hand calculator. You might start out mechanically and compute the top line under the square root. That is, you would compute 12.05x(5,380,000-12.05)=64,829,000. This is a big number! It is getting very close to the largest number that many calculators can handle. If the MEAN had been 120.5, then the corresponding number would have been 648,290,000. With a number this large, many calculators would report an overflow error--the number is too big. You would not be able to complete the calculation. A simple habit avoids the problem. The trick is to compute the part of the formula in the box first, that is, (5,380,000-12.05)/5,380,000. In this case, its value is very close to 1. The calculation is completed by multiplying by 12.05 and taking the square root.

That the part in the box is almost exactly one reveals a remarkable property of the formula for the STANDARD DEVIATION. When the part in the box is almost 1, then the STANDARD DEVIATION can be calculated merely by taking the square root of the MEAN. So, in our last example, the MEAN is 12.05. Its square root is 3.47, which is the STANDARD DEVIATION calculated through the full formula.

This simplification will happen whenever the MEAN is very much less than the total number of trials. It has a practical and theoretical significance. Practically, it provide a quick way of determining the STANDARD DEVIATION. When the mean is very much smaller than the number of trials, you can find the STANDARD DEVIATION just by taking the square root of the MEAN. Many people will be able to do this in their heads! The theoretical importance is that the exact value of the total number of trials does not enter into the calculation or affect the result. This means that we can calculate our confidence intervals without knowing the total number of trials exactly. For example, you may know that the expected number of cases of some rare disease in a country is 100. The total number of trials--the overall population--is very much greater than this MEAN of 100. So we can use the approximation and find that the SD is 10. We know that there is a probability of 0.954 that the actual number of cases of the disease will lie in the interval 100±(2x10), that is, from 80 to 120. Calculations cannot be much easier and faster than this!

American West

Accident rate of 4.3 per million flights. = Highest of all airlines tabulated.

Test hypothesis: American West is just as risky as other airlines.

The probability of an accident on one flight is just the same on American West as on other carriers:
=2.23 in 1,000,000
=0.00000223

Accident rate for carriers excluding American West.

⇓

Fit a bell curve to the probabilities of different numbers of accidents in the 1.16 million flights flown in period reported.

⇓

MEAN = 0.00000223 x 1,160,000 = 2.6

STANDARD DEVIATION $= \sqrt{2.6} = 1.6$

MEAN is very much less than total number of trials. Therefore approximate SD by $\sqrt{\text{MEAN}}$

⇓

Compute MEAN±2SD confidence interval.

⇓

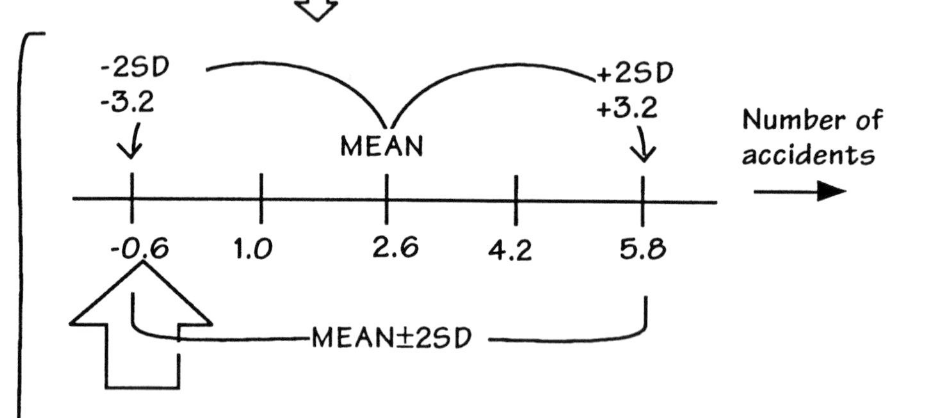

DANGER! A negative value is physically impossible. The bell curve approximation fails!

USAir's record of accidents is well within the range expected if its accident rate is the same as other airlines. This is not surprising since its rate of 2.6 per million flights is closer to the industry average than most other airlines. But what about the other airlines? American West has the highest accident rate: 4.3 per million flights. Is it possible that American West is more risky? We will repeat the test carried out on USAir with American West, first to see if there is a basis for concern in the data and, second, because it will illustrate an important point of methodology.

We will test the hypothesis that American West is just as risky as other airlines. We take this to mean that the probability of an accident on American West is given by the accident rate among carriers other than American West. This figure was calculated in the data table given earlier and is 2.23 in 1,000,000 or 0.00000223. (Note as before that we make the test slightly more strict by assuming that this figure gives the exact probability rather than an estimate of it. A more sophisticated test that corrects for this will be more lenient, since it cannot rule out the possibility that the true probability of an accident in the industry is closer to American West's accident rate.)

If the hypothesis is true, there will be certain definite probabilities for 0, 1, 2, ... accidents. We will fit a bell curve to these probabilities. The Calculation is shown opposite. The MEAN, the expected number of accidents, is 2.6. The STANDARD DEVIATION is computed using the approximation formula discussed above. It is 1.6. We now compute the MEAN±2SD confidence interval, the interval in which the number of accidents will lie with probability 0.95 if the hypothesis is true.

This interval extends from -0.6 to 5.8. At this point we must halt. The confidence interval extends past 0 into negative numbers. This is physically impossible. What can it mean for there to be a negative number of accidents?! What has happened is that the bell curve approximation has failed. If you compute a confidence interval like this that extends into a physically impossible range of values, then you have shown that the bell curve approximation has failed. The probabilities do not conform to the bell curve.

Why did the bell curve approximation fail?

Compute probabilities of different numbers of accidents:

Number of accidents	Probability
0	0.0752603
1	0.194684
2	0.251804
3	0.217123
4	0.140413
5	0.072644
6	0.0313192
7	0.0115738
8	0.00374236
9	0.00107563
10	0.000278243

Total number of flights is 1,160,000. Probability of an accident on one flight is 0.00000223.

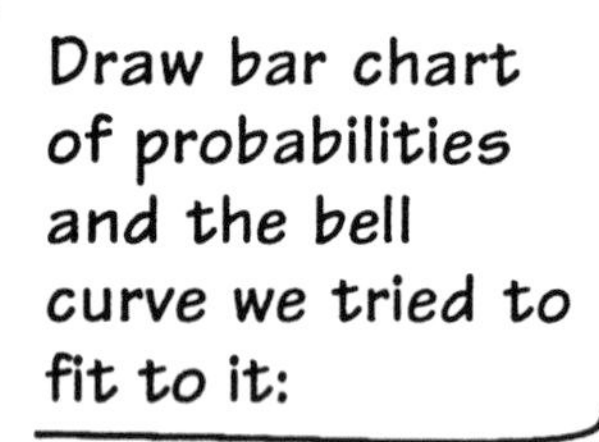
Draw bar chart of probabilities and the bell curve we tried to fit to it:

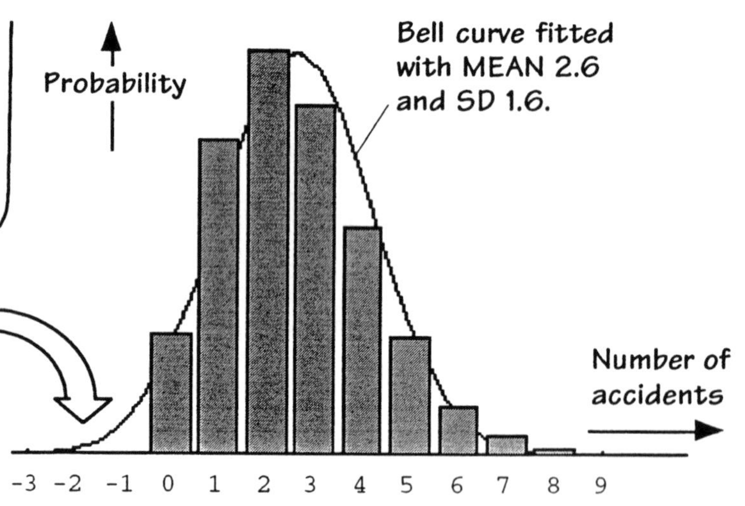

From table: Probability that the number accidents lies in the interval 0 to 5
= 0.0753+0.1947+...+0.0726
= 0.952

⇨ Actual number of accidents 5 *just* lies in this interval.

⇨ Accept the hypothesis that American West is just as risky as other airlines.

Why did the bell curve approximation fail? A few calculations will tell us. We can calculate the probabilities of the various numbers of accidents directly using the tedious methods outlined in the last chapter. These results, calculated with the aid of a computer, are shown opposite. We can then represent these probabilities in a bar chart. Note that there are bars for numbers of accidents ..., 3, 2, 1, 0. But they do not extend below 0, since it makes no sense to talk about negative numbers of accidents. The bar chart also shows the bell curve that was fitted by our earlier calculation. The peak of the curve is located at the MEAN, 2.6. It has a long "tail" that extends to the right and fits the probabilities of higher numbers of accidents, 5, 6, 7, 8, ... The bell curve has the same shape on each side of the mean. Therefore it must have a "tail" of the same shape extending from the peak at 2.6 towards smaller numbers of accidents. This tail fails to fit the probabilities, for they cut off abruptly at 0 accidents.

We are assured that a bell curve will fit probabilities in situations like this, so why did it fail? That assurance includes the requirement that there be very many trials before the bell curve fits. In the case of the repeated die throws, we needed to have 60 or more throws before the probabilities began to conform to the bell curve. In our present example, we do not yet have enough trials. Not enough trials, you say! We have 1,160,000 trials! How can that not be enough? The index that tells us when we have "very many" is the expected number of successes. This number is just the mean, which is calculated by multiplying the probability of success by the number of trials. It must be greater than 5. Thus, for die throws where the probability of success is 1/6, 1,160,000 throws would be very, very many; the expected number of success is 1/6 x 1,160,000 = 193,333. But for American West it is small, since the expected number of success is just 2.6

Finally, we can use our table of probabilities to see if our original hypothesis about American West passes the test. We can calculate the probability that the number of accidents lies in the interval 0 to 5 just by adding the five probabilities on the table. We find that there is a probability of 0.952 that the number of accidents lies in this interval, if the hypothesis is true. The actual number of accidents is 5. It just lies inside this interval. So, working at the 0.05 significance level, we would accept the hypothesis that American West is as risky as other airlines--although on these data, the airline only just passes the test.

Assignment 11: Testing Hypotheses

1. On average, across the nation, 1 in 43 births are twins. That means that the probability of a birth being twins is 1/43 = 0.023=2.3%. In a certain town, there is a concern that there in an abnormal rate of twin births. You have been asked to determine whether the twin birthrate in the town is compatible with the national average. That is, you are to test the truth of the hypothesis: "The twin birth rate in this town is compatible with the national average."

(a) State the hypothesis that will be tested in the form:
"The probability that a birth in the town is a twin birth is ..." --you complete the sentence with a number.

You expect twin births to be distributed probabilistically and you proceed to fit a bell curve to the distribution, assuming that there are 1,000 births overall in the town in one year.

<u>If the hypothesis stated in 1(a) is true:</u>

(b) What is the mean of the bell curve that depicts the probabilities of different numbers of twin births among the 1,000 burths in the town?

(c) What is the standard deviation of this curve?

(d) What are the confidence intervals:

MEAN +/- 1SD
MEAN +/- 2SD
MEAN +/- 3SD

2. Now assume that you learn that there were 35 twin births among the 1,000 births in the town for the year in question. This rate of 35/1000 or 3.5% is 50% higher than the national average of 2.3%.

(a) Is the outcome of 35 twin births compatible with the hypothesis of 1(a) if we expect the outcome to be in the range predicted by the hypothesis with 0.95 probability? (Explain!)

(b) Is the outcome of 35 twin births compatible with the hypothesis of 1(a) if we expect the outcome to be in the range predicted by the hypothesis with 0.997 probability? (Explain!)

(c) What does it mean to say that an outcome is statistically significant at the .05 significance level?

(d) Is the outcome of 35 births statistically significant at the .05 significance level?

(e) What does it mean to say that an outcome is statistically significant at the .003 significance level?

(f) Is the outcome of 35 births statistically significant at the .003 significance level?

3. When possible anomalies are reported in the press, it is quite common for them to be reported too incompletely for statistical analysis. In a case such as this, it would be very common merely to state that the actual rate of twin births is 3.5% and that this is 50% higher than the national average of 2.3%. If the report does not also give the total number of births, one cannot judge whether the rate is within the range to be expected when there are 2.3% twin births on average or whether it is outside that range.

To get a sense of just how critical it is to know the total number, recompute 1(b), (c) and (d) for the case in which the total number of births is 400 so that a 3.5% twin birth rate corresponds to 14 births. That is, still assuming that the hypothesis of 1(a) is true:

(a)What is the mean of the bell curve that depicts the probabilities of different numbers of twin births in the town?

(b) What is the standard deviation of this curve.

(c) What are the confidence intervals:

MEAN +/- 1SD
MEAN +/- 2SD
MEAN +/- 3SD

(d) In this new case, how would your evaluation of the hypothesis differ?

4. These calculations involve approximating the distribution of probability by a bell curve. That approximation can fail if the bell gets too close to the zero mark of the scale. To see this recompute the same quantities as in Question 3. but with the total number of births now set at 86. That is, the twin birth rate is still 3.5%, but it now corresponds to a total number of .035 x 86 = 3 births. Compute:

(a)What is the mean of the bell curve that depicts the probabilities of different numbers of twin births in the town?

(b) What is the standard deviation of this curve?

(c) What are the confidence intervals:

MEAN +/- 1SD
MEAN +/- 2SD
MEAN +/- 3SD

Notice that the confidence intervals now extend past the 0 twin births point which shows that for the small sample size in this example, the approximation is not reliable.

Chapter 12

Applications of the Bell Curve II

Sampling and Estimation

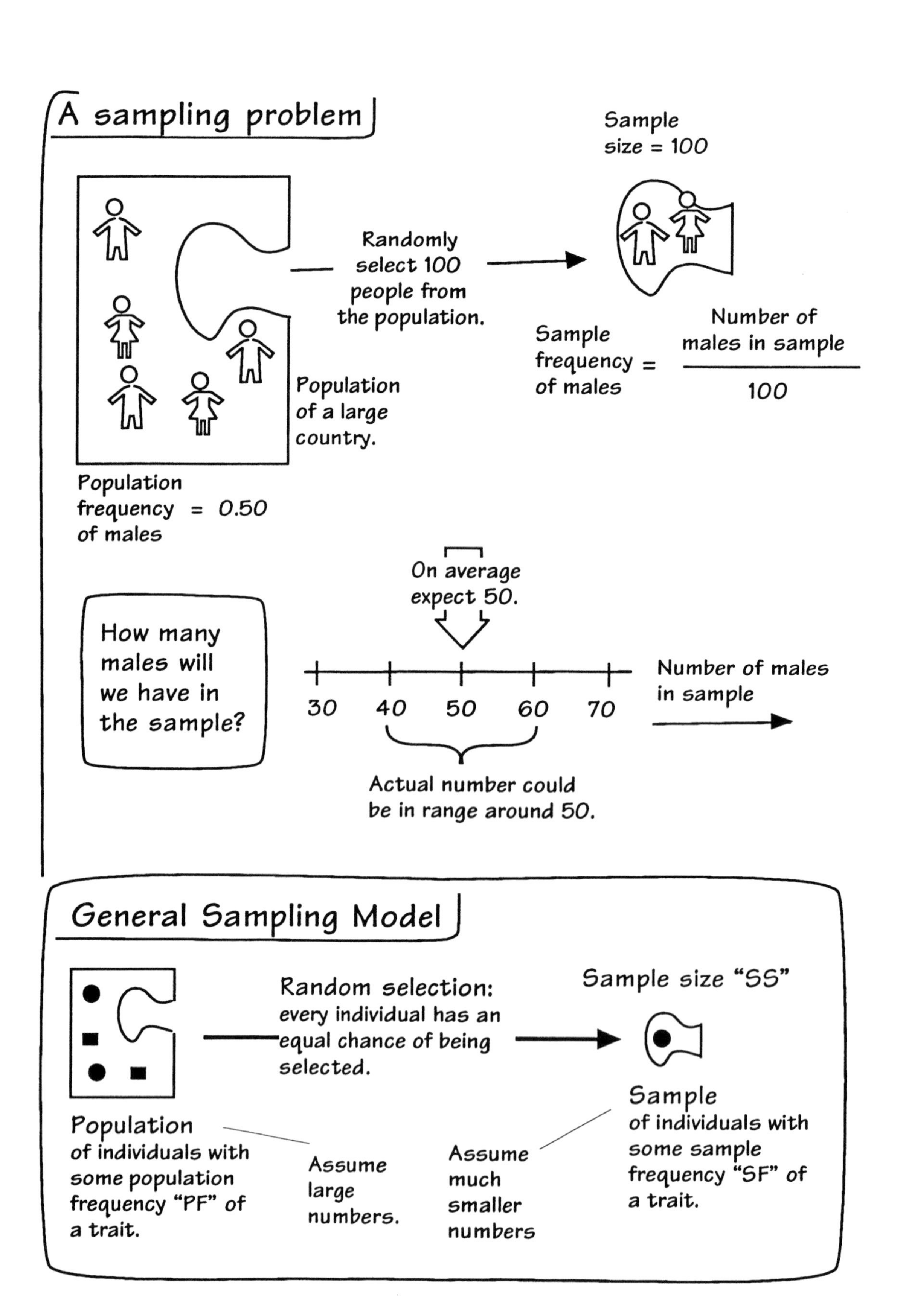

A sampling problem
Sample size = 100
Randomly select 100 people from the population.
Population of a large country.
Sample frequency of males = Number of males in sample / 100
Population frequency of males = 0.50
On average expect 50.
How many males will we have in the sample?
30
40
50
60
70
Number of males in sample
Actual number could be in range around 50.
General Sampling Model
Random selection: every individual has an equal chance of being selected.
Sample size "SS"
Population of individuals with some population frequency "PF" of a trait.
Assume large numbers.
Assume much smaller numbers
Sample of individuals with some sample frequency "SF" of a trait.

The techniques that we have learned with the bell curve have many more applications. In this chapter, we will look at how these techniques can be used in sampling and estimation problems. A typical sampling problem is shown opposite. We imagine a country with a large population and in which 50% of the people are male. We will take a random sample of 100 people. How many of these people will be male?

We expect about 50% of the sample to be male; so we expect about 50 males in the sample of 100. The "about" gives away our focus. For we cannot expect *exactly* 50 males. There could be slightly fewer or slightly more. It seems plausible that we may have 48 or 52. As we stray further from the 50, at what point have we gone too far? That will determine an interval in which the number of males will probably be found. The question and the style of its answer should sound familiar, so you can start to guess how our analysis will proceed.

Before we proceed, we need to fix some terminology. Our examples will conform to a general sampling model. This model will have a population of individuals. There will be some trait of the individuals that interests us. The frequency of that trait in the population is the population frequency "PF". In our example, the population is a population of real people; the trait was being male; and the population frequency was 0.5. The populations need not be human. They could be loaves of bread in a bakery and the trait of interest whether a loaf is under weight. To sample we randomly draw a certain number of individuals from the population. In our model, it is essential that the selection be random. This means that every individual has an equal probability of being selected for the sample. If you like, you could imagine every individual's name on a slip of paper in a huge hat. The sample would be formed by shaking the hat well and blindly taking out slips. The number of individuals in the sample is called the sample size "SS". We will assume that the sample size is always very much smaller than the number of individuals in the population. The sample frequency "SF" is just the frequency of the trait in the sample. In our example, it is the frequency of maleness in the sample of 100. We expect it to have a value somewhere around 0.50.

Note that the population and sample frequencies are always proper fractions so their values lie in the interval 0 to 1.

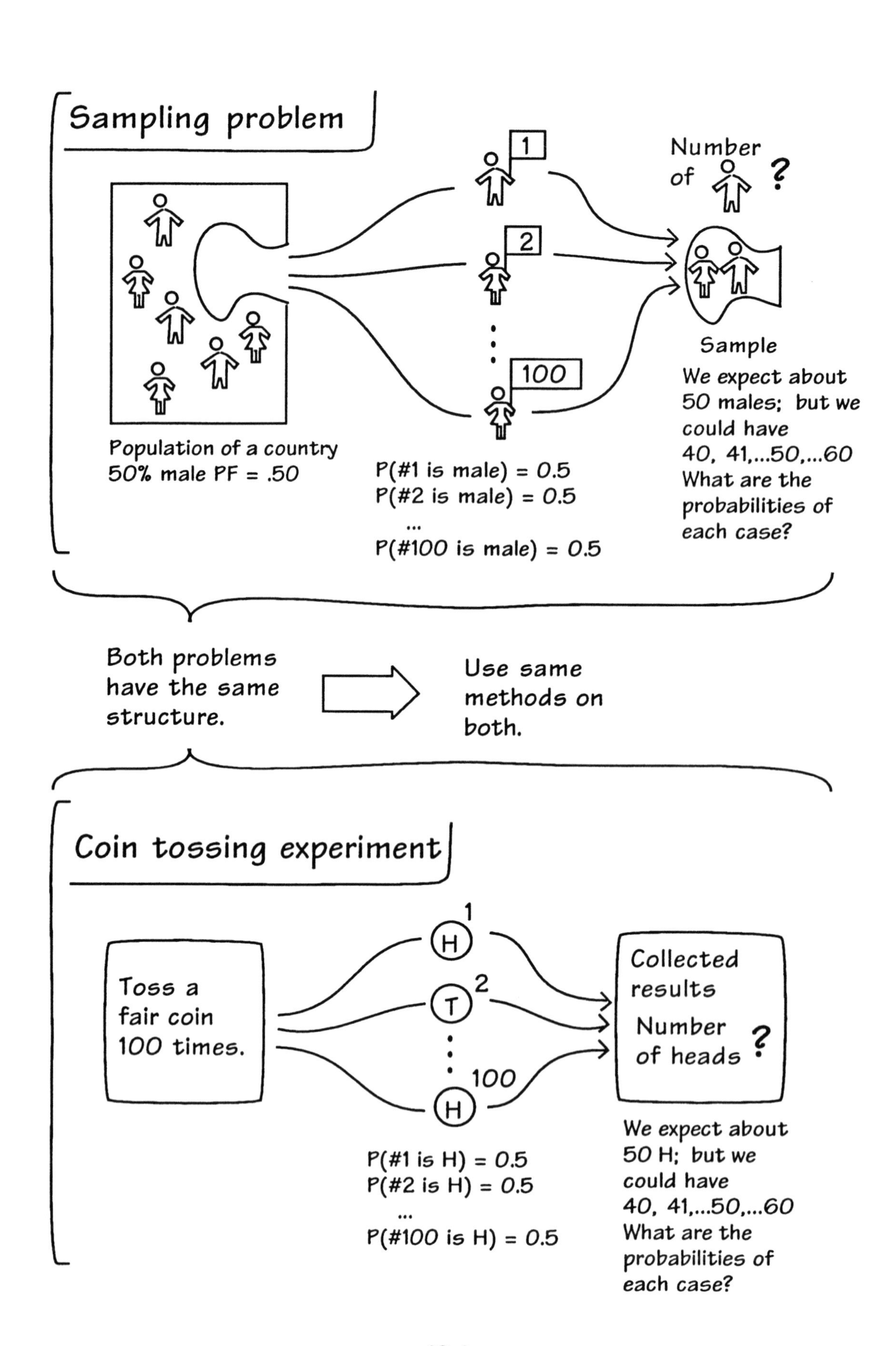
Sampling problem
1
2
100
Number of
?
Population of a country
50% male PF = .50
P(#1 is male) = 0.5
P(#2 is male) = 0.5
...
P(#100 is male) = 0.5
Sample
We expect about 50 males; but we could have 40, 41,...50,...60
What are the probabilities of each case?
Both problems have the same structure.
Use same methods on both.
Coin tossing experiment
Toss a fair coin 100 times.
1
H
2
T
100
H
Collected results
Number of heads ?
P(#1 is H) = 0.5
P(#2 is H) = 0.5
...
P(#100 is H) = 0.5
We expect about 50 H; but we could have 40, 41,...50,...60
What are the probabilities of each case?

How can we use the techniques we already know to treat the sampling problem? All we need to see is that the sampling problem has essentially the same structure as our familiar die throwing experiment. To see this, we will compare the sampling problem with one closely related to the die throwing problem, a coin tossing problem. If we use this coin tossing problem, the analogy becomes very clear since both problems employ the same numbers and probabilities.

In the coin tossing problem, we will throw a fair coin 100 times. There is a probability of 0.5 that each toss yields a head. We will count the number of heads in the 100 throws. We seek the probabilities of the various numbers of heads.

The two problems are closely analogous. The 100 people drawn in the sample are just like the 100 coin tosses. For each person, there is a probability of 0.5 that the person is male; for each coin toss, there is a probability of 0.5 that the coin lands head up. The outcome of a male on each of the individual drawings is independent of the outcomes of the other drawings; correspondingly, the outcome of each coin toss is independent of the other outcomes. After the 100 people of the sample are drawn, we will have some number of males in the sample. The probabilities of these different numbers is fully determined by the mechanism used to form the sample. Correspondingly, after 100 coin tosses, we will have some number of heads in the collected results. The probabilities of these different numbers is fully determined by the mechanism used to generate the results. Since the two mechanisms are mathematically identical, the probabilities of the various numbers of males in the sample will correspond exactly to the probabilities of the various numbers of heads in the collected results of the coin tossing experiment. These last probabilities conform to a bell curve and we can use the formulae of the last chapter to find its MEAN and STANDARD DEVIATION. Therefore the probabilities of different numbers of males in the sample will also conform to a bell curve and we can use the same methods to fit the bell curve.

Note that the analogy between the two problems holds as long as we assume that the sample size is very much smaller than the size of the population. In all the examples to follow, we will assume this. If the assumptions fails, then, whether the analogy holds depends on the precise methods used to take the sample. This problem is discussed further in the appendix.

Compute probabilities of different numbers of males in sample

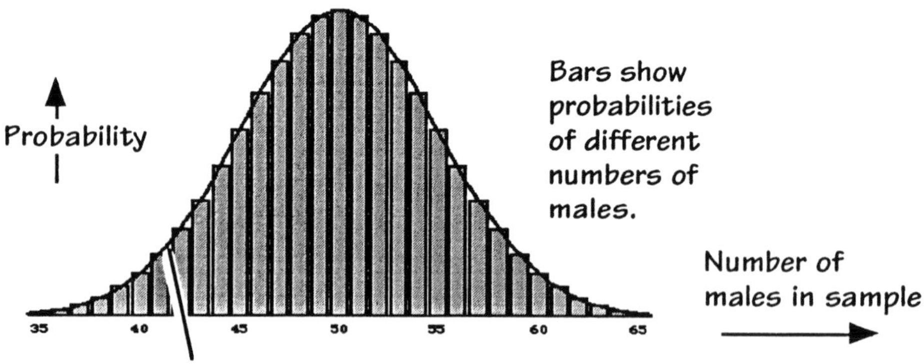

A bell curve is fitted by the values of:

MEAN = Expected number of males in sample = 0.5 x 100 = 50

$$\text{STANDARD DEVIATION} = \sqrt{\frac{\text{MEAN} \times (\text{Sample size} - \text{MEAN})}{\text{Sample size}}} = \sqrt{\frac{50 \times (100 - 50)}{100}} = 5$$

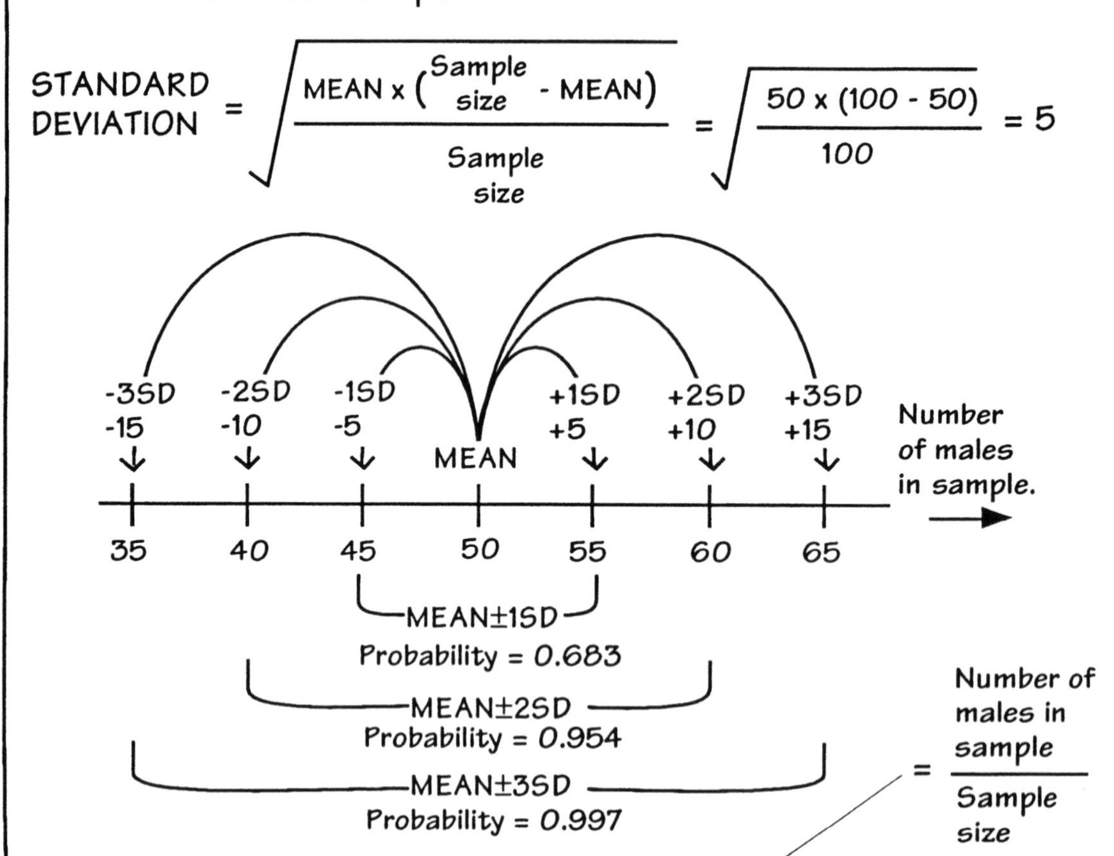

= Number of males in sample / Sample size

Number of males in sample lie in the interval 40 to 60 with probability 0.954.

Sample frequency lies in the interval 0.40 to 0.60 with probability 0.954.

We can now calculate the probabilities of different numbers of males in the sample using familiar techniques. As before we have probabilities for each of the different numbers of males that may be in the sample: ..., 48, 49, 50, 51, 52, ... These probabilities will conform to a bell curve. We fit the bell curve by finding its MEAN and STANDARD DEVIATION. The MEAN is just the expected number of males in the sample, which is the probability of a male 0.5 multiplied by the number of people in the sample 100. We use the familiar formula for STANDARD DEVIATION, except we replace the total number of trials by its analog in sampling, the sample size. The result, as the calculation opposite shows, is 5.

We can now form the MEAN±1SD, MEAN±2SD and MEAN±3SD confidence intervals in the usual way. For example, to find the MEAN±2SD interval, we start at the MEAN=50 and go down and up by 2SD=2x5=10, to arrive at 40 and 60. Thus the three intervals run from 45 to 55, 40 to 60 and from 35 to 65, respectively.

The probabilities 0.683, 0.954 and 0.997 are associated with these intervals as before. For example, there is a probability of 0.954 that the actual number of males in the sample will lie in the MEAN±2SD interval of 40 to 60.

If we take 0.954 as a comfortable level of certainty, we can now answer our original question: how close to the 50 expected will the actual number of males be? Most likely, the actual number will be in the range 40 to 60.

Recall that the sample frequency was the *fraction* of males in the sample. 40 males in the sample of 100 corresponds to a sample frequency of 40/100 = 0.40. Thus our likely range of 40 to 60 males in the sample corresponds to an interval of 40/100 to 60/100, that is 0.40 to 0.60.

General sampling model

Estimating Sample frequency SF from Population frequency PF

Population
with KNOWN population frequency PF of some trait

Sample
with UNKNOWN number of individuals with the trait.

What are the probabilities of the different numbers of individuals with the trait?

Probabilities will conform to a normal curve with:

MEAN = Expected number of individuals with trait = PF x SS

$$\text{STANDARD DEVIATION} = \sqrt{\frac{\text{MEAN} \times (\text{SS} - \text{MEAN})}{\text{SS}}}$$

P(Number in MEAN $\pm$ 1SD) = 0.683
P(Number in MEAN $\pm$ 2SD) = 0.954
P(Number in MEAN $\pm$ 3SD) = 0.997

From these we estimate the probabilities of different sample frequencies SF.

$$\text{Sample frequency SF} = \frac{\text{Number of individuals in sample with trait}}{\text{Sample size}}$$

We have seen how to compute the probabilities of different numbers of males in our example sampling problem. The technique can be used quite generally. The technique is represented as a general recipe opposite.

We have some population of individuals with a known population frequency of some trait of interest. The population frequency is just the fraction of individuals in the population with the trait. We draw a random sample of size SS and ask after the probabilities of different numbers of individuals with that trait in the sample. These probabilities will conform to a bell curve. We find the MEAN of the curve as the expected number of individuals in the sample with the trait. That is given by the population frequency PF multiplied by the sample size SS. The STANDARD DEVIATION is calculated by the usual formula with the number of trials replaced by the sample size SS.

With the bell curve fitted, we can determine the confidence intervals MEAN±1SD, MEAN±2SD and MEAN±3SD in the usual way. There will be a probability of 0.683, 0.954 and 0.997 respectively that the number of individuals in the sample with the trait will lie in these intervals.

Once we know the probabilities of different numbers of individuals with the trait in the sample, we can compute the corresponding intervals for the sample frequency SF.

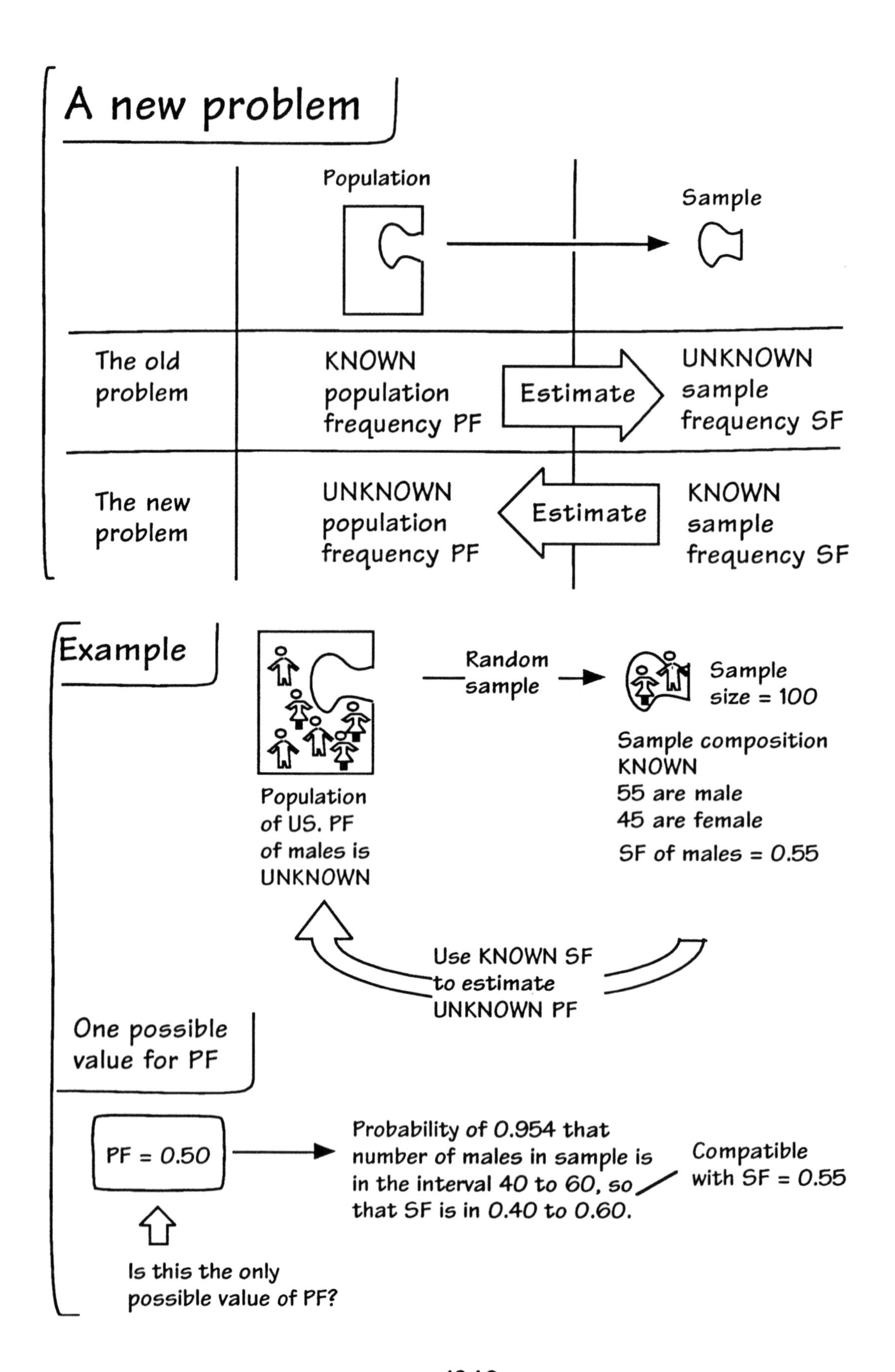
A new problem
Population
Sample
The old problem
KNOWN population frequency PF
Estimate
UNKNOWN sample frequency SF
The new problem
UNKNOWN population frequency PF
Estimate
KNOWN sample frequency SF
Example
Random sample
Sample size = 100
Population of US. PF of males is UNKNOWN
Sample composition KNOWN
55 are male
45 are female
SF of males = 0.55
Use KNOWN SF to estimate UNKNOWN PF
One possible value for PF
PF = 0.50
Probability of 0.954 that number of males in sample is in the interval 40 to 60, so that SF is in 0.40 to 0.60.
Compatible with SF = 0.55
Is this the only possible value of PF?

We have seen how to estimate the composition of a sample from the composition of the population. In posing this problem, we assume that we know the composition of the population and do not know the composition of the sample. There will be occasions when this happens. Far more likely are situations in which we do not know the composition of the overall population. We take a sample of the population in order to find out something about the composition of the population as a whole. Thus we do want to know how to use a known sample frequency to estimate the values of an unknown population frequency. This new estimation problem is the one we now turn to. It is slightly more complicated than the one we started with. Fortunately it will turn out that the final calculations involved are no more complicated.

As an example of this type of problem, imagine that we do not know the frequency of males in the population of the U.S. We seek to estimate it by taking a random sample of 100 people and counting how many are males. Let us say that we find that there are 55 males in the sample. This means that the sample frequency is 55/100 = 0.55.

What compositions of population could lead to a result like this? That is, what population frequencies are compatible with this sample frequency? We already know one value from our earlier sampling problem. If the population frequency were 0.50--that is, 50% of the population is male--then we calculated that we would expect the sample frequency to lie in the interval 0.40 to 0.60 (with probability 0.954). The actual sample frequency of 0.55 lies within this interval. Therefore a population frequency of 0.50 is compatible with actual outcome.

Is the population frequency of 0.50 the only population frequency compatible with the sample frequency? Clearly not. We would expect, for example, that a population frequency of 0.55 would be compatible with it. It would correspond to the case of the compositions of both sample and population agreeing exactly. While it may be improbable for the two compositions to agree *exactly* we certainly cannot rule out its possibility. If both 0.55 and 0.50 are possible population frequencies, then presumably anything in between would also be possible, such as 0.51 and 0.52.

Which population frequencies are compatible with a sample frequency of 0.55?

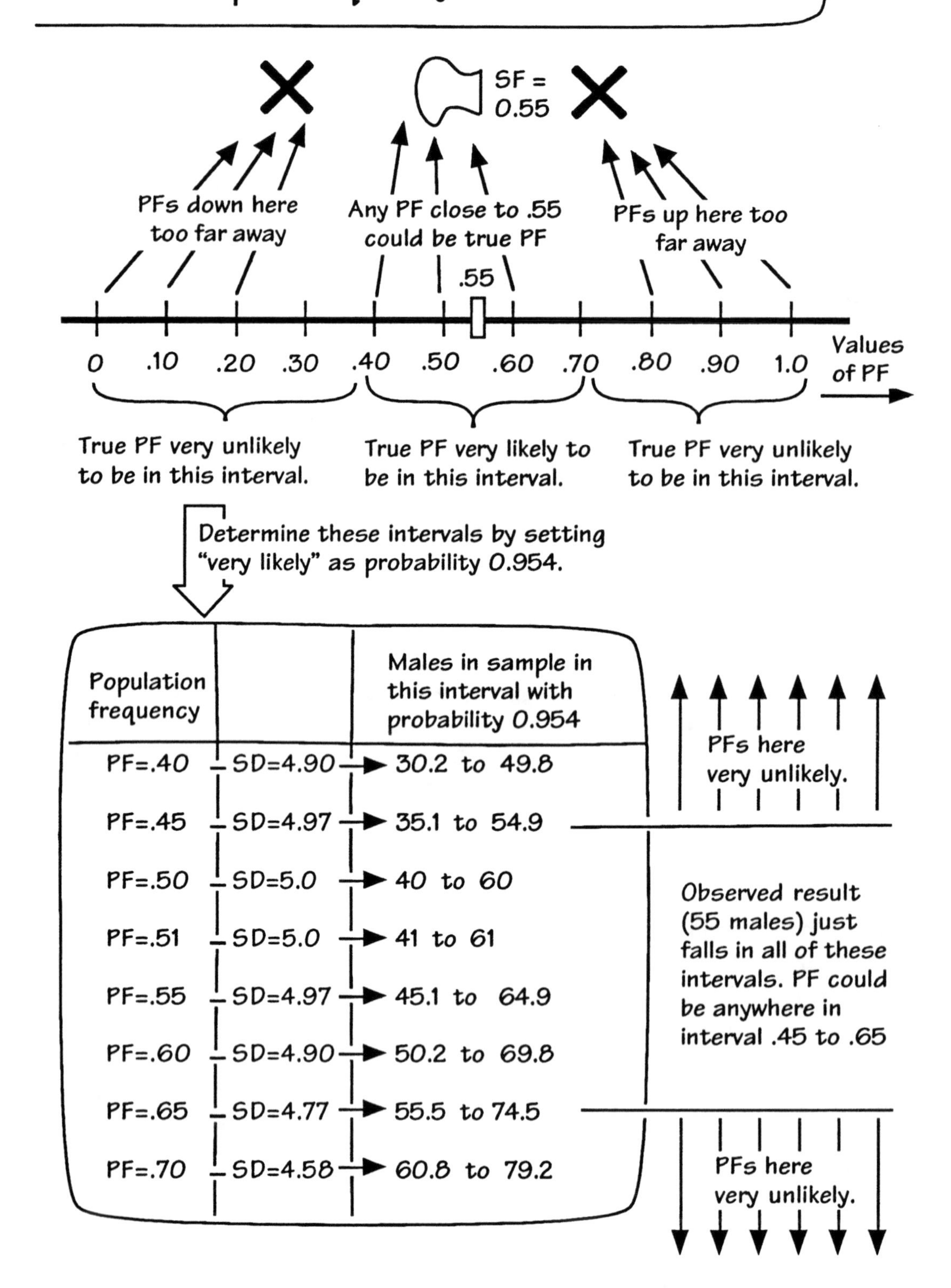

In our example, many population frequencies will be compatible with the actual sample frequency of 0.55. Many will not be compatible. Our job is to delimit those that are. Where do we find them? The whole idea of taking a sample is that the sample will give some approximate picture of the composition of the population as a whole. So, if the sample frequency is 0.55, we would expect the population frequency to be in some interval around this same value of 0.55. Population frequencies that stray too far from 0.55 will become less and less likely.

We can use what we know of sampling to determine the interval in which the population frequency is very likely to lie. Let us take our "very likely" to mean 0.954. We have seen that a population frequency of 0.50 is compatible with a sample frequency at this level of certainty. We had to do a calculation to know this: If the population frequency is 0.50 and the sample size 100, then the number of males in the sample will be probabilistically distributed on a bell curve. Its MEAN, we found, was 50 and STANDARD DEVIATION was 5. Thus, with probability 0.954, we expected to find the number of males in the sample in the interval MEAN±2SD, which is 40 to 60. Therefore at this level of certainty, we find that a population frequency of 0.50 is compatible with the sample result.

Our goal is to find all the population frequencies that are compatible with the sample results at this level of certainty. This will be an interval of population frequencies around the population frequency of 0.55. To find them, we need to repeat the above calculation with many different population frequencies. The results of these repeated calculations are given in the table opposite. In the case of a population frequency of 0.40, for example, we find the STANDARD DEVIATION of 4.90 and then the confidence interval MEAN±2SD of 30.2 to 49.8. Since the sample result of 55 lies outside this interval, the sample result is not compatible with a population frequency of 0.40.

We can read from the table how far the population frequency can stray from 0.55. When it has dropped to 0.45 or when it has risen to 0.65, then the population frequency ceases to be compatible with the sample result of 55, that is, with a sample frequency of 0.55. Therefore, at the level of probability of 0.954, the population frequency will lie in the interval 0.45 to 0.65.

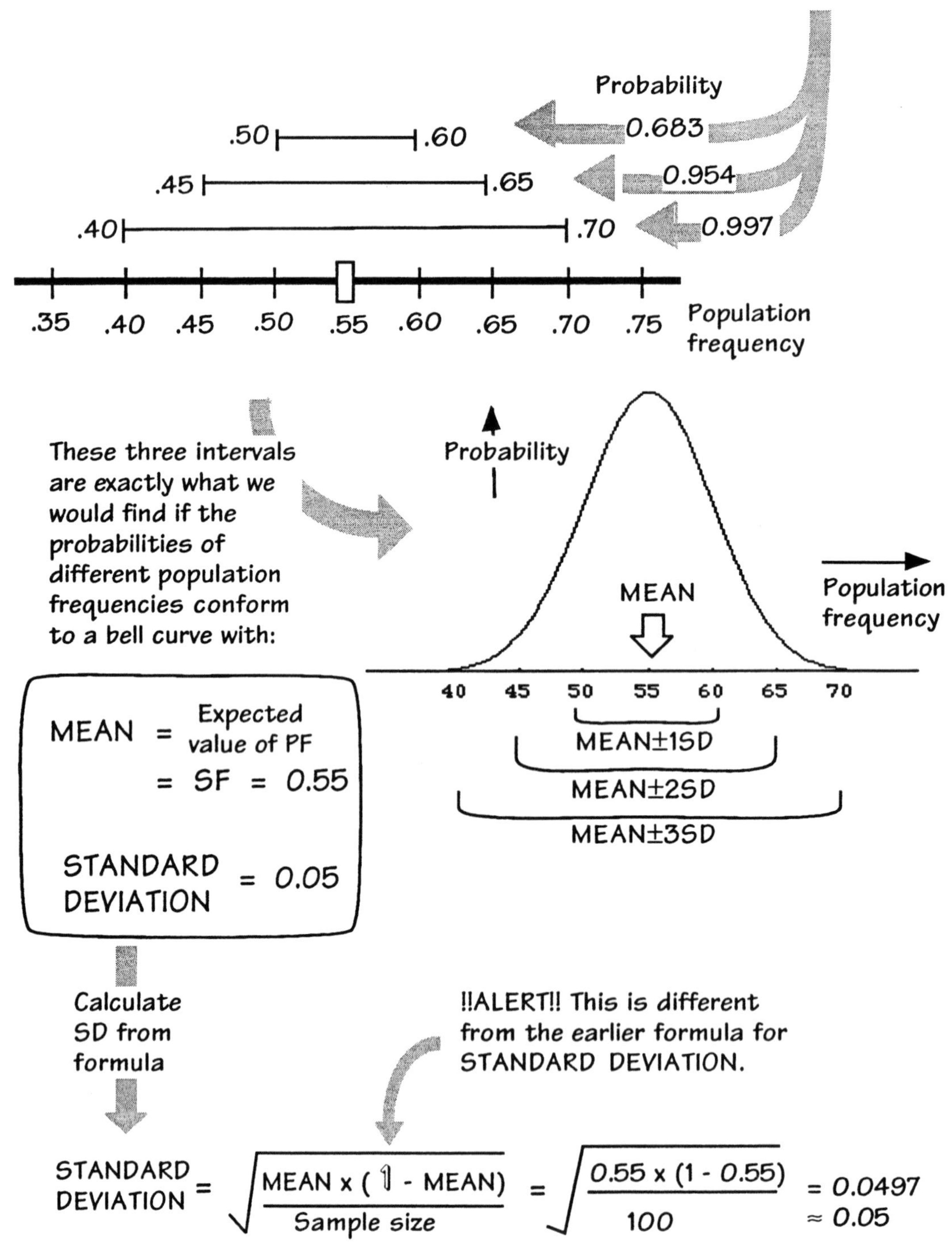

$$\text{STANDARD DEVIATION} = \sqrt{\frac{\text{MEAN} \times (1 - \text{MEAN})}{\text{Sample size}}} = \sqrt{\frac{0.55 \times (1 - 0.55)}{100}} = 0.0497 \approx 0.05$$

We have found that the population frequency will lie in the interval 0.45 to 0.65 with the probability of 0.954. We should be a little cautious in reading the meaning of this probability. The true population frequency is just some fixed value. It has none of the random variation measured by probabilities. The probability of 0.954 actually measures the rate of success of the method we are using to estimate the population frequency. That is, it tells us how strong our degree of belief ought to be in our estimate of the population frequency. This issue is discussed further in the appendix.

Let us seek a broader picture of the probability distribution of the population frequency. We can repeat the above calculation for the two other levels of confidence we are used to using, probability 0.683 and 0.997. The result is that we find the three intervals for the population frequency 0.50 to 0.60, 0.45 to 0.65 and 0.40 to 0.70, corresponding to probabilities 0.683, 0.954 and 0.997.

These three intervals look just like the sort of intervals derived from probabilities that conform to a normal curve. And that is just what these probabilities do. These three intervals arise since the probabilities of different population frequencies conform to a bell curve with a MEAN of 0.55 and a STANDARD DEVIATION of 0.05.

The MEAN and STANDARD DEVIATION can be calculated from the rules shown opposite. The MEAN is given by the obvious rule: it is just the expected (i.e. mean) value of the population frequency. That expected value is just the sample frequency. In this case it is 0.55. The STANDARD DEVIATION is given by a formula very similar to our earlier formula. But--*beware*--it is not exactly the same. The total number of trials has been replaced by the sample size in one place, as you would expect. But in another, it has been replaced by a 1.

That we can still use the curve is a happy outcome. It means that our calculations in estimating the population frequency will be no more complicated than in earlier problems, even though this is an intrinsically more complicated problem. Unfortunately, the method laid out here has its limits. It will work only if the interval of values of the population frequency is small. If that interval starts to become much larger than the one indicated in this problem, then the method will cease to be reliable.

General sampling model

Estimating Population frequency PF from Sample frequency SF

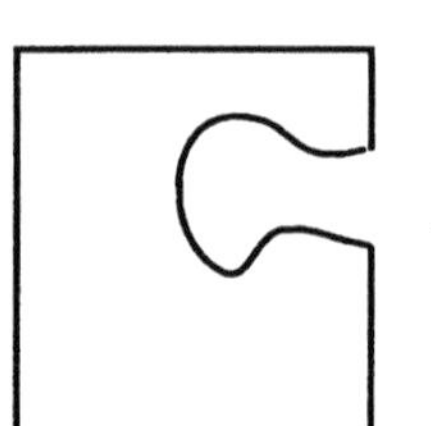

random sample →

Sample size SS

Sample frequency SF = $\frac{\text{Number of individuals in sample with trait}}{\text{Sample size}}$

Population
with UNKNOWN population frequency PF of some trait

Sample
with KNOWN number of individuals with the trait.

What are the probabilities of the different populations frequencies?

Probabilities will conform to a normal curve with:

MEAN = Expected population frequency = Sample frequency SF

$$\text{STANDARD DEVIATION} = \sqrt{\frac{\text{MEAN} \times (1 - \text{MEAN})}{\text{SS}}}$$

⇦ Notice the difference in the formula!

P(PF in MEAN ± 1SD) = 0.683
P(PF in MEAN ± 2SD) = 0.954
P(PF in MEAN ± 3SD) = 0.997

!!BEWARE!! This method only works when the interval of values of PF is small.

We can summarize what we have learned through the above example in a general method for estimating the population frequency from a known sample frequency. We have a population of individuals with some trait of interest. The population frequency, the fraction of individuals in the population with the trait, is unknown. We seek to estimate it by taking a random sample from the population. We known the number of individuals in the sample with the trait. From it we calculate the sample frequency, the fraction of individuals in the sample with the trait.

The known sample frequency is compatible with many population frequencies, but some are more probable than others. The probabilities of the different population frequencies conform to a bell curve. Its MEAN is given by the sample frequency SF. Its STANDARD DEVIATION is given by the formula opposite. The probabilities that the population frequency will be in the intervals MEAN±1SD, MEAN±2SD and MEAN±3SD are 0.683, 0.954 and 0.997 respectively.

There are a number of cautions to be observed. We now have two very similar methods. Be sure that you are using the right one! The choice of method depends on which quantities are known and which are not. So decide which it is that you know: the composition of the population or of the sample; and decide which you need to find out: the composition of the population or the sample. That will determine which of the two methods you use.

The formula for the STANDARD DEVIATION in this new method is different from the one we have used so far. (In fact it is not that different. It is really still the old formula modified to allow for the fact that the MEAN no longer varies in value from 0 to the number of trials; it now varies between 0 and 1.) Again, be careful to use the right formula in the right place.

Finally the method summarized opposite will only be reliable as long as the size of the intervals of the population frequency are small, such as in the above example.

Example

Study by Department of Justice (reported July 1994):
In 1988, of 258 instances of parents killing their children 55% = 142 were cases in which mothers were the killers.

Does this datum mean that mothers are *more* likely than fathers to kill their children?

Source: John M. Dawson and Patrick A. Langan, "Murder in Families," Bureau of Justice Statistics: Special Report, July 1994.

Analysis

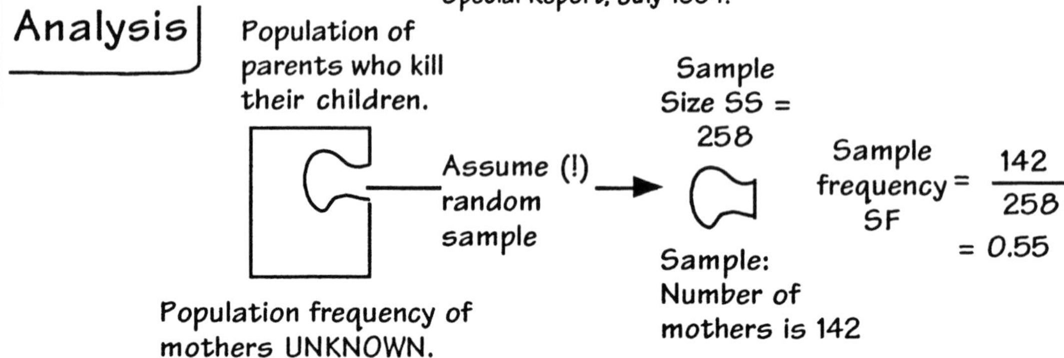

Probabilities of different population frequencies conform to a bell curve with:

$$\text{MEAN} = \text{Sample frequency SF} = 0.55$$

$$\text{STANDARD DEVIATION} = \sqrt{\frac{\text{MEAN} \times (1 - \text{MEAN})}{\text{Sample size}}} = \sqrt{\frac{0.55 \times (1 - 0.55)}{258}} = 0.031$$

Compute confidence interval:

$$\text{MEAN} \pm 2\text{SD} = 0.55 \pm 2 \times 0.031 = 0.55 \pm 0.062 = 0.49 \text{ to } 0.61$$

$$\text{Probability}\left(\begin{array}{c}\text{Population} \\ \text{frequency lies} \\ \text{in interval} \\ 0.49 \text{ to } 0.61\end{array}\right) = 0.954$$

On the basis of the sample data, we conclude with probability 0.954 that 49% to 61% of parents who kill their children may be the mothers. This interval just includes the 50% point at which mothers and fathers are equally likely to be the killers.

Here is another example of the estimation of a population frequency from the sample frequency. We hold a special horror for parents who kill their children, especially so if the killer is the mother. Therefore a study by the Department of Justice seemed especially disturbing. It showed that 55% of parents who killed their children were mothers. Does this mean that mothers are more likely to kill their children than fathers?

This is a sampling problem. We have a sample of size 258 from the population of parents who kill their children. In 142 cases the killer was the mother. So the sample frequency is 142/258 = 0.55. What does this tell us about the probabilities of different population frequencies? These probabilities are normally distributed. The MEAN and STANDARD DEVIATION are computed opposite. From them we find the MEAN±2SD confidence interval to be 0.49 to 0.61. So, with probability 0.954, the population frequency will lie in this interval; we expect 49% to 61% of parents who kill their children to be mothers.

This interval just contains the 0.50 population frequency at which mother and fathers are equally likely to be the killers. Thus we cannot rule out this possibility. Notice also that we could have conducted the analysis as a test of the hypothesis that the population frequency is 0.50. We would have found that the hypothesis just passes the test. Analyzing the problem as an estimation problem, however, is far more informative since, in addition, it gives us an indication of the possible range of the population frequency. Since that range extends so far into the higher values, we are left with the impression that the sample data favors the view that mothers are more likely to be the killers.

As always, we must be very careful in interpreting results such as these. Mothers may be more likely to be the killers. But that is *not* the same as saying that mothers have a greater intrinsic propensity to harm their children. We must look at other factors. For example, 27% of families are single parent families and in 87% of these the single parent is the mother. Data such as these suggest that mothers have far greater contact with their children than fathers. Thus, even if mothers have less of an intrinsic tendency to harm their children, this may be outweighed by their greater exposure to the children. An analysis that seeks to draw conclusions about intrinsic tendencies to violence will need to determine how to correct for this factor.

Tossing a biased coin

A coin is tossed 200 times. It gives us 150 heads. We are sure that the coin is not fair. On the basis of this datum, what is our estimate of the chance of a head on each toss?

Analysis

To use our techniques, we will make the problem look like a sampling problem.

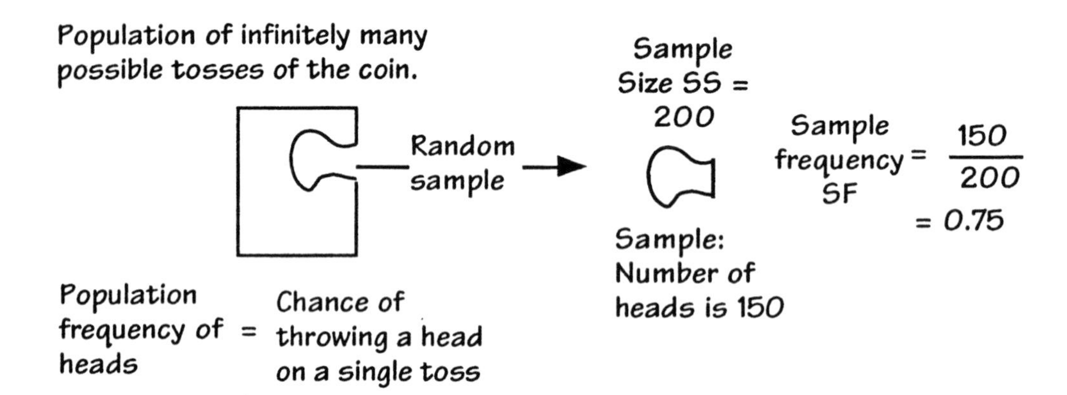

Probabilities of different population frequencies conform to a bell curve with:

$$\text{MEAN} = \text{Sample frequency SF} = 0.75$$

$$\text{STANDARD DEVIATION} = \sqrt{\frac{\text{MEAN} \times (1 - \text{MEAN})}{\text{Sample size}}} = \sqrt{\frac{0.75 \times (1 - 0.75)}{200}} = 0.031$$

Compute confidence interval:

$$\text{MEAN} \pm 2\text{SD} = 0.75 \pm 2 \times 0.031 = 0.75 \pm 0.062 = 0.69 \text{ to } 0.81$$

$$\text{Probability}\left(\begin{array}{c}\text{Population} \\ \text{frequency lies} \\ \text{in interval} \\ 0.69 \text{ to } 0.81\end{array}\right) = 0.954$$

We conclude that the chance of tossing a head on a single toss lies in the interval 0.69 to 0.81 (with probability 0.954).

These same techniques can be used in cases that do not appear to be sampling problems at all. Consider, for example, the case of a coin tossing experiment. It gives us 150 heads in 200 throws. We are sure that the coin is not a fair coin; that is, we are sure that there is not an equal chance on each toss of a head or a tail. What is the chance of a head on a single toss?

The datum tells us that 75% of the tosses came up heads. So our best guess of the chance of a head on a single toss would be 0.75. But we cannot be sure that it will be exactly 0.75. Both higher and lower chances could still give us 150 heads in 200 tosses. These acceptable values will lie in an interval around 0.75. We can use the technique developed here to find that interval.

To do this, we make the problem look like a sampling problem. Recall that the chance of a head on a coin toss equals the frequency of heads that would result if the coin were tossed infinitely often. In other words, the chance we seek is actually a population frequency in the populations of infinitely many coin tosses. We can think of the 200 tosses observed as a tiny sample of these infinitely many throws.

We proceed as before. The probability of different population frequencies will conform to a bell curve. We compute the MEAN and STANDARD DEVIATION of the curve and form a confidence interval. We conclude (with probability 0.954) that the chance of tossing a head on any one toss will lie in the interval 0.69 to 0.81.

This last result also shows that we were right to believe that the coin was not fair. If it were fair, the outcome of our 200 tosses must be compatible with a 0.5 chance of tossing a head. But this 0.5 does not lie in the interval.

What can make this example perplexing is the two distinct uses of the notion of probability. One is the chance of a head or tail on a single toss of the coin. The other is the degree of belief that we have in the different chances compatible with the 150 heads in 200 tosses. To help you keep the two uses distinct, I have used the word "chance" only for the first usage and "probability" only for the second.

Appendix Sampling with and without replacement

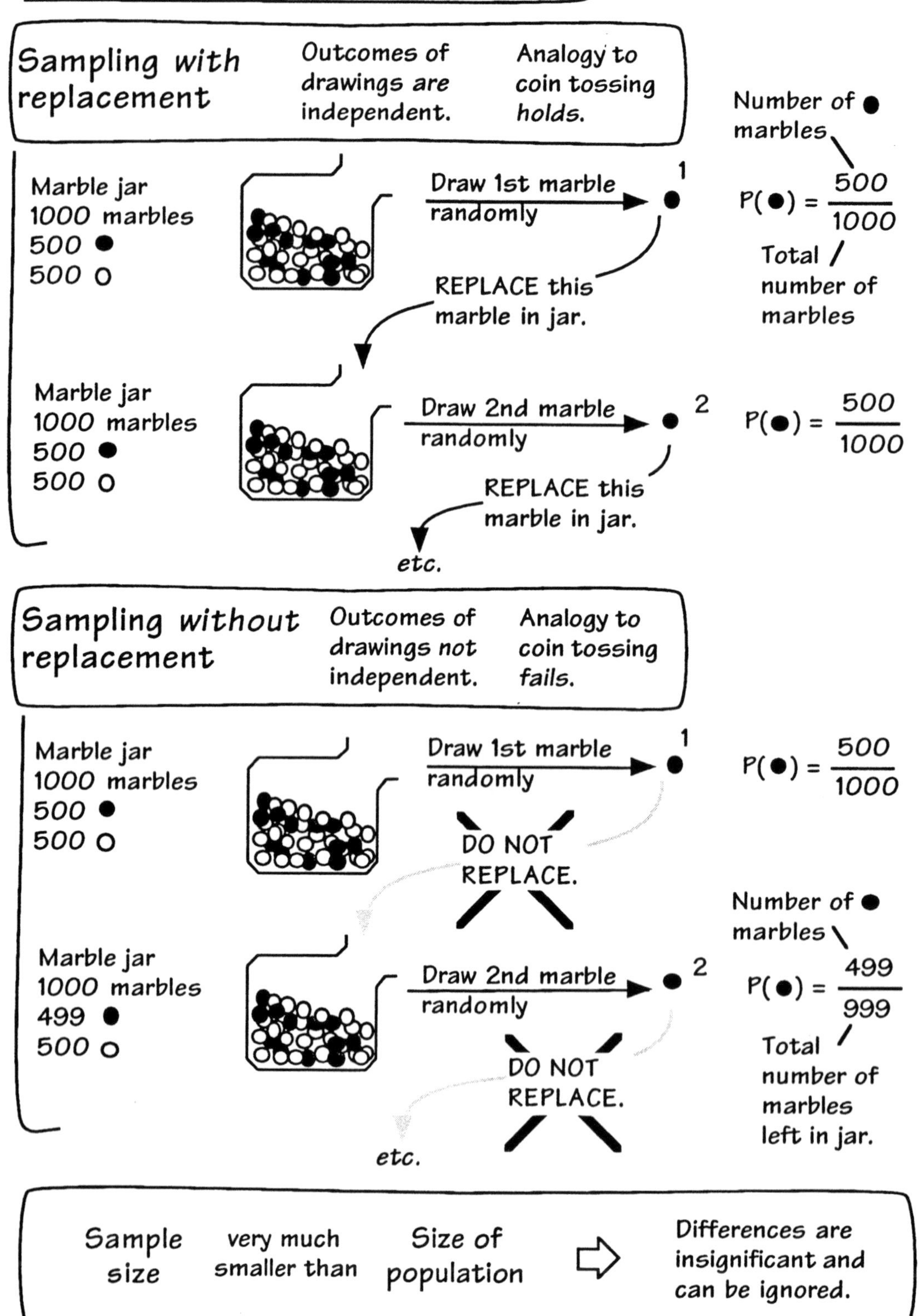

There are two schemes for sampling, sampling with and without replacement. It is easy to see the difference if we imagine that we are sampling from a jar of black and white marbles. In sampling with replacement, we draw the marbles one at a time and *replace* them in the jar after each drawing. In sampling without replacement, we *do not* replace each marble after it is drawn.

As long as the sample size is very much smaller than the size of the population, it does not matter for our purposes which scheme is used. If the sample size becomes large, then we must be more careful. In the case of sampling without replacement, the outcomes of the individual drawings cease to be independent. The analogy to the coin tossing problem required this independence. For, with coin tosses, what one throws on the first toss does not affect what one throws on the second. In sampling with replacement, the outcomes remain independent.

This effect is illustrated opposite through sampling by both schemes from a jar with 500 each of black and white marbles. In sampling with replacement, if we draw a black marble in the first drawing, that marble is replaced. So the composition of the jar is the same on the second drawing. The probability of a black marble on both first and second drawing remain the same.

When we sample without replacement, things are different. If the first drawing is a black marble and it is not replaced, then the composition of the jar has changed for the second drawing. There are now only 499 black marbles, so the chance of a black marble on the second drawing is reduced. Eventually this dependence of outcomes will completely alter the probabilistic behavior of the system. If we draw 1,000 marbles without replacement, we have drawn the entire jar. There is now only *one* outcome possible: 500 black marbles and 500 white marbles, which must happen with a probability of one. This is quite unlike what would happen with 1,000 coin tosses. The outcome of 500 heads and 500 tails would be just one of many outcomes that could happen.

Puzzle

We draw a sample of 100 people from the U.S. population and find 55 are males. We estimate that the frequency of males in the U.S. population is 0.55 *with probability 0.954.*

But the frequency of males in the U.S. population is some definite, fixed number (=0.49)!

How can a population frequency of 0.49 have a probability of anything other than 0 (wrong number) or 1 (right number)?

Answer

The probability 0.954 measures how justified we are in believing that the population frequency lies in 0.45 to 0.65 on the evidence of a sample frequency of 0.55.

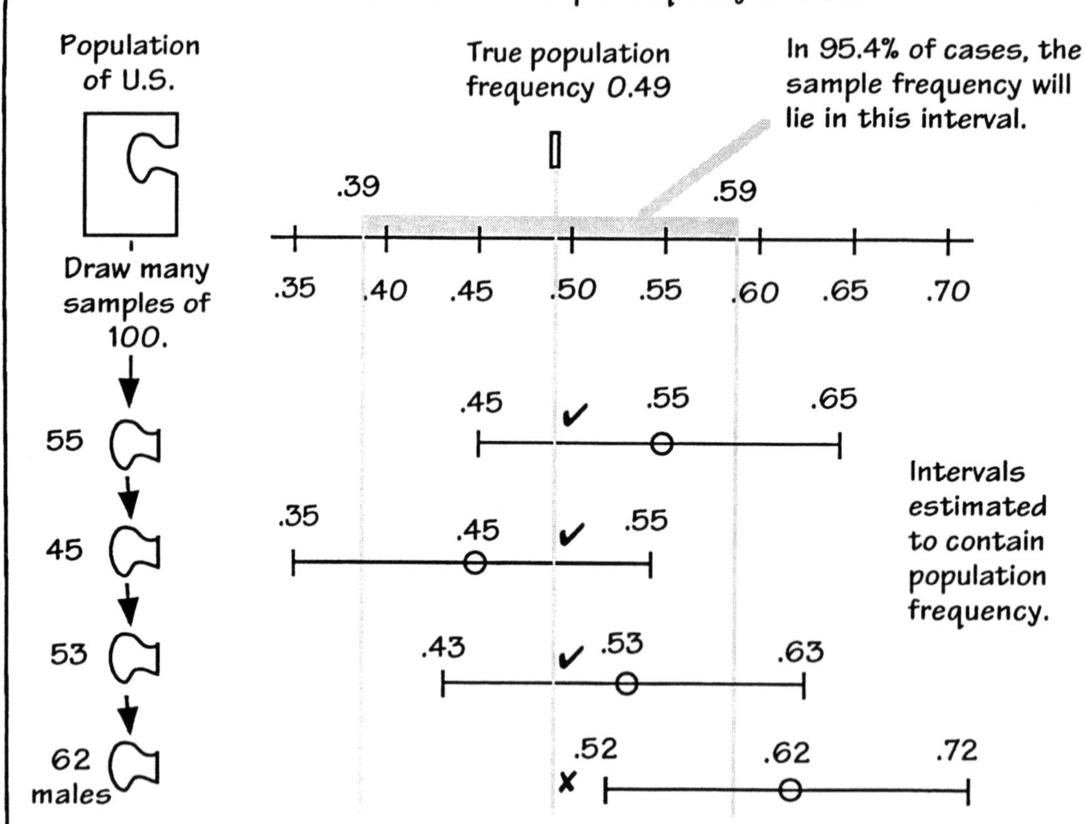

Method succeeds in 95.4% of cases.  Probability that method succeeds in any one case is 0.954.

The meaning of "probability" is slightly different in the two methods that we have used to relate sample and population frequency. When we sample from some particular population, its population frequency is fixed. The frequency of males in the U.S. population is fixed at 0.49. If we draw a sample of 100, the sample frequency of males will vary randomly in some interval. We describe this randomness by means of probability statements. When we are given a known sample frequency, we also use probability statements to describe intervals in which the population frequency may lie. On its face, this doesn't make sense. The population frequency either lies in the interval we name or it does not. This calls for a probability of 1 or 0. How can we resolve this puzzle?

When we estimate the population frequency from a sample frequency, the probabilities refer to the chances of success of our method of estimation. They do not indicate any random variability in the population frequency; they merely indicate the degree of belief we are warranted to have in our estimates.

To see how this works, imagine that we take many samples of 100 from the U.S. population. In 95.4% of these cases, they will have sample frequencies within the interval 0.39 to 0.59. Now imagine that we do not know the true population frequency of 0.49. We will use these sample frequencies to estimate the population frequency. In each case, we use the method described in this chapter to form the MEAN±2SD confidence interval. The results of such calculations are shown opposite for sample frequencies of 0.55, 0.45, 0.53 and 0.62. With most of the samples, the true population frequency of 0.49 will lie within the interval calculated. This happens, for example, with the sample frequencies of 0.55, 0.45 and 0.53. In some cases, however, the interval will fail to contain the true population frequency of 0.49. This happens with the sample frequency of 0.62.

You can see from the figure opposite that the method succeeds if the sample frequency lies within the interval 0.39 to 0.59. It does this 95.4% of the time. Therefore the method will succeed in 95.4% of cases. We restate this as probability claim: the probability of success of the method of estimating population frequencies with MEAN±2SD intervals is 0.954. Since our degree of belief ought to match the probability of success of our methods, the probabilities are also degrees of belief.

Assignment 12: Sampling and Estimation

1. Roughly ten percent of people are left handed. Assume that you are designing a study of the frequency of left handedness. You plan to sample 200 people randomly from the general population and count how many are left handed. In the following you will compute the range of left handers you expect to find in your sample.

(a) Under the assumptions given above, what is the population frequency "PF" of left handers?

(b) What is the sample size "SS"?

The number of left handers in the sample ("sample number") will be normally distributed:

(c) What is the MEAN of the sample number?

(d) What is the STANDARD DEVIATION ("SD")?

(e) What is the interval MEAN +/- SD? What is the probability that the sample number lies in this interval?

(f) What is the interval MEAN +/- 2SD? What is the probability that the sample number lies in this interval?

(g) What is the interval MEAN +/- 3SD? What is the probability that the sample number lies in this interval?

2. After the survey is complete, you find that there are 15 left handers in your sample of 200. For various reasons, you are no longer so confident that 10% of the population is left handed. So you decide to treat the relevant population frequency (=the frequency of left handedness in the population) as an unknown. You will now determine what ranges of values of this unknown population frequency are compatible with the sample number of 15 from your survey of 200. The unknown population distribution is normally distributed.

(a) What is the sample frequency (=frequency of left handers in your sample)?

(b) What is the MEAN of the population frequency?

(c) What is the STANDARD DEVIATION?

(d) What is the interval MEAN +/- SD? What is the probability that the population frequency lies in this interval?

(e) What is the interval MEAN +/- 2SD? What is the probability that the population frequency lies in this interval?

(f) What is the interval MEAN +/- 3SD? What is the probability that the population frequency lies in this interval?

In October 1994, researcher James Fries and co-workers reported the results of an investigation into whether running exercise is of benefit to the elderly. (James F. Fries, "Running and the Development of Disability with Age," *Annals of Internal Medicine*, **121** (1994), pp.502-509.) The study compared a test group of elderly runners with a control group of elderly non-runners. One result was far lower mortality amongst the runners. The relevant statistics for an eight year period were:

Test group of elderly runners:	537 individuals; 8 deaths (death rate 1.5%)
Control group of elderly non-runners:	423 individuals; 30 deaths (death rate 7.1%)

One worry about a study with such small numbers of outcomes (8 deaths, 30 deaths) is that the result is just a statistical fluctuation. That is, we might wonder if the population death rate amongst elderly runners and non-runners is the same but chance alone has led to us getting different death rates in the two groups. In questions 3, 4 and 5, you will assure yourself that this is not the case by computing the possible ranges of population frequencies compatible with the above statistics.

3. Consider the population of elderly runners. Assume the test group of 537 runners is a random sample of them. What is the range of death rates in the population of elderly runners that is compatible with the 8 deaths in the sample group? Calculate this interval at 95.4% confidence, that is, there should be a probability of .954 that the population death rate lies in this interval.

(Hint: Your calculation should follow the steps laid out in question 2.)

4. Consider the population of elderly non-runners. Assume the test group of 423 non-runners is a random sample of them. What is the range of death rates in the population of elderly non-runners that is compatible with the 30 deaths in the sample group? Calculate this interval at 95.4% confidence, that is, there should be a probability of .954 that the population death rate lies in this interval.

(Hint: Your calculation should follow the steps laid out in question 2.)

5. Use the results of 3. and 4. to argue that the study results are incompatible with the hypothesis that the death rate is the same in the populations of elderly runners and elderly non-runners.

Chapter 13

Science and the Bell Curve

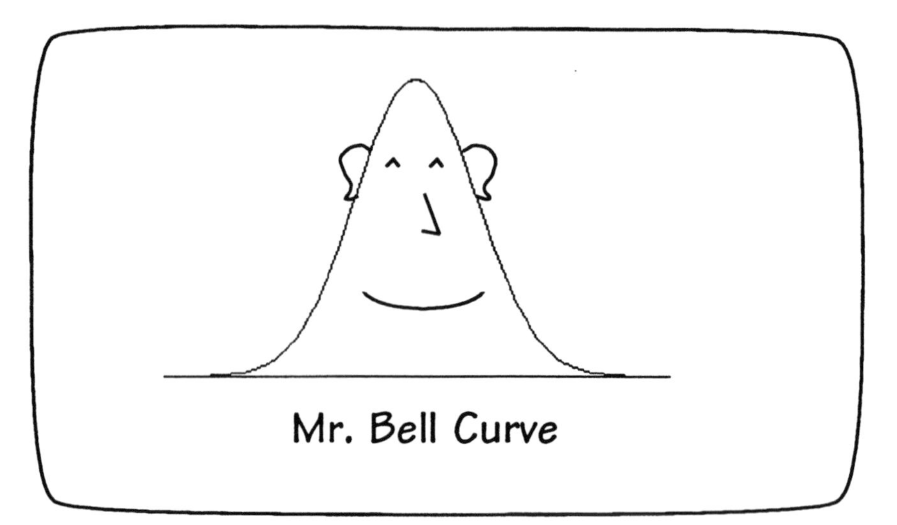

The bell curve is the most important curve in probability and statistical analysis.

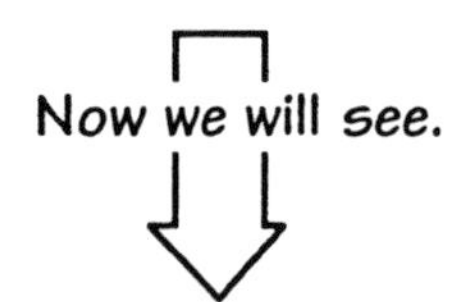

In general science, the bell curve is one of the most common and important of all curves.

* Demographics: distribution of heights, I.Q.'s, ages...
* Population genetics: distribution of genes
* Physics of molecules:
  - random motion of individual molecules
  - distribution of velocities of molecules
* Theory of errors

.
.
.

A bell curve appears whenever one deals with any sort of random process with many individual parts.

In the preceding chapters, we have seen how the bell curve can be used in probability theory and statistical analysis. Of all mathematical curves, the bell curve is the most important for probability theory and statistics. In this chapter we will broaden our horizons. We will see that the bell curve has an importance that goes well beyond probability and statistics. In science as a whole, the presence of bell curves is so pervasive that it may well count as the most important of all curves in science.

When we study demographics, we find that the distributions of various traits over a population conforms to bell curves. We have already seen this with the distribution of IQs. It holds also for heights and the many other characteristics of individuals that actuaries pursue in order to form insurance rates and government policies. In biology, in population genetics, when we look to the distribution of various genes in some species, we find bell curves. In the physical sciences, we look at matter and the molecules that form it. We find bell curves in the random motions of the individual molecules and in the distribution of their velocities. And the theory of errors that first gave us the bell curve requires this curve as its core.

Why do we hear bells so often in science? All of these and many more applications have one thing in common: they deal with systems that have some sort of random process (probability, statistics, theory of errors) or systems that involve many individuals (the huge populations of people in demographics, genes in population genetics and molecules in physical science). Whenever we have some sort of randomness in a process or whenever we have a system that has many individuals, we will find a bell curve somewhere.

One of the most important mathematical results for science:

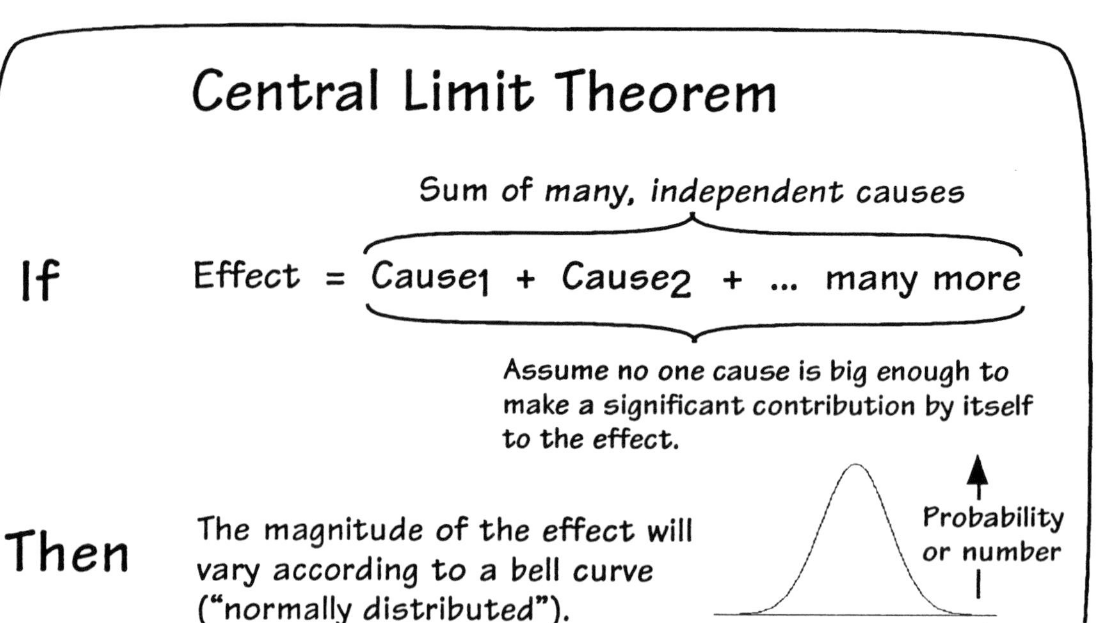

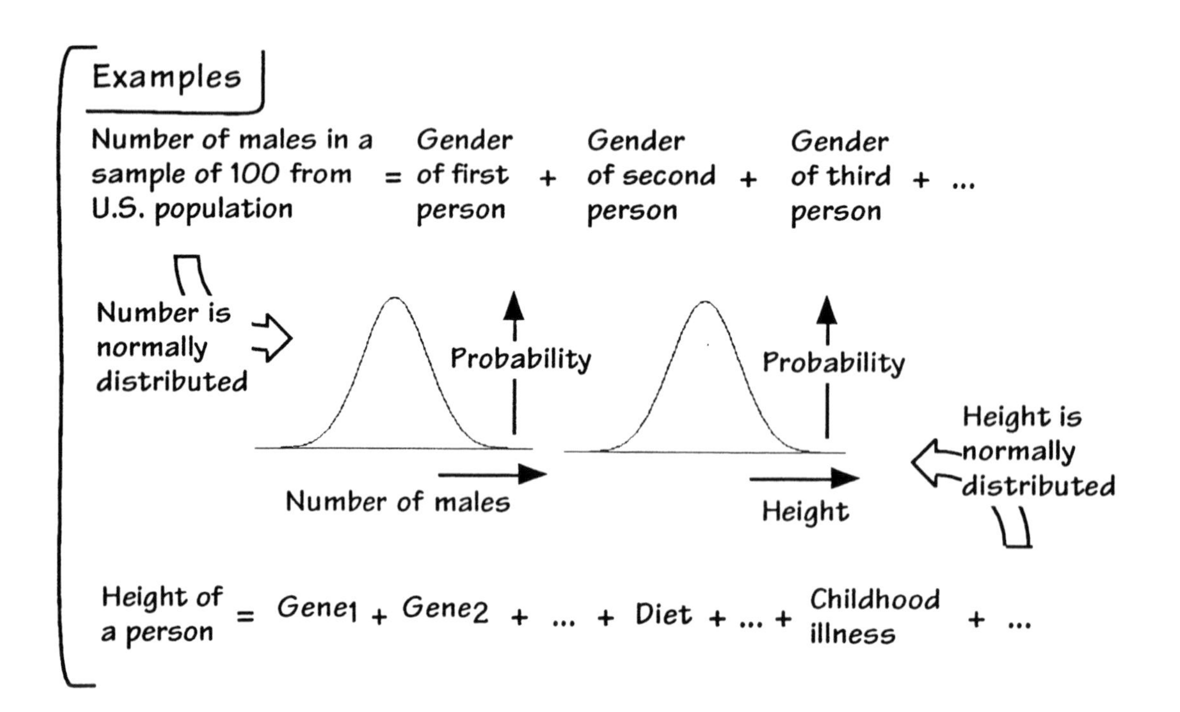

The pervasiveness of the bell curve in science is due to a small piece of mathematics of big importance, the central limit theorem. By itself, the theorem only tells us about numbers. It tells us nothing directly about the world. It is not a law like the laws of thermodynamics. However, if we go to a science that uses numbers to describe nature and ask what the theorem tells us of those number, it usually tells us that we will see a lot of bell curves.

The theorem is stated in a somewhat intuitive form opposite. It deals with cases in which we have an effect that is the combined result of the many causes. Note the "many causes" must satisfy several conditions: There must many of them. They must be independent in the sense of probability theory. That is, the character of one cause must not affect the character of another. And none of the causes must be enough to make a significant contribution to the final effect on its own. If these conditions are satisfied, then the distribution of the effect will conform to a bell curve.

The bell curve in sampling arises from the theorem. The effect is the number of males, for example, in a sample of 100 drawn from the U.S. population. Many causes fix this effect: the gender of the first person sample, that of the second, and so on. The causes are independent. The gender of the first person sampled does not affect that of the second, and so on. And the gender of no one person plays a significant in determining the final result.

The height of a person is also normally distributed. For young women, for example, the mean height is around 5'4" to 5'5" with a standard deviation of several inches. This arises again through the central limit theorem. A person's height is fixed by a huge array of smaller causes. Some are listed opposite. There will be the contributions of diverse genes. Variations in diet affects height--improvements in people's diet in the last century have led to a population that is taller overall. Childhood illnesses and accidents can also affect adult height. The combined action of these causes will give us a bell curve if the conditions of the theorem are met. They largely are met, but they can fail in many ways. For example, a single gene may overwhelm all other factors and produce an exceptionally short person. Or independence may fail. A child living in poverty may be more likely to have a poor diet and childhood illness.

The central limit theorem also fixes the mean and standard deviation of the bell curve.

$$\text{MEAN of Effect} = \text{MEAN of Cause}_1 + \text{MEAN of Cause}_2 + \dots$$

$$\left(\text{SD of Effect}\right)^2 = \left(\text{SD of Cause}_1\right)^2 + \left(\text{SD of Cause}_2\right)^2 + \dots$$

Example

Draw a sample of 100 people from the U.S. population. How many males in the sample?

$$\text{Number of males} = \text{Gender of first person} + \text{Gender of second person} + \dots + \text{Gender of 100th person}$$

Count as: female = 0; male = 1

$$\text{MEAN of number} = \text{MEAN of first person} + \text{MEAN of second person} + \dots + \text{MEAN of 100th person}$$

$$= 0.49 + 0.49 + \dots + 0.49$$

$$= 49$$

$$\left(\text{SD of number}\right)^2 = \left(\text{SD of first person}\right)^2 + \left(\text{SD of second person}\right)^2 + \dots + \left(\text{SD of 100th person}\right)^2$$

$$5^2 = 0.5^2 + 0.5^2 + \dots + 0.5^2$$

The central limit theorem says a little more and that little more should be mentioned here, even though we will only explicitly use this material once in the remainder of the chapter.

We pick out a particular bell curve by specifying its MEAN and STANDARD DEVIATION, just as we pick out a particular circle by specifying its center and radius. The central limit theorem also picks out the particular bell curve for the effect by telling us how to find its MEAN and STANDARD DEVIATION.

The MEAN is computed by the simplest of rules. We find the MEAN of the effect by adding up all the MEANs of the individual causes. Here the MEAN of a cause is just the average amount that it contributes to the effect. The simplest rule for the STANDARD DEVIATION would be to add up the STANDARD DEVIATIONs of the individual causes. The theorem gives a slightly more complicated rule. We find the STANDARD DEVIATION *squared* by summing the STANDARD DEVIATIONs *squared* of the individual causes. (It is possible to define a STANDARD DEVIATION of a cause even if the cause itself is not associated with a bell curve. This STANDARD DEVIATION measures how much the individual cause contributes to the spread of the final effect.) The result of using an addition of squares rule rather than straight addition is that the STANDARD DEVIATION does not grow as fast when extra terms are added; that is, the effect does not spread as much as it would with straight addition.

We can see our familiar sampling example opposite how the results we know from the earlier chapter fit with these formulae. In assessing the contribution of each person sampled to the total number of males in the sample, we count each female as a 0 and each male as a 1. Since 49% of people in the U.S. are male, on average each person sampled contributes 0.49 to the final sum. This 0.49 is the MEAN. We add 100 of these to get the MEAN of the effect, 49. Similarly, it turns out that the SD of each individual person sampled is 0.5. If we take these, square them and add all 100 together, we get 25. This equals the 5^2, so the SD of the effect is 5, which corresponds with our earlier result. Notice that, if we had found the SD of the effect by adding without squaring, we would have found an SD ten times bigger, 50. We see how the use of the addition of squares rule reduces growth of the SD of the effect.

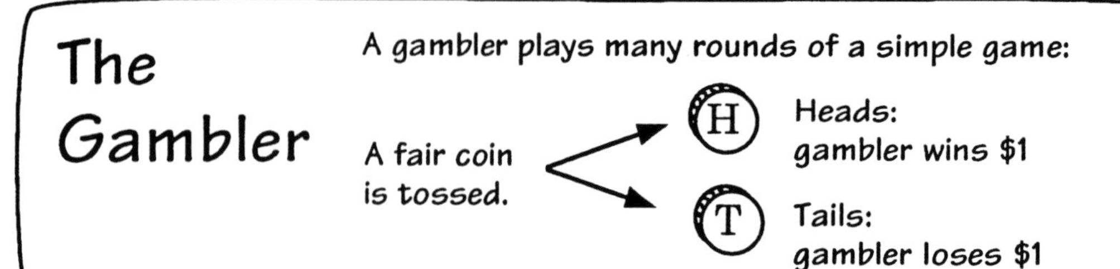
The Gambler
A gambler plays many rounds of a simple game:
A fair coin is tossed.
H
Heads: gambler wins $1
T
Tails: gambler loses $1

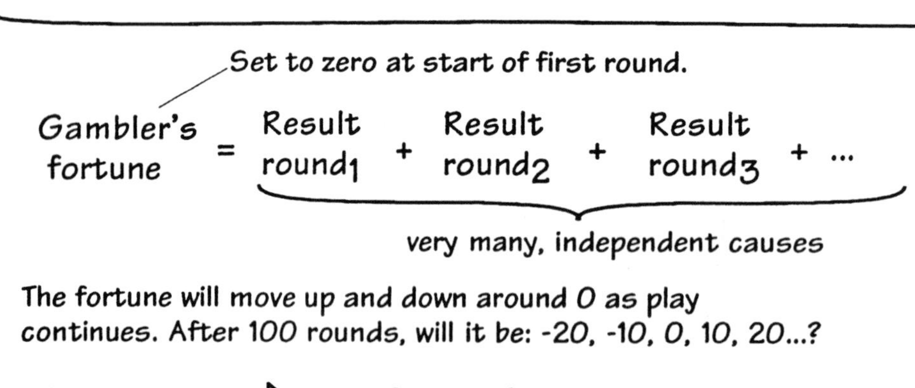
Set to zero at start of first round.
Gambler's fortune = Result round1 + Result round2 + Result round3 + ...
very many, independent causes
The fortune will move up and down around 0 as play continues. After 100 rounds, will it be: -20, -10, 0, 10, 20...?

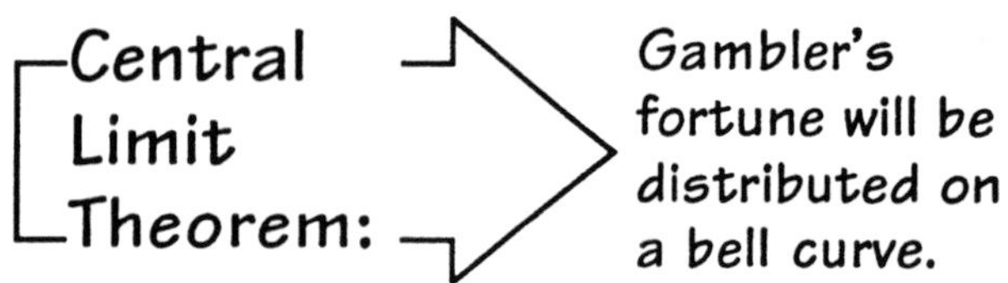
Central Limit Theorem:
Gambler's fortune will be distributed on a bell curve.
Mean will be 0 = expected fortune. On each round it is equally likely that fortune goes up or down by $1.

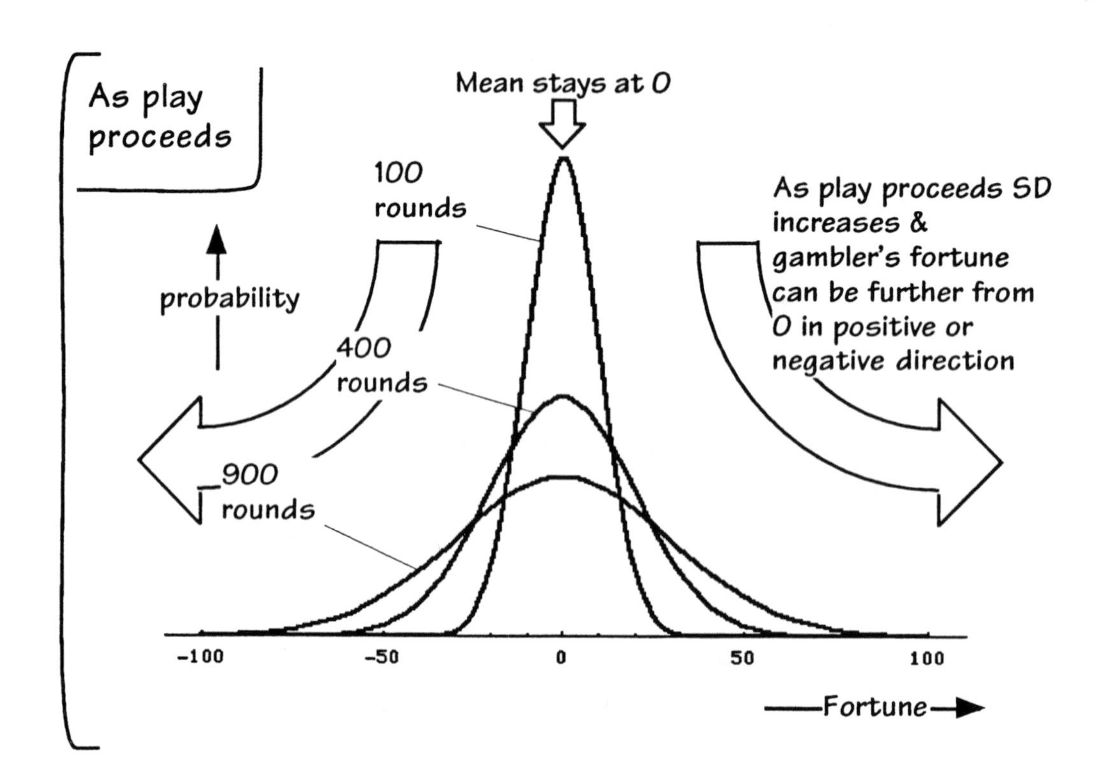
As play proceeds
Mean stays at 0
100 rounds
400 rounds
900 rounds
probability
As play proceeds SD increases & gambler's fortune can be further from 0 in positive or negative direction
-100
-50
0
50
100
Fortune

As a bridge to more serious applications in science, let us consider what the central limit theorem tells us about a simple gambling game. A fair coin is tossed repeatedly. The gambler wins or loses one dollar according to whether the coin comes up heads of tails. We will set the amount of money the gambler starts with--the gambler's fortune--at zero dollars. As play proceeds, wins or losses will appear as a positive or negative fortune.

At any time, the gambler's fortune will lie in some interval around zero. Since the fortune is subject to the chance outcome of coin tosses, we cannot say after 100 rounds exactly what the fortune will be. The best we can give are probabilities for different fortunes. The central limit theorem tells us that these probabilities conform to a bell curve, since the conditions for the theorem are met. Once 100 or more rounds have been played, the gambler's fortune is the resultant of many causes: the wins or loss on any individual round. The causes are independent: whether the gambler wins or loses on round one does affect a win or loss on round two; and so on. Finally, no one round makes a significant contribution, for each round contributes at most one dollar to the gambler's fortune.

Opposite we see the bell curves associated with the gambler's fortune at various stages of play: after 100, 400 and 900 rounds. Notice that the mean of each curve remains fixed at a fortune of 0. It is easy to understand why this is so. On each round, the gambler has an equal chance of winning or losing one dollar. So *on average* each round will not alter the gambler's fortune. Therefore *on average* the gambler's fortune will stay at zero; that is, over time its mean will remain at zero.

What is more interesting is the spread of the bell curve. The spread for the 100 round curve it relatively small. This means that the gamblers fortune is likely to be close to zero. The spread increases for the curves for 400 and 900 rounds. This shows that the gambler's fortune can be further and further away from the starting value of zero as play proceeds. Look closely at how the bell curves spread over time, for this is the phenomenon we shall be examining in several sciences.

Use what we know of coin tosses to compute how quickly the bell curve for gambler's fortune spreads out.

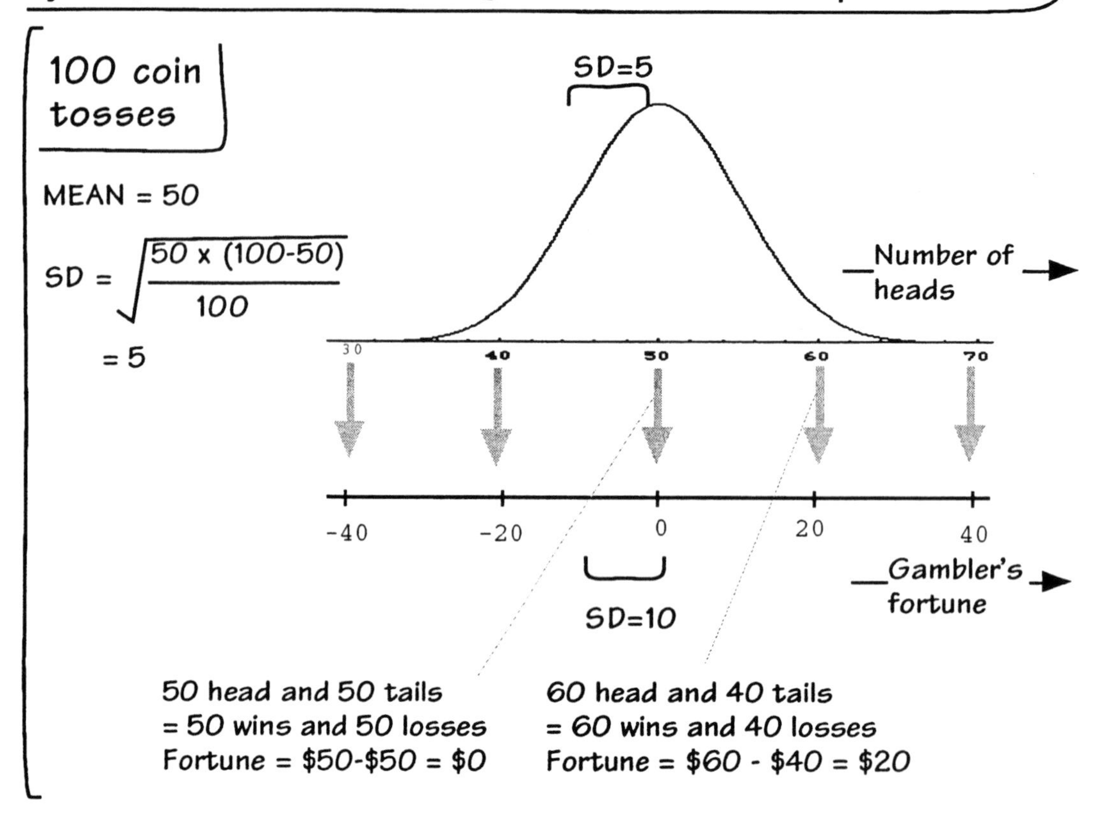

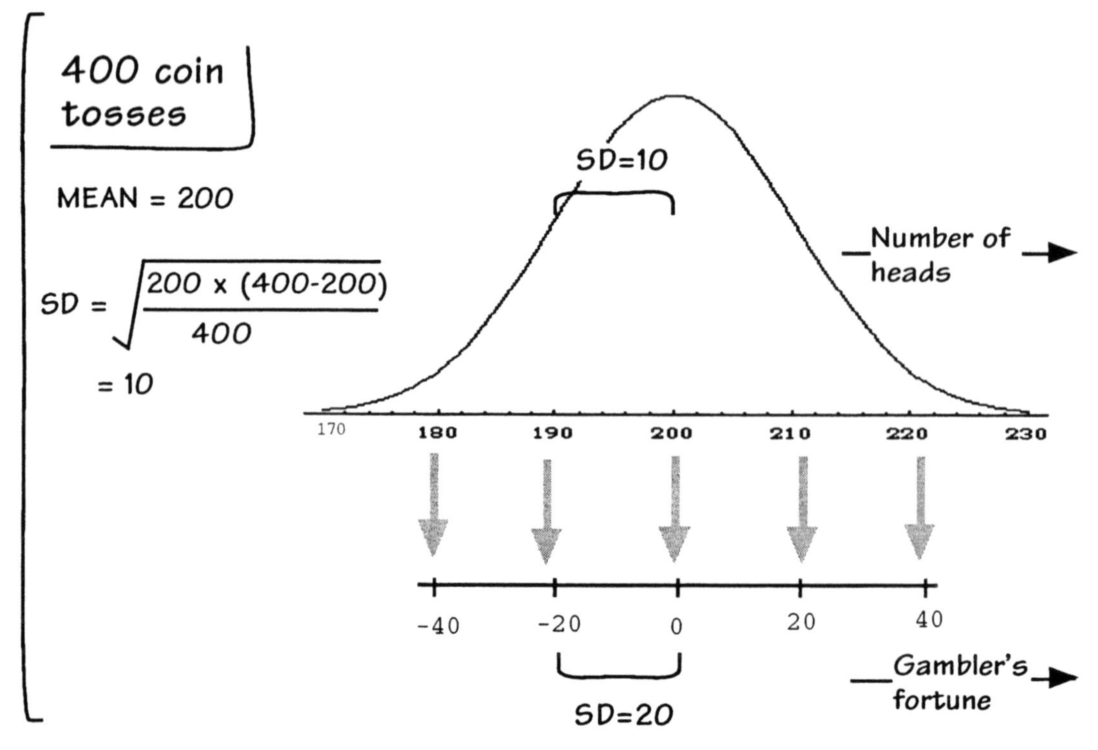

Since this spreading of the bell curve with time will concern us, let us compute exactly how fast it spreads with time. We can do this using what we already know about the probabilities of tossing different numbers of heads and tails on many tosses of a fair coin.

Take the case of 100 coin tosses. How many heads will there be? We expect some number around 50. The probabilities of the different numbers will conform to a bell curve. Following our usual calculations, we find that its MEAN will be 50 and its STANDARD DEVIATION will be 5. This bell curve is drawn opposite. Now, the number of heads thrown relates directly to the amount won by the gambler, that is, to the gambler's fortune. 50 heads corresponds to the fortune staying at $0. It is the case of 50 heads and 50 tails, which is $50 won and $50 lost. Similarly 60 heads corresponds to a fortune of $20. It is the case of 60 heads and 40 tails, which is $60 won and $40 lost--a net fortune of $60-$40=$20. Thus the cases of 30, 40, 50, 60, 70 heads are the cases of fortunes -$40, -$20, 0, $20, $40. Therefore the probabilities of 30, 40, 50, 60, 70 heads are the probabilities of fortunes -$40, -$20, 0, $20, $40. So, if we draw our scales carefully as opposite, the same bell curve can be used for both the number of heads and the gambler's fortune.

This joint bell curve shown opposite is massed over the interval 40 to 60 heads, that is, a fortune of -$20 to $20. Therefore we read that the number of heads will be roughly in the interval 40 to 60 and the gambler's fortune will lie roughly in the interval -$20 to $20. A more precise measure of the spread of the curve is the standard deviation. On the number of heads scale, the standard deviation is 5. So one SD extends from 50 to 55 or to 45 heads. This corresponds to an interval that extends from $0 to -$10 or to $10. Thus one SD on the gambler's fortune scale is $10.

The same calculation is repeated for the case of 400 coin tosses opposite. Notice, in this case, that the bell curve has spread further on the gambler's fortune scale. The gambler's fortune will now most likely lie in the interval -$40 to $40. Similarly we find that the SD that measures this spread more accurately has grown to $20.

Continuing these calculations:

Number of rounds played	MEAN of gambler's fortune	STANDARD DEVIATION of gambler's fortune
100 = 1 x 100	0	10 = 1 x 10
400 = 4 x 100	0	20 = 2 x 10
900 = 9 x 100	0	30 = 3 x 10
1600 = 16 x 100	0	40 = 4 x 10
. . .	. . .	. . .

Pattern: STANDARD DEVIATION grows in direct proportion to $\sqrt{\text{Number of rounds played}}$

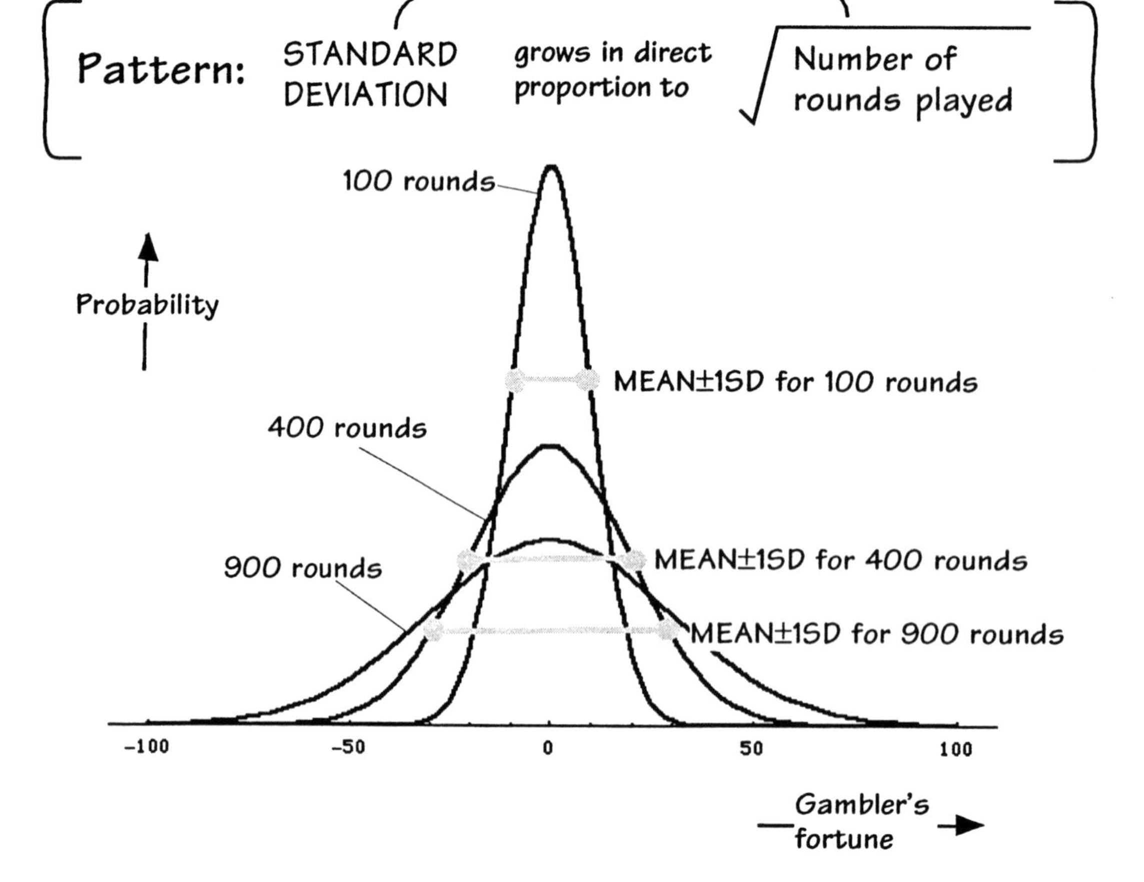

The point of the calculation of the previous pages is to find exactly how much the bell curve of the gambler's fortune spreads with repeated plays. That spread is represented most precisely by the standard deviation of the curve. After 100 rounds, the standard deviation is 10. After 400 rounds, the standard deviation is 20. If we continue the calculation, the result is the table shown opposite. As the number of rounds increases through 100, 400, 900, 1600, ... , the standard deviation grows more slowly: 10, 20 30, 40, ...

This rate of growth is governed by a simple pattern. The standard deviation grows in direct proportion to the square root of the number of rounds played. At 100 rounds, the standard deviation is 10. Now let the number of rounds grow by a factor of 4 to 4x100=400. The standard deviation grows by a factor of the (square root of 4) = 2; so it grows to 2x10=20. Similarly, if the number of rounds grows by a factor of 9 to 9x100=900, the standard deviation grows by a factor of the (square root of 9) = 3; so it grows to 3x10=30.

This growth is shown in the drawing of the bell curves for the three cases of 100, 400 and 900 rounds. The three bars indicate the growth of the MEAN±1SD interval for each curve, showing how the growth of this interval measures the growth in spread of the curves.

Important special case of the

Central Limit Theorem

SAME cause repeated many times — e.g. the gambler makes the same wager many times.

Effect = Cause1 + Cause2 + Cause3 + ...

STANDARD DEVIATION of Effect grows in direct proportion to $\sqrt{\text{Number of causes}}$

A quick way to arrive at this special case:

$$\left(\text{SD of Effect}\right)^2 = \left(\text{SD of Cause}_1\right)^2 + \left(\text{SD of Cause}_2\right)^2 + \ldots$$

All of these are the same. In the case of the gambler, each one equals 1.

⇩

$$\left(\text{SD of Effect}\right)^2 = \text{Number of Causes} \times \left(\text{SD of individual Causes}\right)^2$$

⇩

$$\text{SD of Effect} = \sqrt{\text{Number of Causes}} \times \text{SD of individual Causes}$$

In the case of the gambler:

$$\text{SD of Effect} = \sqrt{\text{Number of Rounds}}$$

The pattern seen in the case of the gambler illustrates an important special case of the central limit theorem. This special case arises whenever the many causes that act are the same. They are the same in the case of the gambler, for example, since the gambler makes the same wager over and over again. In this special case, we will always have the result that the standard deviation of the bell curve of the effect grows in direct proportion to the square root of the number of causes.

In the remainder of the chapter, we will see how useful this special case can be. It will help us understand any random process that involves the repeated action of the same cause through time. In that case, the number of causes grows in direct proportion to time. Therefore such processes will be characterized by an effect whose spread (as measured by the standard deviation) grows in direct proportion to the square root of time.

There is a simple way to see how the special case of the theorem arises--for the brave who do not mind following a few steps in a proof. Recall that the standard deviation of the effect is given by the addition of squares rule shown opposite. The square of the SD of the effect is equal to the sum of the squares of the SDs of the individual causes. If all the individual causes are the same, then all their squares are the same. So the SD of the effect squared is found by adding together the SDs squared of an individual cause as many times as there are causes. That is, the SD squared of the effect is equal to the number of causes times the SD squared of one individual cause. Taking square roots, it follows immediately that the SD of the effect equals the square root of the number of causes times the SD of one cause.

In the case of the gambler, the SD of one cause--one round of play--is just equal to 1. Therefore the SD of the effect equals the square root of the number of rounds played. This simple result fits with the table given above for the relationship between the SD of the gambler's fortune and the number of rounds played.

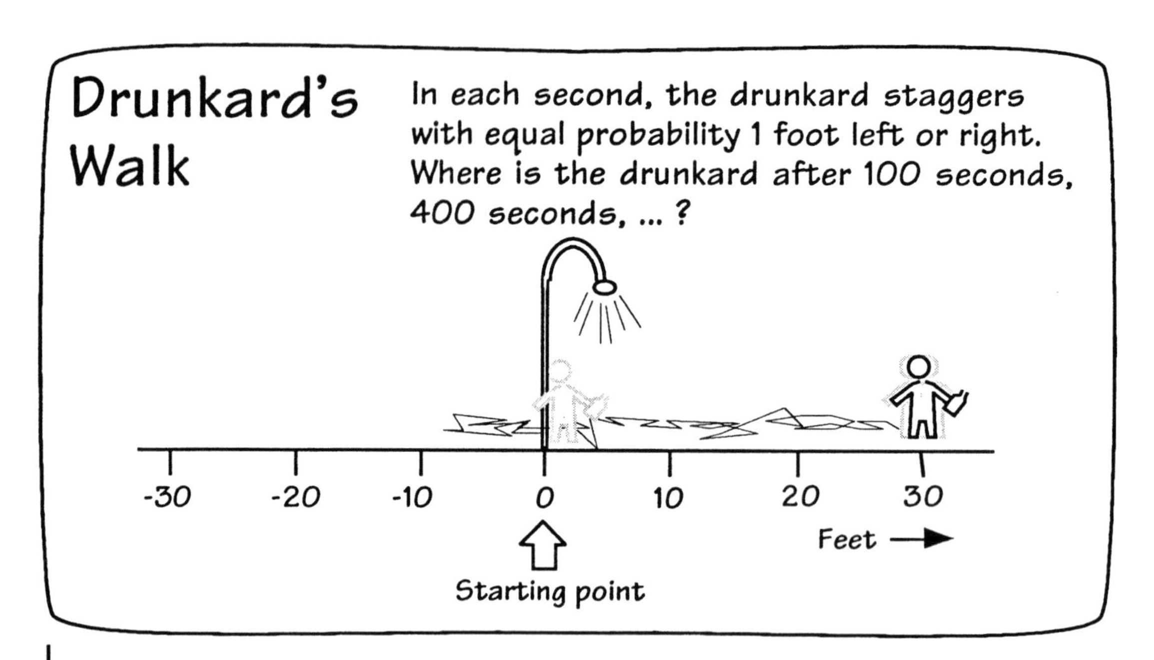

Answer

Drunkard's position on the street shifts left and right by exactly the same probabilistic mechanism as the gambler's fortune moves up and down. Therefore the same results apply to the drunkard.

After 100 seconds

SD = 10

Drunkard is within 20 feet of start with probability 0.95.

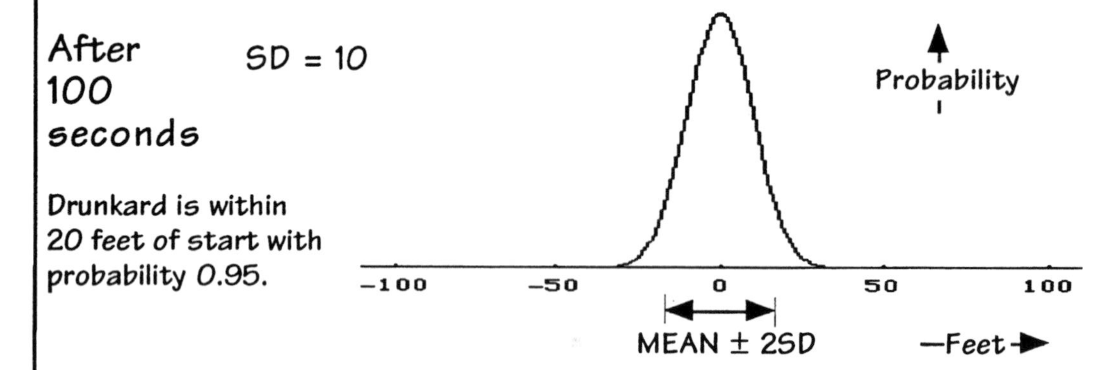

After 400 seconds

SD = 20

Drunkard is within 40 feet of start with probability 0.95.

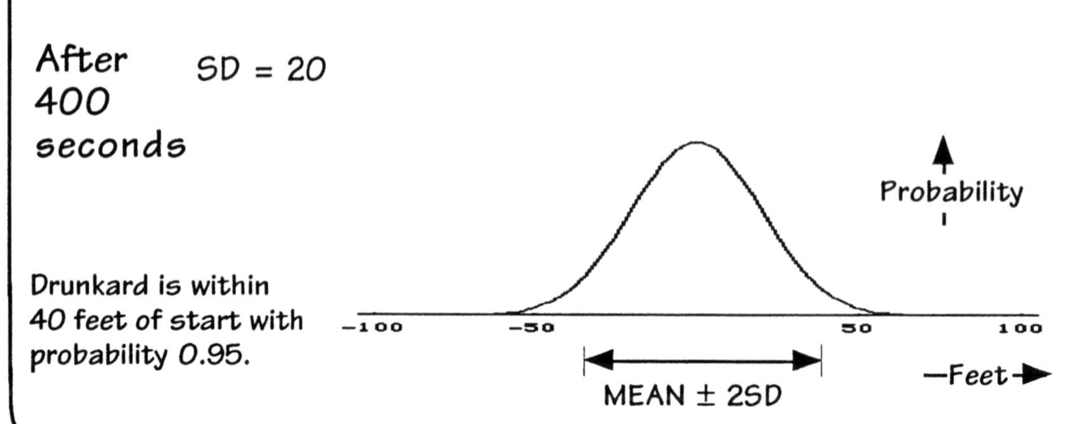

Another example of this special case of the central limit theorem is the drunkard's walk. The drunkard starts at a lamp post on the street and staggers left and right. In each second there is an equal probability that the drunkard staggers one foot left or one foot right. Where will the drunk be after some nominated time?

The central limit theorem applies. The drunkard's position is the accumulated effect of many, independent causes, the many staggers. So the probabilities of the different positions will conform to a bell curve. Each stagger has an equal chance of being to the left or right, so, on average, the drunkard's position will remain centered on the lamp post. Therefore the mean of the bell curve will remain at the lamp post, the zero feet position. The curve will spread over time, indicating the greater probability of finding the drunkard further away from the lamp post. The special case of the central limit theorem applies. The effect--the position of the drunkard--is the sum of many of staggers, all the same in the sense that each has an equal chance of being a step left or right. Therefore we know how rapidly the bell curve will spread with time: its standard deviation will grow in direct proportion to the square root of the number of staggers. Since there is one stagger per second, this means that the standard deviation will grow in direct proportion to the square root of the time elapsed.

The analysis of the drunkard's walk is made even easier by noticing that it is mathematically identical with the gambler. The gambler's fortune moves up or down by one dollar each round with equal chance. Correspondingly, the drunkard's position moves left or right by one foot each second with equal chance. The random element in the case of the gambler is supplied by the toss of a coin and for the drunkard by intoxication. But we would have the same walk if a sober walker chose to take a step left or right according to the outcome of a coin toss.

Because the two problems are the same, we can reuse the results of the gambler, as shown opposite. For example, after 100 rounds, the standard deviation of the gambler's bell curve will be \$10. So, after 100 seconds, the standard deviation of the drunkard's bell curve will be 10 feet. Thus the drunkard will be within $\pm$2SD=20 feet of the lamp post with probability 0.95. After 400 seconds, that distance will have grown only to $\pm$40 feet. Such is the penalty of public intoxication: after the effort of 400 staggers, the drunkard has probably advanced less than 40 feet in an undetermined direction!

Brownian motion and diffusion

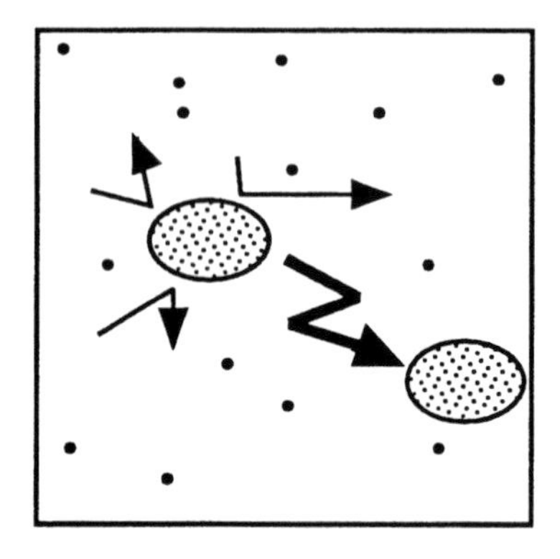

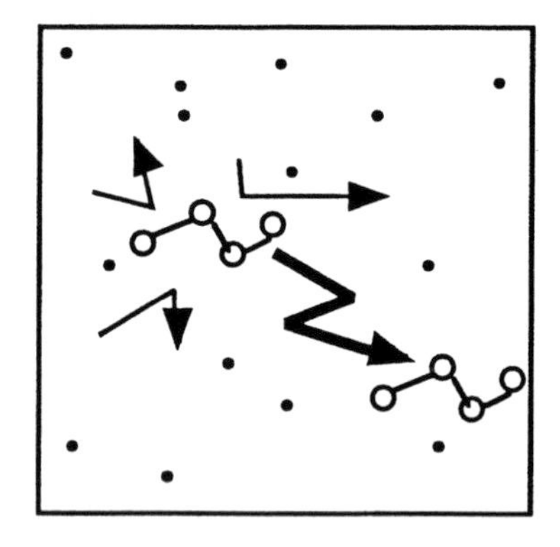

Pollen grain in water jiggled by lots of little collisions with water molecules.

Perfume molecule in air jiggled by lots of little collisions with air molecules.

Both executed a slighter fancier "drunkard's walk" and are governed by the "same cause" special case of the central limit theorem.

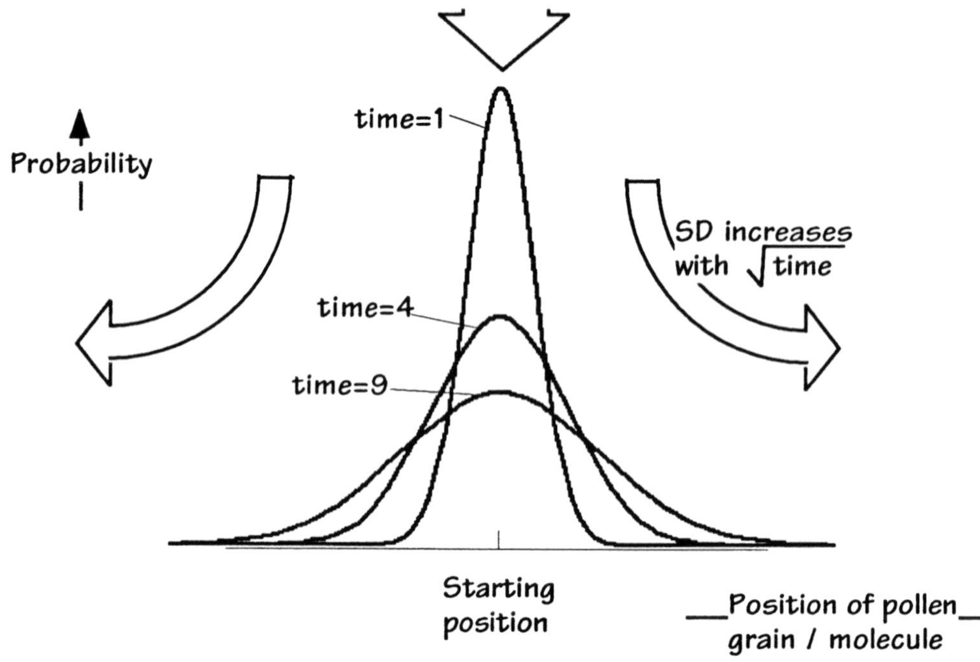

We can now use our excursions into the dark vices of gambling and public drunkenness to illuminate more scientific applications of the central limit theorem. In 1828, the botanist Robert Brown reported his observations from the previous year of many tiny particles suspended in water. The particles included pollen grains, as well as dust, soot and even pulverized fragments from the Sphinx! Under his microscope, he found them all to jiggle back and forth in an endless, random dance. After much debate, the origin of this Brownian motion was decided by Albert Einstein in 1905. Water molecules are in constant motion; the sustained rain of their impacts jiggles all tiny bodies suspended in water. This gave us visually some of the most powerful evidence for the reality of the atoms that make up molecules. (Their reality was still a matter of debate as recently as 1905!)

The Brownian motion of pollen grains is governed by the central limit theorem and the "same cause" special case. In fact it is just a slightly fancier version of the drunkard's walk. The effect is the position of the pollen grain. The same causes that are repeated to produce it are the repeated collisions with water molecules. The process is slightly more complicated than the drunkard's walk in so far as the collisions do not happen quite so regularly and the size and direction of the displacement of the grain will vary from collision to collision. Nevertheless, the conditions of the theorem are still met. If we consider displacement of the pollen grain in just one direction, the probabilities of different positions will conform to a bell curve. The mean will be the original position. From the special case of the central limit theorem, we have the law that governs how far the pollen grain can stray from its starting point: the standard deviation of its bell curve will grow in direct proportion to the square root of time.

Another example is almost exactly the same. Consider a perfume molecule floating in air. It will be bombarded by rapidly moving molecules of nitrogen and oxygen in the air in the same what that the pollen grain is bombarded by water molecules. The perfume molecule will execute a motion like Brownian motion, but this one will be invisible to a microscope, because individual perfume molecules are too tiny for it to see. The perfume molecules overall motion will be just like the pollen grain. Its average position--the mean of its bell curve--will remain in its starting position. Its bell curve will spread out in all directions with time. The standard deviation of the bell curve will grow in direct proportion to the square root of time.

Diffusion of a cloud of perfume.

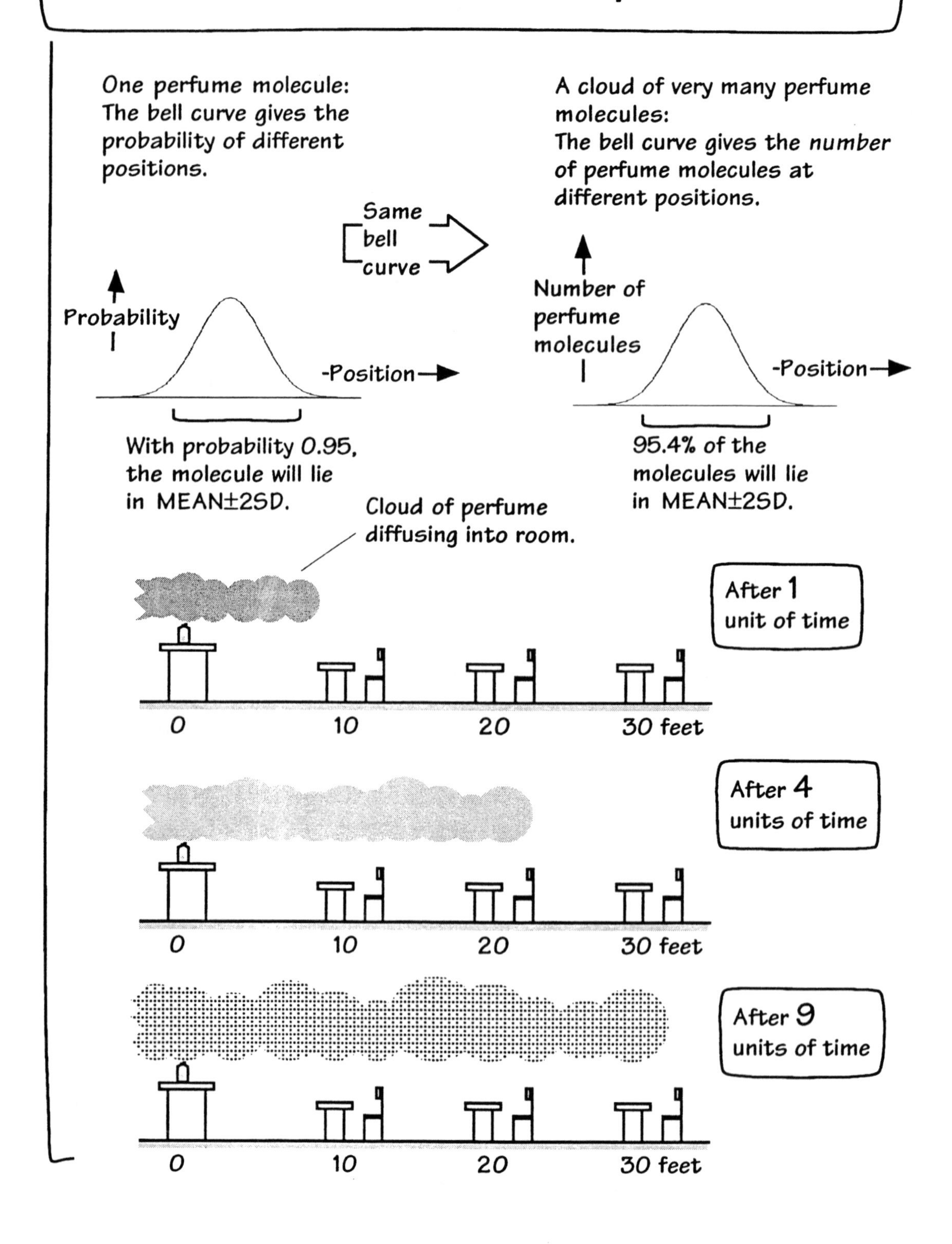

Once we know how a single molecule of perfume diffuses, we can find out how a cloud of perfume will diffuse. The connection between the two is simple. The bell curve for a single molecule tells us the probability that the molecule will have diffused to this or that position. Pick some interval of positions at some time. Imagine that the bell curve tells us that the molecule will diffuse there with a probability of 1/100. If we released 100,000 molecules from the same starting point at the same time, we then know that there is a 1/100 chance that each one will end up in that interval. That is, on average 1/100 x 100,000 = 1,000 of these molecules will end up in that position.

In short, the bell curve tells us the probability that an individual molecule will be in this or that position; the same bell curve tells us how many molecules from a cloud will diffuse to that position on average.

We can now use what we know about the speed of spread of the bell curve to infer how fast a perfume cloud will diffuse. Imagine that we have a room with very still air and that we release a cloud of perfume molecules at one end. (In plain language, we pull the stopper out of the bottle!) The numbers of molecules will distribute themselves in a cloud that conforms to the bell curve. Its MEAN will be zero so that the cloud will be centered on the starting point. 68.3% of the molecules will lie within MEAN ±1SD; 95.4% of the molecules will lie within MEAN±2SD, and so on. So, as the standard deviation grows with time, the cloud will spread out.

From the special case of the central limit theorem and our considerations of one molecule, we know that the standard deviation will grow in direct proportion to the square root of time. Therefore we can make inferences of this type: Assume that the cloud spreads to 10 feet in one unit of time. (A time unit may be one minute, for example.) Then after 4 time units, it will have spread to (square root of 4)=2 times as far; that is, to 20 feet. Similarly after 9 units of time, it will have spread to 30 feet, and so on.

Thus the central limit theorem gives us remarkably detailed knowledge of the rate of spreading of perfume by diffusion--even though we have used scarcely any information about the detailed mechanisms of the molecular collisions that spread the perfume!

Genetic drift and Natural selection

in evolutionary biology

Random fluctuation is the driving force of change that natural selection molds.

Example

Moth: *Biston betularia*

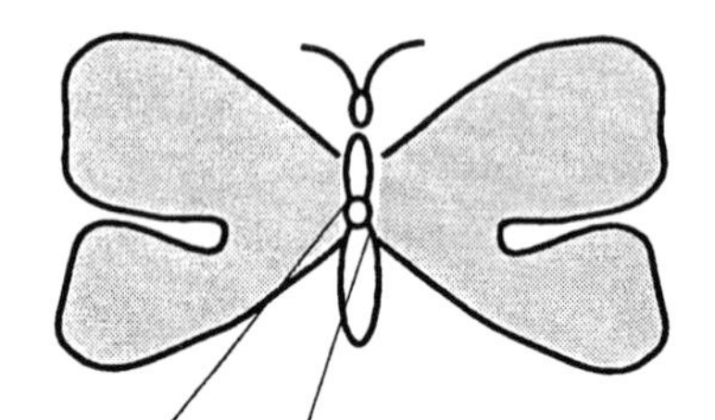

Each moth has two genes "G" = gray
"B" = Black
that determine color.

GG
Gray moth

GB, BG
Dark moth

BB
Black moth

B is not quite dominant over G, so GB, BG moths are dark, but not black.

The cloud of perfume moves through the air of the still room. But no breezes carry it. Rather what drives its spread is the accumulated effect of many small, random collisions between air molecules and perfume molecules. The accumulated effect of many random occurrences can exert a very powerful driving force. We shall now look at an example of this in evolutionary biology. We shall see that genetic drift, a kind of diffusion in the gene pool of a population, is the force that generates change in the composition of the population. That force is undirected, just as the diffusion of perfume is indiscriminate. However, once natural selection is allowed to mold the diffusing effects of genetic drift, adaptive evolution occurs.

To make the discussion concrete, we will consider a famous example, the case of the moth, *biston betularia*. These moths exist in several shades: a light gray, a dark color and black. The color of the moth is determined by its genes. Each moth has two genes, one from each parent, that combine to give its color. Each of these genes may be a "G" gray gene or a "B" black gene. If the moth had two G genes--we will write this as GG--then it will be gray. If the moth has two B genes--BB--then it will be black. But what of the moth with one of each, BG or GB? In the simplest cases in genetics, one gene--say B--would be dominant over the other. This means that the color of a BG moth would be fixed by the B gene; the G gene would have no effect on the color. These moths are close to having B dominant over G. But B is not quite dominant. Thus a BG or a GB moth will be dark, but not black.

A small population of moths in a forest.

Assume:

- The forest supports a constant population of 500 moths; that is, the population has 1000 color genes (2 per moth) in its gene pool.
- The gene pool of the initial "0th" generation has 100 B genes and 900 G genes.

How will the composition of the gene pool change through the generations of moths?

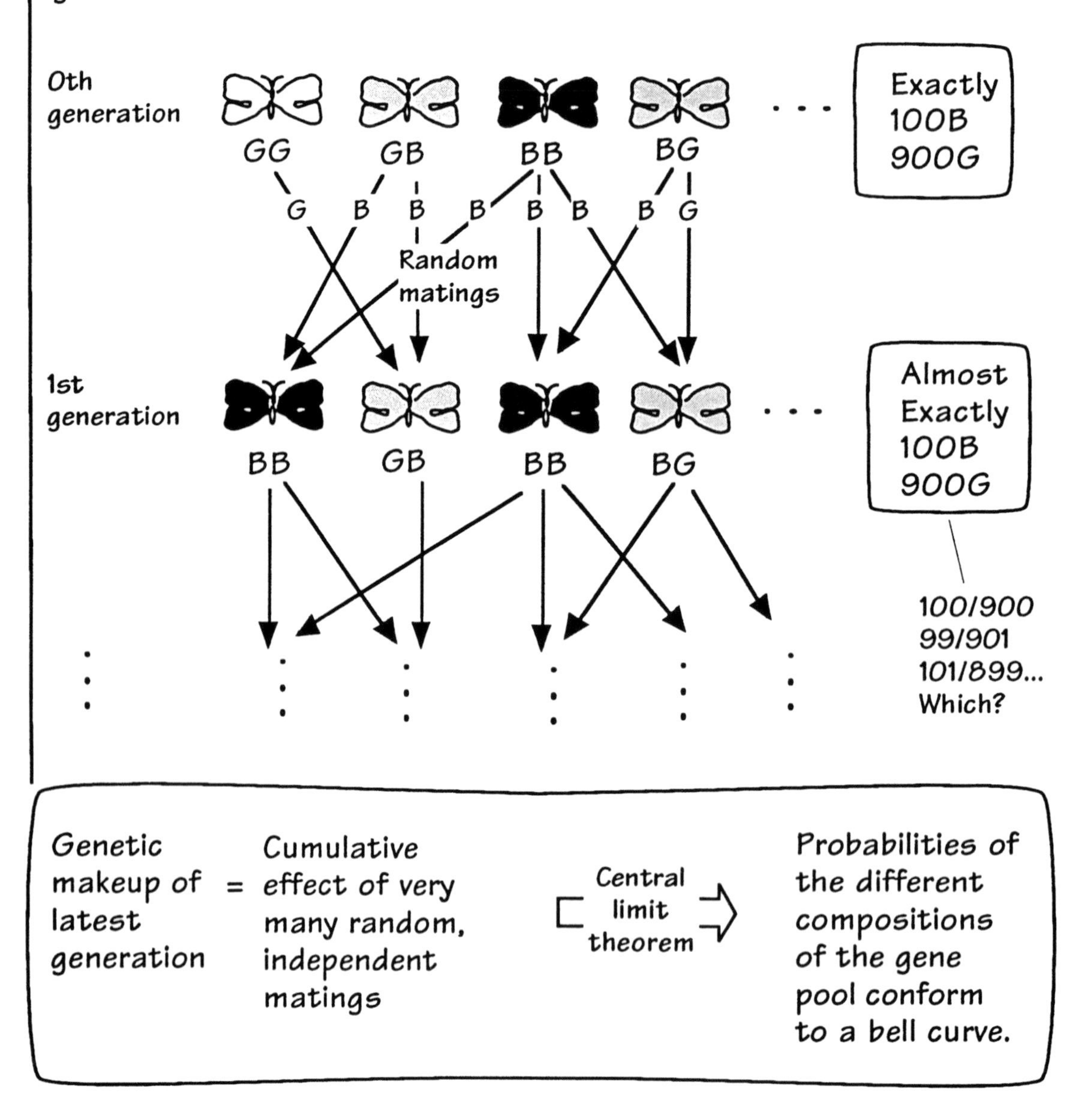

Consider a small forest that has sufficient food to support a population of 500 moths. We will assume that the moth population remains stable at 500 moths. Since each moth carries two color genes, this represents a gene pool of 1,000 color genes. Our focus will be how this pool of 1,000 genes varies in composition through the succeeding generations of moths. To give us a concrete starting place, let us assume that the initial "zeroth" generation has exactly 100B genes and 900G genes. Thus most of the moths will be gray.

The composition of the next generation will arise through many random matings between the moths of the present generation. Consider some individual of the next generation. Several random factors contribute to that individual's final genetic makeup. First we have the random pairing of male and female parent. Then each parent randomly contributes one of its two color genes to the offspring. For example, imagine that a GB and GG moth mate. The GB moth will contribute one of its genes, possibly a G or a B. The GG moth will contribute one of its genes, in either case, a G. Thus the offspring may be either GG or BG.

In the zeroth generation, 90% of the genes are G. We would expect the first generation to reflect the genetic composition of the zeroth fairly closely. Therefore, in the first generation, we would expect *about* 90% of the genes to be G. But it may be a little more or a little less according to the vagaries of chance matings. Thus we can at best specify probabilities for the different possible compositions of the first generation: ..., 98B/902G, 99B/901G, 100B/900G, ... As the generations pass, we expect the possible compositions to drift further from the initial 100B/900G.

The central limit theorem applies to this system, at least for the early stages of the drift. The effect is the composition of the latest generation. It represents the sum of the many, independent random matings in the history of the gene pool--the many causes required by the theorem. Therefore the probabilities of the different compositions of the gene pool will conform to a bell curve. The random process that leads to a new generation is not so far from a simple sampling process. Pick some individual of new generation. Its two color genes are close to a random sample of the gene pool of the previous generation. It is not quite a random sample, however. The individual cannot inherit both genes from the same parent, for example.

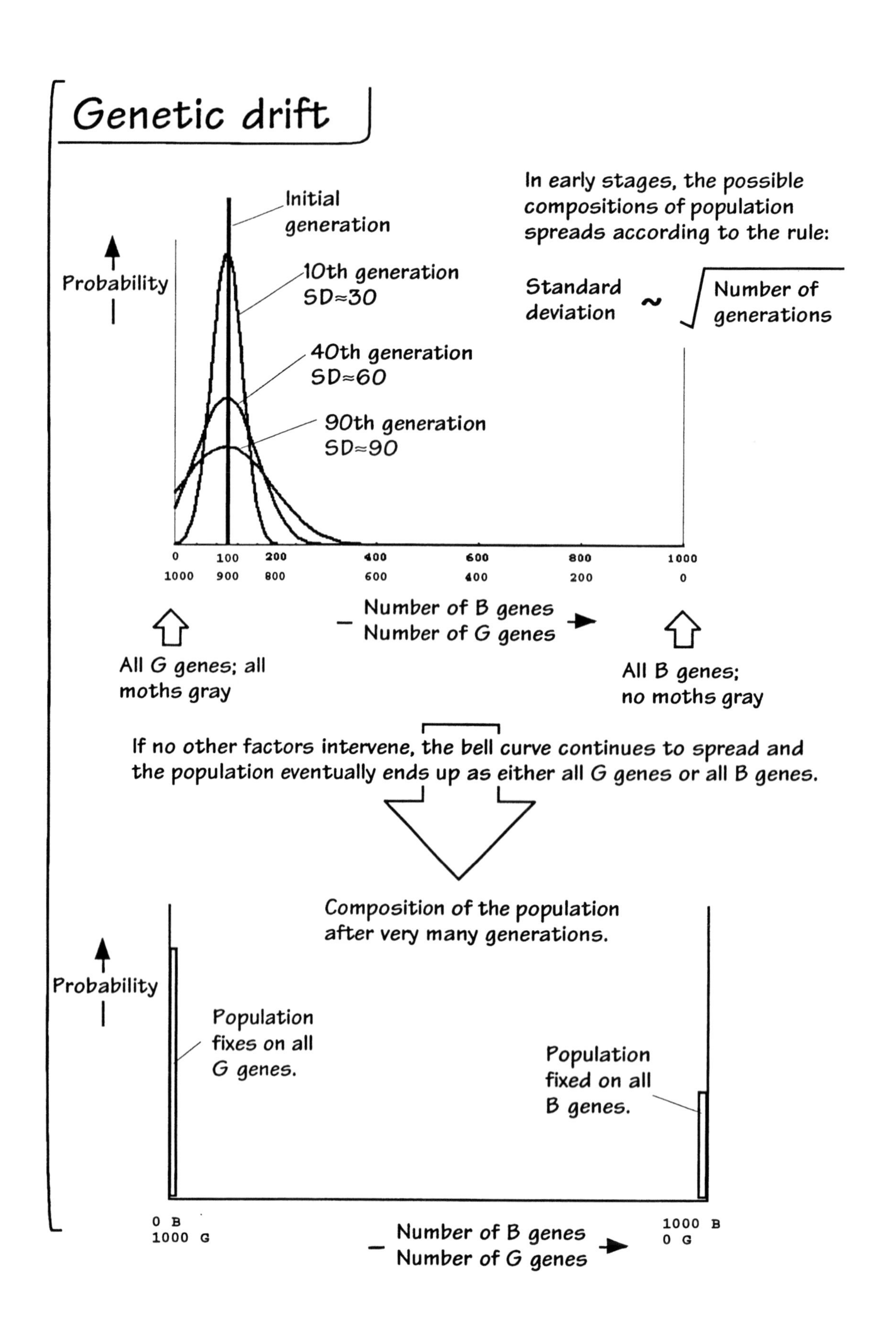
Genetic drift
Initial generation
Probability
10th generation SD≈30
40th generation SD≈60
90th generation SD≈90
In early stages, the possible compositions of population spreads according to the rule:
Standard deviation ~ √(Number of generations)
0 100 200 400 600 800 1000
1000 900 800 600 400 200 0
Number of B genes
Number of G genes
All G genes; all moths gray
All B genes; no moths gray
If no other factors intervene, the bell curve continues to spread and the population eventually ends up as either all G genes or all B genes.
Composition of the population after very many generations.
Probability
Population fixes on all G genes.
Population fixed on all B genes.
0 B
1000 G
Number of B genes
Number of G genes
1000 B
0 G

The central limit theorem tells us that the probabilities of different compositions will conform to a bell curve. These bell curves are illustrated opposite. The bottom axis of the graph corresponds to the different compositions the population may have. At the extreme left, we have the case of zero B genes; that is all 1,000 genes in the population are G and all moths are gray. At the other extreme, all 1,000 genes in the population are B; there are no G genes and all the moths are black. The intermediate numbers represent different possible compositions. 500B and 500G genes for example--the midpoint of the axis--represents the case in which there are exactly as many G and B genes. The moths in this population will be a mix of gray moths (with GG genes), dark moths (with BG and GB genes) and black moths (with BB genes).

The curves drawn give us the probabilities of the different compositions of the populations. We know for sure--probability 1--that the initial generation has 100B genes. This is represented by a spike at 100B--the only possibility for the initial generation. As the generations pass, it is more and more possible that the actual composition of the population has strayed from the original of 100B and 900G genes. The probabilities of different compositions are represented by the bell curves. Over time, they spread into ever flatter curves. This spreading is called "genetic drift".

A more exact calculation tells us how fast the bell curves spread. In the early stages, as shown opposite, the bell curves spread in accord with our familiar rule: the standard deviation grows in direct proportion to the square root of the number of generations. This rule is exemplified by the standard deviations of 30, 60 and 90 associated with the number of generations 10, 40, and 90.

The central limit theorem and the bell curve only prevail for the early stages of genetic drift. If ever the population reaches a state of all G or all B genes, genetic drift ceases. For once the population is all one gene, the same is true of all future generations. Whatever random mating may occur, the composition of the color genes in the population cannot alter. Thus the central limit theorem ceases to apply. The bell curves give way to a massing of the populations at the 0B or 1,000B compositions. This breakdown of the bell curve is already visible in the graph. The curves for 40 and 90 generations are cut off noticeably at the 0B point.

Natural selection checks genetic drift

Pre-Industrial England	G gene gives moths camouflage against light tree bark. Black moths seen and eaten by birds.	⇨	Post-Industrial England	B gene gives better camouflage against soot blackened bark. Gray moths seen and eaten by birds.
	Population is 1% black moths. Gray moths favored by natural selection.	— Natural selection ⇨		Population is 90% black moths. Black moths favored by natural selection.

Directional selection

Natural selection discourages drift towards more G genes.

Natural selection encourages drift towards more B genes.

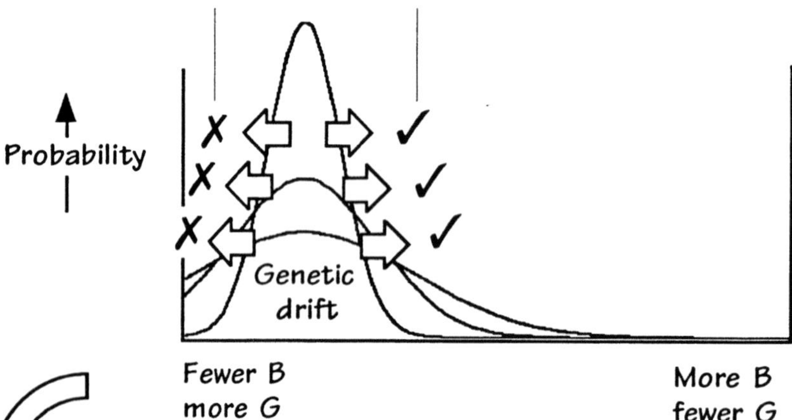

Fewer B more G

More B fewer G

Population migrates towards more B.

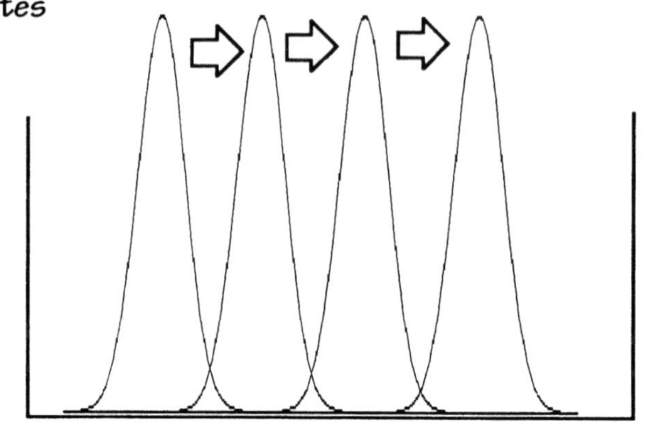

Natural selection molds undirected spread of random selection to yield migration towards better adapted population.

In real populations, genetic drift does not continue unchecked. For in addition to the random forces producing genetic drift are the forces of natural selection. In the case of moths, their color has a great deal to do with their survival. In preindustrial England, the gray color of these moths provided a decided advantage. It camouflaged them against the light tree bark. Darker moths stood out and were easy prey for hungry birds. Thus nature selected in favor of the lighter moths. Genetic drift towards more B genes is discouraged.

What makes this case so interesting is that we could watch what happened when this situation changed. With industrialization in England, the soot of the factories darkened the tree bark. The light moths now stood out as easy prey for the birds; the dark moths were camouflaged. Natural selection now favored the dark moths. From generation to generation, the dark moths survived better than the light ones. The population of black moths grew from 1% to 90%, in some areas in less than a century. The gene pool moved away from more G genes to more B genes.

This adaptation of the moth population is an example of directional selection. The driving force of the adaptation is the random spreading of the bell curve--the same spreading that has been the subject of this chapter. That spreading is molded by natural selection. It discourages motion towards populations with more G genes and encourages motion towards populations with more B genes. The net effect is that the bell curve does not spread and flatten in all directions. Instead it migrates towards the populations with more B genes. This migration is the adaptation of the moth population to the darker tree bark.

If no other selective forces acted, the population would migrate towards all B genes and the gray moths would disappear completely from the population. But there are other forces--and that is a story for another day.

Why does water evaporate?

A simple molecular model explains why water is liquid.

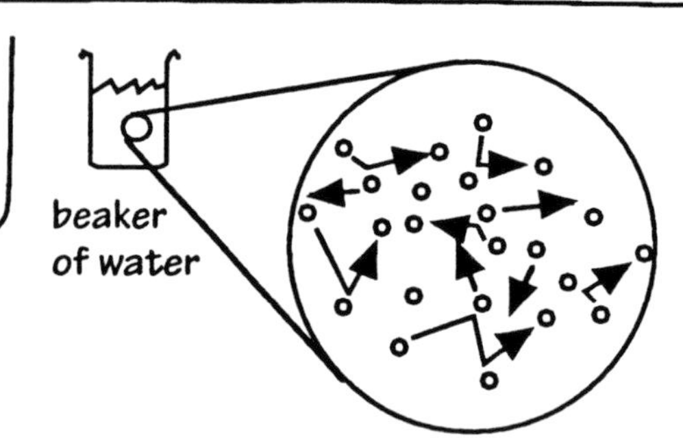

Many water molecules moving at different speeds colliding with one another.

The water molecules attract one another with a weak force.

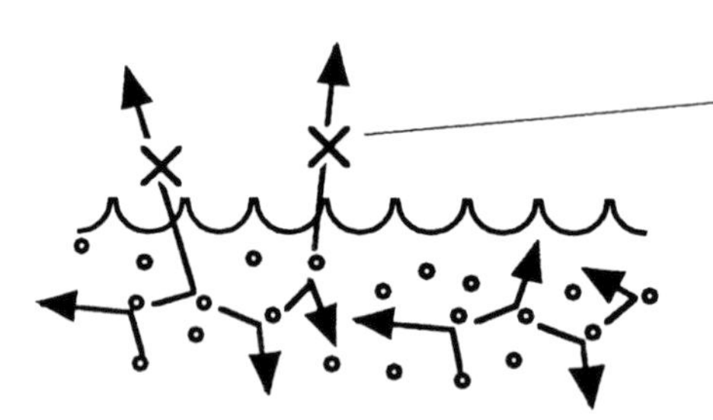

The forces are strong enough to prevent water molecules at the surface flying into the air space above.

Therefore, water stays in the beaker as a liquid.

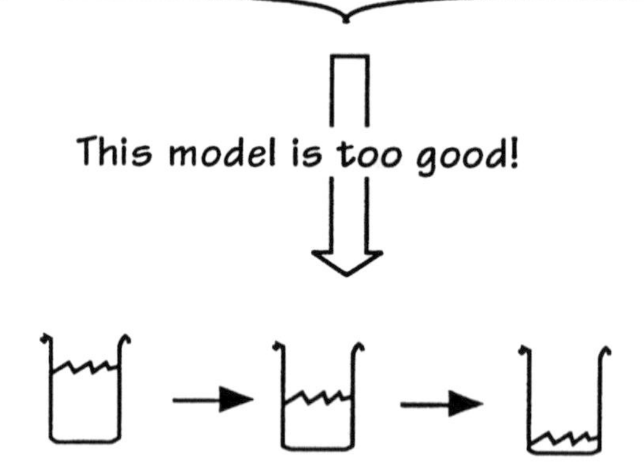

The water in the beaker does slowly evaporate.

Somehow ALL the water molecules eventually do get enough velocity to escape.

HOW DOES THIS HAPPEN?

Our final example of the use of the bell curve comes from the physics of liquids. We are all familiar with a simple model of liquids. Water, in this model, consists of very, very many water molecules rushing to and fro, smashing into each other. This model can explain a lot. It explains, for example, how water can conform rapidly to any shape container. But it does not explain everything about a liquid like water.

Consider some water in a beaker. If its molecules are rushing about so frantically, why don't they jump out of the beaker? They don't. Water remains politely in a liquid puddle at the bottom of the beaker. Why does the water remain in the liquid form? We need a new element in the model to explain this. The water molecules are attracted weakly to each other. These attractive forces tend to pull the water molecules together. Consider the surface of water. A molecule within the liquid may head towards the surface. Once it reaches the surface, all the attractive forces of the other water molecules pull it back towards the body of the water. Thus the water molecules are trapped at the liquid surface and cannot escape.

This model explains why water molecules are trapped in a liquid. But it explains it too well! For our experience is that water remains liquid only temporarily if left in an open beaker. Over a period of a day or two, water in a beaker evaporates. That is, eventually *every single water molecule* somehow has enough velocity to escape. How is this possible?

An application of the central limit theorem.

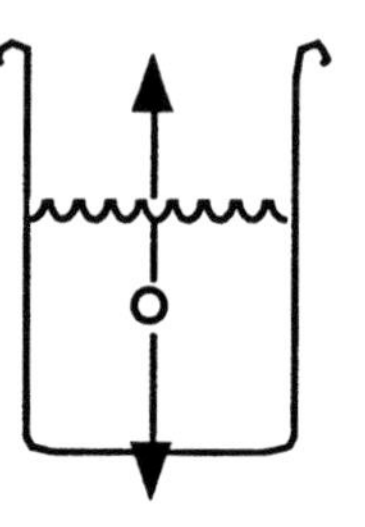

Consider the velocities of water molecules in the vertical direction.

Velocity of a water molecule in the vertical direction = Collision with first molecule + Collision with second molecule + Collision with third molecule + ...

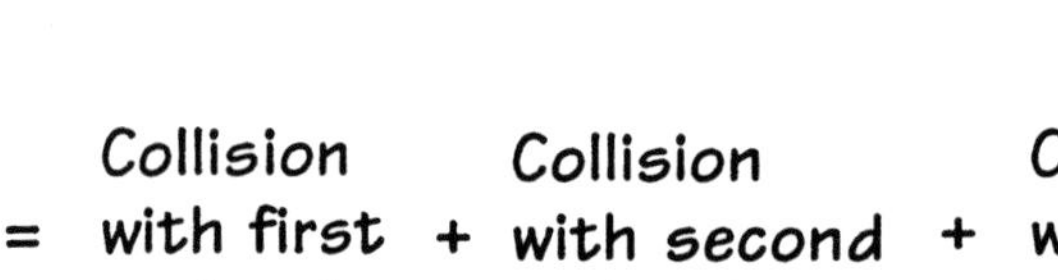

Very many, independent causes

Central limit theorem

The vertical component of velocity is normally distributed:

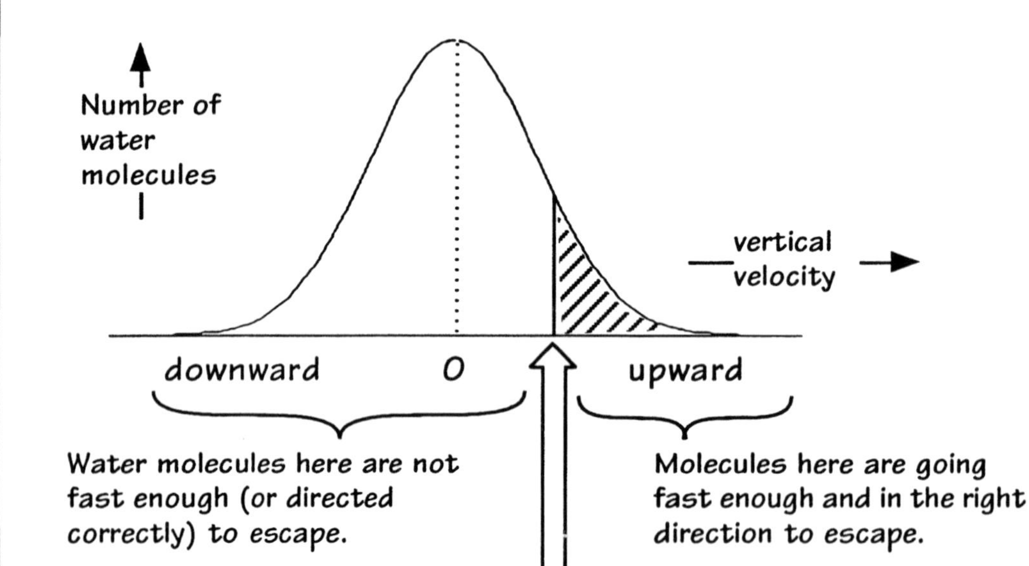

Minimum velocity needed to escape attractive forces of other molecules.

To understand how water evaporates we need a better understanding of how different velocities are distributed over the water molecules. Water molecules move in all directions in water. The direction that interests us is the vertical direction, for motion in this direction carries a water molecule out of the liquid.

What determines how fast a water molecule is moving in the vertical direction? That is determined by the molecule's history of many collisions with other water molecules. The outcome of these collisions depends in part on the velocity of these other molecules. And that depends on their own history of collisions.

We now have some of the conditions of the central limit theorem. The vertical component of a water molecule's velocity is the effect of very many causes, the individual collisions. Although it cannot be obvious at this superficial level of discussion, the other conditions also hold. Thus we have the vertical velocity of the molecule distributed according to a bell curve.

In the curve shown, the axis across the page shows the vertical component of velocity of a water molecule. The bell curve shows the number of molecules in the water that has each vertical velocity. The mean of the curve is zero; that is, the average vertical velocity is zero. This is as we would would expect, since, on average in the liquid, there is no bulk motion either up or down. A randomly picked molecule is as likely to be going up as down. (We assume here that the water is in a sealed container so that no evaporation occurs.) We also see that most of the molecules have vertical velocities that cluster around zero.

As we proceed towards greater upward vertical velocities, there will come a point at which a molecule does have enough vertical velocity to escape the water's surface--if the surface is exposed to air so that escape is possible. Thus any molecule with a velocity greater than this can also escape the water's surface. Since water evaporates slowly, we must presume that under normal conditions, only few molecules will be in the shaded regions of the curve opposite, the region where the water molecules have enough velocity to escape.

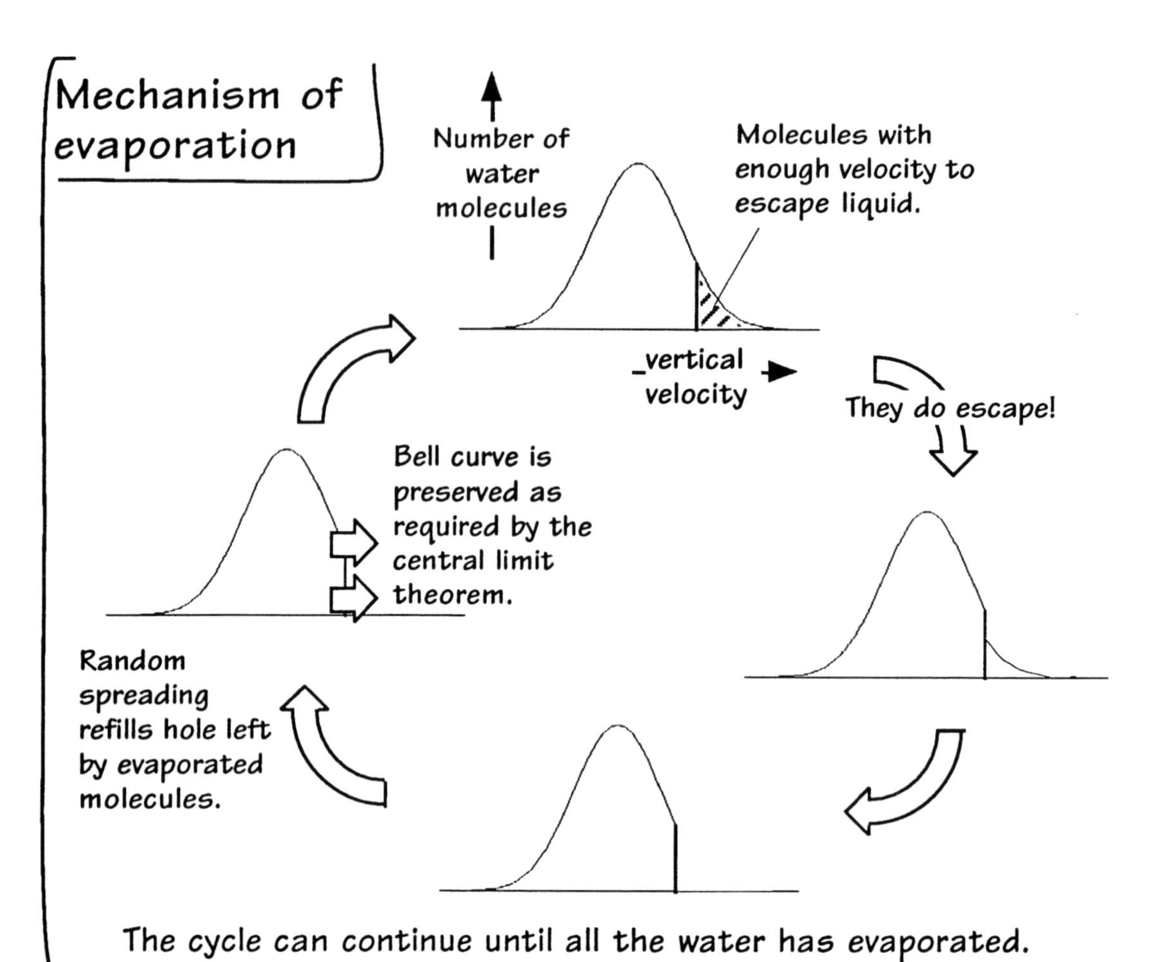

The cycle can continue until all the water has evaporated.

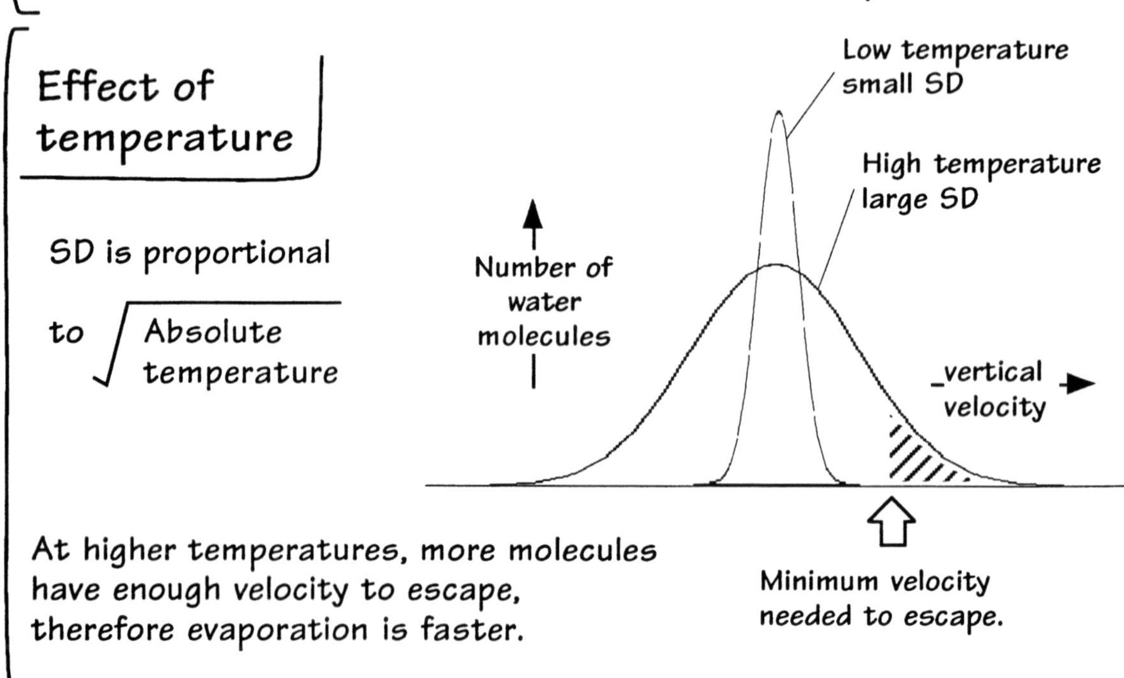

At higher temperatures, more molecules have enough velocity to escape, therefore evaporation is faster.

Now consider a body of water exposed to dry air. Some portion of the molecules will have sufficient vertical velocity to escape the surface of the water into the air. They do escape. Random collisions then boost more, slower moving molecules up to a higher velocity. These molecules can then also escape. We have entered a cycle that can continue indefinitely until all the water is evaporated.

The drawings of the bell curves opposite help us visualize this process and to see the role of the central limit theorem in it. The bell curve tells us how many molecules there are at each of the velocities. When some escape, the effect is to remove a portion of bell curve. More precisely, if the molecules with velocities in the shaded portion of the graph do escape, then that means this piece of the curve drops to a zero value. Once this has happened, the resulting curve is no longer a bell curve. But we know from the central limit theorem that the many random collisions of the molecules will produce a bell curve. These collisions fill out the curve and restore it. The cycle is complete. The overall effect is that a portion of the molecules have escaped into the air; that is, they have evaporated.

For simplicity this cycle has been represented as a discontinuous two step process. Of course in real water the two steps proceed at the same time and the process is quite smooth.

How is it that all the molecules eventually get a high enough velocity to escape? Random collisions are assured to produce a full range of velocities. We selectively remove just the molecules that have a high vertical velocity. Random collisions then boost more, slower moving molecules up to a higher velocity. Eventually all the molecules go through the process.

There is much more that we can learn from the bell curve of velocities. Why is it that hot water evaporates faster? The key is the standard deviation of the bell curve. A more detailed analysis shows that the standard deviation of the bell curve increases in direct proportion to the square root of absolute temperature. As the temperature increases, the mean of the bell curve stays at zero and the bell curve spread out further. As it spreads, more molecules enter the region in which they have enough velocity to escape and the evaporation proceeds faster. Therefore heating water promotes evaporation.

Concluding morals

- Random processes are by their nature unpredictable.
- The cumulative effect of many random processes is regular and predictable.
- The notions of the bell curve, the statistical spread and the standard deviation that measures it, are very powerful tools for understanding random processes.
- Outside science, these tools are best known for their use in statistics. They also are core methods for most sciences including physics, chemistry and biology.

Random processes are by their nature unpredictable. We cannot know in advance whether a fair coin will land heads up or tails up; whether a person sampled randomly from the population is male or female; whether the offspring of two moths with will be dark or black. But that does not mean that we can know nothing about these random processes. Probability theory gives us powerful tools for finding the order that lies behind the confusion of the random processes. In particular, if we allow the effects of many random processes to accumulate, we arrive at results that display a reliable consistency. We may not know if a person randomly sampled from the population will be male or female. But, if the population is 50% female, we can be virtually certain that, if we sample 100 people, between 85 and 115 will be female.

The bell curve is the tool that allow us to compute these regularities with great ease. Once we have a random process whose probabilities conform to a bell curve, we need only know the mean and standard deviation of the curve to enable us to list off an array of conclusions about the process. In the systems we have looked at, these essential quantities are computed with a few simple formulae.

When do we have a random process whose probabilities conform to a bell curve? The central limit theorem lays out those conditions. They are very general. Essentially all we need is an effect produced by many, independent causes. These conditions were satisfied in the sampling and estimation problems we considered. Throughout the diverse branches of science, one finds effects that result from the combined action of many, independent causes: the genetic composition of biological populations and the velocities of molecules are only two examples. This makes the bell curve one of the most important curves in science.

Assignment 13: A Practical Exercise

Your assignment is to scan any current media (current newspapers, magazines etc.) in search of a statistical claim backed up with the statistical data. The claim must be one whose analysis requires the use of a bell curve and the types of techniques developed in this book. (An article that merely presents some numerical information is usually not of this type.)

You are then to check whether the claim is properly supported by the statistics offered. (e.g. are the results statistically significant in the precise sense, could the result depend on a chance fluctuation *etc.*)

As a model consider questions 3, 4 and 5 of Assignment 12. The data analysed were reported in an article in the *Pittsburgh Post Gazette* which claimed that exercise is beneficial to the elderly. That claim was well supported by the data, as we found when we analysed them. ("Death, disability lower in elderly runners," *Pittsburgh Post Gazette* , October 1, 1994, p.A8.)

Notice that very commonly reports of statistical results do not give enough of the data to enable analysis. Often the sample size is missing. If this is the case, you should make some of realistic guess of the value of the missing number and then complete the calculation.

You assignment should contain

(a) A clear citation to your source.

(b) A statement of the claim advanced in your source.

(c) A clear statement of the statistics that the source uses to back up the claim.

(d) A statement of whether the data reported is sufficient to enable analysis. If it is not, state clearly which number or numbers you are guessing to enable completion of the calculation.

(e) Your analysis (which uses some of the methods developed in this class).

!!ALERT!!

Don't leave this assignment to the day on which it is due. To find an item that you can work with takes some patience. You will need to scan the media for a week to get something that you can analyze . Start now! If you are not finding suitable material in your usual reading, try reading a little more widely.

Further Assignments

1. Find a second example of a statistical claim in the current media and analyze it in the same way as in Asssignment 13.

OR

2. Find an example of a controlled study in the current media. Describe it indicating (in so far as the information is available):
(a) The hypothesis under test.
(b) How the test and control groups were formed.
(c) The outcome and how it was used to decide the hypothesis of (a).
(d) An analysis of the statistics of the experiment.
To see how to do (d), look at the analysis given in questions 3, 4 and 5 of Assignment 12, which gives a statistical analysis of the outcome of a controlled experiment. If your source does not include all the data needed for the analysis, guess the missing data. Be sure to indicate clearly which data were guessed.

OR

3. Find some process which is restricted strongly by its energy needs (and has not been discussed already in earlier chapters). Give an analysis complete with relevant calculations of how these energy needs limit the process.
(a) Describe the process clearly.
(b) Indicate the sources of energy available for the process and give numerical estimates of the quantities of energy available.
(c) Indicate the energy needs of the process and give numerical estimates of these needs.
(d) Show how the limitations on the energy supply affect the process. Use the numbers presented in parts (b) and (c) to compute the relevant limit.
As a model for this assignment, consider the analysis given in Chapter 5 for the prospects of an electric car. We compared the energy needed to drive a car with the energy a battery can supply. We found that the range and performance of an electric car are severely limited by the energy the battery can supply. That is, if we kept the weight of the battery down to a reasonable size, then we found that batteries of traditional design were hard pressed to deliver quantities of energy required for the range and performance desired.

OR

4. Find an example of an effect which is the result of many applications of the same cause.
(a) Describe the process that yields the effect.
(b) Explain how the central limit theorem applies to it.
(c) Indicate how the relationship between the standard deviation of the effect's spread and the number of causes helps us to understand how the process develops.

Questions for Study and Review

1. Controlled Studies

You have been asked to analyze through a controlled study whether passive smoking is harmful to the health. (Passive smokers do not smoke but are in smoky environments where they inhale the smoke generated by smokers. The non-smoking spouse of a smoker is usually a passive smoker.)

(a) State clearly the hypothesis to be tested.

There are obvious ambiguities in the notion of harmful effects of passive smoking that make testing hard. e.g. How much smoke must a non-smoker inhale to be a passive smoker? Exactly what is meant by "harmful to the health?" Explain how your statement of the hypothesis avoids these and other problems of ambiguity.

(b) Indicate how you will form your test and control group.

(c) Is your study a prospective or retrospective study?

(d) What are the advantages and disadvantages of a prospective study and of a retrospective study?

(e) Is your study a double blind study? If not, why not? How might this omission affect the results of your study?

2. Conservation of Energy

Science fiction writers love to write about transporters. Their use is standard in much science fiction. The unit sits in an orbiting spaceship. The travelers step in. The dials are turned, the lights flash, the travelers twinkle out and reappear below on the surface of the planet. Their return journey is effected in the same way. Obviously we have no idea in current science of how to build such a machine. However, if some scientific breakthrough does permit construction of such a machine, it will be subject to the law of conservation of energy. So it is possible for us to estimate at least a lower limit on the energy needs of such a machine.

For simplicity, we shall assume that the transporter "dematerializes" the traveler in one location, "rematerializes" that same traveler in another and that this is its sole effect.

Imagine that the transporter is used to elevate a 90 kilogram (approximately 200 lb.) traveler up a 1000 meter cliff on the surface of a planet with the same gravity as earth.

(a) If 9.81 Joules of energy are released when a one kilogram mass is lowered by one meter, how many Joules are released when one kilogram is lowered 1000 meters.

(b) Express your answer to (a) in kiloJoules. (1000 Joules = 1 kiloJoule)

(c) How many kiloJoules of energy are released in lowering a 90kg traveler down the 1000 meter cliff?

(d) Assume that the transporter can elevate the 90 kg traveler up the 1000 meter cliff with little energy supplied in fuel ("little" here means much less than your answer to (c)). How could this transporter be used in conjunction with the cliff to create a perpetual motion machine?

(e) Since the law of conservation of energy tells us that perpetual motion machines are impossible, we must conclude that the transporter of (d) cannot use little energy. What is the minimum energy that must be supplied to the transporter in some type of fuel each time it elevates a 90 kg traveler up a 1000 meter cliff?

(f) Imagine that we use the machine to elevate two travelers up the cliff in each second. What is the minimum rate at which energy must be supplied to the transporter in kiloJoules per second?

(g) If a power supply of one kiloJoule per second is equivalent to one kiloWatt of power, express your answer to (f) in kilowatts.

(h) If a large steam locomotive engine running at full power supplies 850 kiloWatts, how many steam locomotive engines are needed to meet the power needs of (g)?

(i) Where on the StarShip Enterprise would one keep a steam locomotive?

3. Heat Engines

(a) The efficiency of a heat engine is measured as the fraction of heat supplied to it that it converts to work. What type of heat engine is the most efficient?

(b) What formula gives us the efficiency of such a machine?

(c) What temperature scale is used in the formula?

(d) Describe the operation of a heat engine that has zero efficiency.

(e) Could we ever have a heat engine that is 100% efficient? Why not?

(f) If we supply heat at 177°C (=450K) to a heat engine and it exhausts the heat at 77°C(=350K). What is the maximum efficiency of such an engine? For each 100 cal of heat supplied, what is the maximum work a heat engine can recover?

(g) If a refrigerator of maximum performance is used to return the heat to the 177°C heat source from the heat sink at 77°C, how much work would it need to return 100 cal to the heat source? In doing this, how much heat would the refrigerator extract from the heat sink at 77°C?

4. Probability

In the game of roulette, a wheel with the numbers 1, 2, .., 35, 36 is spun and the fall of a ball randomly selects one of these numbers. Bets are placed on the possible outcomes. One may bet on individual numbers as well as a huge range of sets of numbers (e.g. even or odd, 1-18 or 19-36 etc.)

(a) What is the outcome space.

(b) What is the probability of the number 15 as an outcome? What is the probability of the number 17 as an outcome?

(c) If the wheel is spun many times, on average how often do we expect to see the number 15 as an outcome?

(d) Consider the probability P(1 to 18) = P(1 OR 2 OR 3 OR ... OR 18). What condition must the outcomes 1, 2, ..., 18 satisfy for the OR rule to be used? Do they satisfy it?

(e) If it is satisfied, use it to compute P(1 OR 2 OR 3 OR ... OR 18).

(f) Consider the probability P(1 to 18 OR 1 to 12). What condition must the outcomes (1 to 18) and (1 to 12) satisfy if the OR rule is to be used? Do they satisfy it?

(g) Consider the probability P(1 to 18 AND even)? These outcomes are independent, so that the AND rule can be used. Use it to compute the probability.
(Hint: P(even) = 1/2)

(h) Consider the probability P(15 AND 17). What condition must these outcomes satisfy if the AND rule is to be used to compute the probability? Do they satisfy it?

(i) What is P(15 AND 17)?
(Hint: do not use the and rule.)

5. Hypothesis Testing

A small car dealer sells somewhere around 50 cars a year. In one year the numbers of cars sold drops to 39--over a 20% drop from the 50 cars sold the previous year. The dealer panics, fears the dealership's market share is disappearing and fires the sales manager (his brother). The sales manager is eager to recover his job and has also studied statistics. He decides to determine whether the drop in sales is clear evidence of loss of market share or if it may plausibly result from a random fluctuation in an otherwise stable market share. To decide this, he proceeds as follows.

He guesses that there are 1,000 cars (of the type his dealership trades) sold each year in his town. If the dealership ends up selling 50 of these, then the chance that the dealership makes any particular one of these thousand sales is 0.05 = 50/1000. This chance measures the dealership's market share. If the market share is stable, this chance will be stable. If the market share is falling, this chance will be falling. To decide bewteen these cases, the sales manager sets out to test the following hypothesis:

There is a chance of 0.05 that the dealership makes each of the1,000 sales in one year.

(a) If this hypothesis is true, how many cars on average should the dealership expect to sell in one year?

(b) The actual number of cars sold will be distributed normally around this value. What is the MEAN of the distribution? What is the STANDARD DEVIATION?

(c) In what interval can the former sales manager be sure that this number will lie with probability 0.95?

(d) Does the actual result of 39 lie in this interval?

(e) What does it mean to say that an outcome is statistically significant at the 0.05 significance level?

(f) Is the actual result of 39 statistically significant at the 0.05 level?

(g) The entire analysis depends on the figure of 1,000 guessed by the sales manages in (a). What would happen to the analysis if he had guessed 2,000 instead? or 10,000?
(Hint: The outcome of the analysis remains essentially unchanged when these numbers are changed.)

(h) Can the car dealer use the reduced sales as proof that his market share has fallen, that is, as evidence that the hypothesis under test is false?

(i) Can the former sales manager use the results of this statistical analysis to rule out the possibility that market share has been lost, that is, that the hypothesis under test is true?

(j) Will the former sales manager get his job back (i.e. when it comes to it, is statistics of any use?)

6. Estimation

The City of Pittsburgh decides that it will give a tax rebate to households with children under 18 years. $1,000,000 has been set aside in the budget for this purpose. As treasurer you have been asked to determine how much should be given to each household with children so that the $1,000,000 is not exceeded. You know that there are 140,000 households in Pittsburgh. But there is no direct data on the fraction of these that have children. Fortunately, a recent survey gives you some guidance. Of 100 households surveyed, 70 had children under 18. You project these figures onto the entire population of 140,000 households and conclude that there are about .70 x 140,000 = 98,000 households with children under 18 in Pittsburgh. If you allow a rebate of $10 per household with children, that will consume $10 x 98,000 = $980,000 of the budgeted $1,000,000 and leave just a little bit of room in case there are more households with children than you thought. You send your recommendation to the Mayor.

As you travel home from work that evening, you start to worry about the "little bit of room" that you left yourself. In the survey 70% of the 100 households looked at had children. That means that **about** 70% of households in the overall population will have children. That means it could be 71% or 72% or...well how high could it get? If it gets up to 72%, you realize to your horror, your rebate will already exceed the $1,000,000 allowed. (72% of 140,000 = 100,800, which would lead to a total cost of $1,008,000). As panic sets in, you recall one of your favorite classes as a student was one that covered some statistical methods. Fortunately you had kept your precious lecture notes in a mink lined box along with your other most prized possessions, such as the calculator that could take square roots. You arrive home , leap into those notes and begin to calculate.

(a) If 70 out of the 100 households sampled have children, what is the frequency of households with children in the sample, the "sample frequency"? (Hint: the answer will be a number between 0 and 1.)

(b) The "population frequency"--the frequency of households with children in the overall population will be normally distributed. What is its mean?

(c) What is the standard deviation of the population frequency?

(d) Compute the interval in which the population frequency will lie with probability 0.954.

(e) Each population frequency corresponds to a definite cost through the tax rebate (e.g. a population frequency of 0.69 or 69% corresponds to 0.69 x 140,000 = 96,600 households with children and a total rebate cost of $10 x 96,600 = $966,000). To what range of costs does the interval of (d) correspond?

(f) You notice that the interval calculated in (e) allows for a considerable overrun of the $1,000,000 budget. If you get into your office early the following morning, you might be able to intercept your memo to the Mayor in the Mayor's mail. Do you succeed?

7. The Central Limit Theorem

You are employed in the quality control section of a pipe manufacturing plant. The machine that cuts the pipe into 10 foot lengths is old and unreliable. Customers have been complaining that the pipes you sell as 10 feet long can be out by several inches. You have been asked to investigate. After a lengthy inspection of the cutting machine you decide that repair is hopeless. There is no one part that is worn or defective. The entire mechanism is worn and wobbly so it cannot cut to a reliable length. The actual length cut depends on the random, independent jiggling of the very many parts of the machine. In an astonishing feat of memory, you recall that this is precisely the circumstance to which the central limit theorem applies.

(a) What does the central limit theorem tell us in this case? (Take the "effect" to be the pipe length and the many causes to be the random jiggles of the many parts of the machine.)

Assume that you measure the pipes in the warehouse and discover that 68% of the pipes are within 1" of the desired length of 10 feet.

(b) How close to the 10 feet will 95.4% of the pipes be?

(c) How close to the 10 feet will 99.7% of the pipes be?

(d) Does this display of mathematical brilliance appease your customers and help you keep your job?

8. The Ultimate Test

You have just shown the three card swindle of Chapter 9 to your friends. It worked perfectly. They all thought that the probability of the hidden side being white was 1/2. You know that it is 2/3. You have even won a little money from them, but your conscience compels you to return it. You now try to convince them that the correct probability is 2/3 and not 1/2. After a while it becomes clear that your arguments and calculations in the probability theory do not impress them. Finally, in desperation, you propose the Ultimate Test. You will take the three cards, sit with your friends, shuffle them repeatedly and count outcomes. In particular, you will wait for every time that the shuffled deck ends up with a white face showing. In those cases you will check how often the other side of the top face is white. Your probabilistic analysis tells you it will happen 2 in 3 times; everyone else expects it to happen 1 in 2 times.

You expect the hidden face to be white 2 in 3 times *in the long run* and this will convince your friends. But in the short run, it may well not work out that way. In ten trials, you could easily get 5 or 6 hidden white faces. That would prove nothing. For that could well happen if your friends were right and the chances of a hidden white face were just 1 in 2. The question is, how many trials are needed in order to separate cleanly between the two hypotheses? (By "trial" we mean a shuffling that results in a white face uppermost.) In other words, how long is the "long run." At this point you might like to hazard a guess, recalling that our hunches in probability and statistics are notoriously unreliable.

(a) Assume that the probability of a hidden white face is 2/3 = .666. Assume that you run 50, 100, 150 and 200 trials. In each case, the number of hidden white faces will be distributed on a bell curve. What is the mean of the distribution in each of the four cases?

(b) For each of the four cases of 50, 100, 150 and 200 trials, what is the standard deviation?

(c) For each of the four cases, in what interval are we 95% sure that the actual outcome will lie?

(d) Repeat the calculation of (a) - (c) using the assumption that the probability of a hidden face is 1/2 = 0.5.

(e) Compare the intervals of (c) and (d). When there is significant overlap between the two intervals, we cannot be sure that the test will distinguish the two hypotheses. For then there is a range of outcomes that will be compatible with either. If the actual outcome lies in this common range, then our test will be inconclusive. When the intervals no longer overlap, then we can be confident that the test will succeed. For then, when we have an outcome, it will point unequivocally to only one of the two hypotheses. Roughly at what point do the intervals cease to overlap? How many trials should you be prepared to make to ensure an outcome that can distinguish the two hypotheses?

(f) Once you have located the number of trials needed to separate these two intervals, you know how many are needed for a test that would satisfy you. But notice that it probably would not satisfy your friends. For example, if you run 200 trial and find 120 hidden white faces, you would conclude from your earlier analysis that this was incompatible with the hypothesis of probability = 1/2 and compatible with the hypothesis of probability = 2/3. But your friends would look at the 120 and say it was not close enough to the 133 expected. Put another way, the 120 corresponds to a frequency of 120/200 = 0.6 and that, they would complain, does not establish the frequency you hypothesize of 0.666. The actual result is too far away. It differs from the 0.666 you hypothesize by .066/.666 = 0.10 (i.e. 10%). "OK, "you say, "let us just continue the trials. Eventually the outcomes will stabilize at the 66.6% I hypothesize." The moment you say this, you regret it, for you start to suspect that a huge number of trials will be needed to bring the frequency close to 66.6%. To get a sense of just how large is this number of trials, start to compute the interval in which the number of outcomes will lie for larger number of trials: say 1,000, 5,000, 10,000. How many trials are needed to be 95% confident that the outcome will lie within 1% of the 66.6% you expect?

Moral: a little learning is a dangerous thing. Had you known only the probability theory, you would have known that the correct answer is 2 in 3. But, without some knowledge of statistics, you would have been quite unable to estimate how many trials were needed to make a clean separation.

Further Questions for Study and Review

1. Repetition Again

Generally saying the same thing twice is just a waste of time, boring and even offensive. Yet scientists put much importance on the repeating of experiments. Indeed they may well refuse to believe an experimental result until it has been repeated successfully.

(a) Why is recovering the same result again so important?

(b) Find some examples of experiments that have been repeated and thereby enter into the mainstream of science.

(c) Find some examples of experiments that could not be repeated and thus failed to enter into the mainstream of science.

(d) Can you think of some science in which it is unreasonable to demand replication of results? Which is it and why is the demand unreasonable there?

2. Secret Messages

We have seen many claims of the presence of hidden messages in texts, but we have concluded that they are bogus. The methods used to find them are so powerful that they can find messages hidden in any sufficiently large body of text. Now let us reverse the problem. Imagine that you wish to hide a message in a text. How would you go about hiding it so that someone who finds it can be sure that it is present by your design and not merely an artefact of overzealous pattern finding?

3. Energy and Other Invisibles

Energy is an odd entity in so far as it is something of which we have no direct experience. We do not see or sense energy directly; we sense it through its consequences. We do not see the energy of motion of a moving baseball; we see the motion of the baseball. We do not see the heat energy of boiling water; we sense its high temperature. It was urged in Chapter 4 that we still have good reason to believe that there really is an entity energy because of the massive success of our theory of energy in accounting for experience and because it was possible to infer directly from different sets of experiences to the existence and properties of energy.

(a) There are many entities in science that we cannot sense directly but which we have little doubt are (or were) really there. Find several examples of these and lay out the reasons we have for believing that they are real. Here are some suggestions. Feel free to find your own:
Radio waves, electrons, helium in the core of the sun, iron in the core of the earth, the big bang.

(b) Outside science we often find entities postulated that we cannot sense directly. Some are postulated illegitimately; others legitimately. Give some examples of each indicating why you judge them to be illegitimate or legitimate. Here is a list of candidate entities to get you started. Feel free to add your own:
Juius Caesar, hobgoblins, telepathic thought waves, bad luck, the will of the people, the nutritive power of food.

4. Conservation of Energy

Might we use compressed air to run an automobile? The idea is attractive. Compressed air can be stored easily in the familiar cylinders used to hold compressed gases. It is easy to procure; all one needs is a compressor to make it. The engines that use it need only a simple design; a familiar piston engine will work well enough. A compressed air engine emits no pollutants at all, although pollution may arise from the generation of the power used by the compressor that supplies the compressed air. Let us calculate if a compressed air car can match the performance of a gasoline powered automobile.

The problem will be whether compressed air can store enough energy in a portable form. Since the weight of the storage cylinders will be our major concern, we will consider lightweight aluminum storage cylinders. One commercial supplier provides 1.04 cubic feet of compressed air at 2,000 pounds per square inch pressure in aluminum cylinders (8in. diameter x 48 in. height), weighing 48 pounds. If the expansion of the air is used with perfect efficiency 2,000 kJ of energy can be extracted from the compressed air.

(a) If we design a car that can support 500 pounds of air cylinders, how many cylinders will we install in the car? (Round off your answer to a whole number!)

(b) If we are able to extract energy from the air at perfect efficiency, how much energy can the cylinders in (a) supply.

(c) If 1650kJ of energy are needed for each mile of city driving, how far can our air powered car run on one charge of compressed air in the city?

(d) If 660 kJ of energy are needed for each mile of freeway driving, how far can our air powered car run on one charge of compressed air on the freeway?

The answers to (c) and (d) do not compare well with gasoline powered cars, but they are not so bad. Given the simplicity and cleanliness of compressed air technology, we might wish to pursue it. We can surely move to lighter air storage at higher pressures. One ingenious possibility was realized by the experimental car LN2000 at the University of Washington. Air is 80% nitrogen. Nitrogen, liquefied by making it extremely cold, is 600 times as dense as nitrogen at ordinary pressure and temperatures, but needs no pressurization. (Nitrogen at 2,000 pounds per square inch pressure is only increased by a factor of 137 in density.) The LN2000 car runs on the pressure of nitrogen gas produced by vaporizing nitrogen using the heat of the surrounding air.

5. Heat Engines

(a) The liquid nitrogen powered car described above uses heat from the air to boil liquid nitrogen to make high pressure nitrogen gas to run the car. Thus, in its operation, heat is taken from one place and no heat is discharged to a cooler place. Does this represent a violation of the second law of thermodynamics?

(b) The compressed air car described above has a similar design feature. While the air is drawn from the aluminum cylinders, these cylinders cool. To recover the maximum energy to drive the car, we must endeavor to keep the cylinders at normal air temperatures. That is, we must ensure that enough heat passes from the surroundings to maintain its temperature. If the cylinder were fully insulated, the cooling would so reduce the pressure of the air contained that we would not recover 2,000 kJ of energy but merely 874 kJ, just 44%. (In practice complete insulation would be very hard to realize since the cylinder seeks to cool itself to temperatures well below freezing.) So in a sense heat drawn from the suroundings is being used to drive the car, but we are not discharging heat to a colder place. Does this violate the second law of thermodynamics?

Hint for both: review all of Carnot's three core notions.

6. Quadruplet Births

In following we will see a quite astonishing example of just how unreliable our raw intuitions about chance can be. The example deals with the occurrence of quadruplet births, "quads." These births are very rare, happening on average once in half a million births. If we consider all of the U.S., Canada and Mexico, on average there will roughly one quad birth per month, that is, roughly twelve each year. We assume that the occurrence of quads are independent, so that one quad birth does not make others any more or less likely.

Since we are expecting roughly twelve quads a year, how likely is it that we have 12 quads in the course of the year distributed exactly one per month: that is, one in January, one in February, one in March, ... and one in December? Would we expect this to happen

--every other year on average?
--every ten years on average?
--every hundred years on average?
--every thousand years on average?

Most people do not expect the uniform distribution to happen as often as once every other year. Many find once in ten years to be plausible. Virtually everyone expects the frequency to be greater than once in a thousand years. But even that estimate proves to be <u>far</u> too high, as we will now see. Finally, however unlikely 12 uniformly distributed births in a year may be, virtually everyone agrees that it is a great deal more likely that a year with no quad births at all. That supposition proves to be wrong too!

To begin, we need to know the probability of just one quad birth in a month. We expect one quad birth in a given month <u>on average</u>. So it is quite possible that we will have no quad births in that month or one quad birth or more than one. Using more advanced techniques not discussed in this text, one can arrive at the following probabilities:

P(0 quads in one month) = .3678
P(1 quad in one month) = .3678

(a) For some month, are the outcomes (0 quad) and (1 quad) mutually exclusive? If so, use the OR rule to compute P(0 quad OR 1 quad).

(b) Consider the outcome (more that 1 quad) for some month and notice that it is the same as NOT- (0 quad OR 1 quad). Use the NOT rule to compute P(more that 1 quad).

(c) Now consider the first outcome that interests us

(1 quad in January AND 1 quad in February AND ... AND 1 quad in December)

Are these ANDed outcomes independent? If so, use the AND rule to compute the probability of a year with 12 quad births uniformly distributed over its 12 months.

(d) Use your answer to (c) to determine how frequently such a uniform outcome year will occur. (Recall: if the probability of an outcome is 0.05, then in the long run that outcome will occur once every 1/0.05=20 trials.)

(e) Now consider the second outcome that interests us

(0 quads in January AND 0 quads in February AND ... AND 0 quads in December)

Are these ANDed outcomes independent? If so, use the AND rule to compute the probability of this outcome of no quad births over the 12 months of the year.

(f) Use your answer to (e) to compute how frequently we will have year with no quad births.

7. Therapeutic Touch

Practitioners of therapeutic touch claim to be able to heal or improve many medical conditions by manipulating what they call the "human energy field." In part of the therapy, the practitioner sweeps his or her hands over the patient's body, two to four inches above the skin. During this sweep, the practitioner claim they can sense the patient's human energy field. Beneficial effects are supposed to derive from the resulting manipulation of the field. While there are many practitioners of therapeutic touch, there has been considerable controversy over it. Critics charge that the procedure offers no intrinsic benefits other than those accrued through the placebo effect; that is, if patients expect a therapy to be effective, that expectation alone can yield benefit.

A study reported in the April 1, 1998, Journal of the American Medical Association (L. Rosa, E. Rosa, L. Sarner and S. Barrett, "A Close Look at Therapeutic Touch") set out to test whether the practitioners really could sense the human energy fields that they claimed to sense. (One of the authors, Emily Rosa, was nine years old when she commenced conducting the tests.) The test was simple. A therapeutic touch practitioner placed his or her hands palm uppermost on a table. One of the experimenters then placed her hand three to four inches above one of the practitioner's hands--the target hand. Whether the target hand would be the right or left was decided by a coin toss each time. The practitioner was then to determine which the target hand was by detecting the energy field of the experimenter's hand. A tall screen with hand holes cut out and towel were used to ensure that the practitioner could not see which hand had been selected as the target hand.

If the practitioners can detect a human energy field reliably, one would expect 100% accuracy in detecting the target hand. If there is no detection at all, one would expect the practitioners to do no better than if they guessed which was the target hand. That is, they would name the target hand correctly (a "hit") in roughly 50% of the trials, since a guess has a probability of 1/2 of success. The results strongly favored this second possibility, that there is no detection of the target hand.

A total of 21 practitioners participated in two series of trials. In the first series, the practitioners scored 70 hits in 150 trials. In the second series, the practitioners scored 53 hits in 130 trials.

We will use these data to test the hypothesis.

> The successes of the practitioners arise from chance guessing alone; that is, the probability of a practitioner correctly identifying the target hand is 1/2.

(a) Consider the first series of 150 trials and assume that the hypothesis is true.
What is the mean of the bell curve that depicts the probabilities of the different numbers of possible hits? What is its standard deviation?
What are the confidence intervals MEAN +/-1SD, MEAN+/-2SD, MEAN+/-3SD?
The actual number of hits in the first series was 70.
Is this outcome in the range of outcomes expected with probability 0.683? With probability 0.95? With probability 0.997?
Is the outcome statistically significant at the 0.05 level? At the 0.003 level?
What is the upshot of the test: do we accept or reject the hypothesis?

(b) Repeat the analysis of (a) for the second series of 130 trials in which 53 hits were scored.

(c) If we combine the two series into one, we have 280 trials with 123 hits. Repeat the analysis of (a) for this combined series.

8. SAT Coaching Services

An SAT coaching service promises an average improvement in SAT scores of 100 points (out of the total of 800 on each test). This claim is scrutinized by comparing the improvement in scores of students who are coached with those who are not coached when they all retake the test. 2000 students who were not coached retake the test and show no improvement on average. Another 2000 students take the test, are coached and then retake the test. These coached students show an improvement of 13 points on average, well below the 100 claimed.

There is some variability in the scores of any student taking the SAT. For example, a 600 point student will probably not score exactly 600 point each time the student takes the test. The scores will vary around 600. It is known that the standard deviation of this spread is 30 points. This fact leads to the suggestion that there is no statistically significant improvement at all for students who are coached: the average improvement is less than the individual variability.

Having learned some statistics, you are suspicious of this suggestion. You would not quarrel with the suggestion if the study had looked at just one student. But it did not. It looked at 2,000 and the 13 point improvement was the average improvement. You recall that averages tend to suppress variations. For example, on repeated coin tosses we expect about 50% to be heads. But if a coin is tossed 2 times, the percentage of heads can readily vary from 0% to 100%. If the same coin is tossed 100 times, the percentage of heads will most likely lie in the interval 40% to 60%. To evaluate your suspicions you proceed with the following calculation. You decide to test :

Hypothesis: Coaching produces no improvement in SAT scores on average.

(a) Consider the total SAT scores of all 2,000 coached students on retaking the test. This total score is the sum of vary many independent causes: the score of the first student + the score of the second student + ... + the score of the 2,000th student. Does the central limit theorem apply to this total score so that the total will be normally distributed?

(b) Use the central limit theorem to compute the MEAN of that normal distribution.

(c) Use the central limit theorem to compute the STANDARD DEVIATION of that normal distribution. Hint: for this special case, the individual causes summed are all alike so you can use the simple rule

STANDARD DEVIATION

$= \sqrt{\text{number of causes}}$ (STANDARD DEVIATION of individual cause)

(d) In what interval will the total score lie with probability 0.954? In what interval will the total score lie with probability 0.997?

(e) If the average improvement of the 2,000 students is 13 points, then what is their total score?

(f) Does this total lie in either of the intervals computed in (d)? Is the result statistically significant at the 0.05 or 0.003 level?

(g) Should the hypothesis be accepted or rejected?